T0139288

# Practical Numerical and Scientific Computing with MATLAB® and Python

# Practical Numerical and Scientific Computing with MATLAB® and Python

Eihab B. M. Bashier

CRC Press
Taylor & Francis Group
Boca Raton  London  New York

CRC Press is an imprint of the
Taylor & Francis Group, an **informa** business

A CHAPMAN & HALL BOOK

CRC Press
Taylor & Francis Group
6000 Broken Sound Parkway NW, Suite 300
Boca Raton, FL 33487-2742

© 2020 by Taylor & Francis Group, LLC
CRC Press is an imprint of Taylor & Francis Group, an Informa business

No claim to original U.S. Government works

Printed on acid-free paper

International Standard Book Number-13: 978-0-367-07669-6 (Hardback)

### Library of Congress Cataloging-in-Publication Data

Names: Bashier, Eihab Bashier Mohammed, author.
Title: Practical Numerical and Scientific Computing with MATLAB®
and Python / Eihab B.M. Bashier.
Description: Boca Raton : CRC Press, 2020. | Includes bibliographical
references and index.
Identifiers: LCCN 2019052363 | ISBN 9780367076696 (hardback) | ISBN
9780429021985 (ebook)
Subjects: LCSH: Science--Data processing. | MATLAB. | Python (Computer
program language)
Classification: LCC Q183.9 B375 2020 | DDC 502.85/53--dc23
LC record available at https://lccn.loc.gov/2019052363

**Visit the Taylor & Francis Web site at**
**http://www.taylorandfrancis.com**

**and the CRC Press Web site at**
**http://www.crcpress.com**

*To my parents, family and friends.*

# Contents

# *Preface*

The past few decades have witnessed tremendous development in the manufacture of computers and software, and scientific computing has become an important tool for finding solutions to scientific problems that come from various branches of science and engineering. Nowadays, scientific computing has become one of the most important means of research and learning in the fields of science and engineering, which are indispensable to any researcher, teacher, or student in the fields of science and engineering.

One of the most important branches of scientific computing is a numerical analysis which deals with the issues of finding approximate numerical solutions to such problems and analyzing errors related to such approximate methods. Both the MATLAB® and Python programming languages provide many libraries that can be used to find solutions of scientific problems visualizing them. The ease of use of these two languages became the most languages that most scientists who use computers to solve scientific problems care about.

The idea of this book came after I taught courses of scientific computing for physics students, introductory and advanced courses in mathematical software and mathematical computer applications in many Universities in Africa and the gulf area. I also conducted some workshops for mathematics and science students who are interested in computational mathematics in some Sudanese Universities. In these courses and workshops, MATLAB and Python were used for the implementation of the numerical approximation algorithms. Hence, the purpose of introducing this book is to provide the student with a practical guide to solve mathematical problems using MATLAB and Python software without the need for third-party assistance. Since numerical analysis is concerned with the problems of approximation and analysis of errors of numerical methods associated with approximation methods, this book is more concerned with how these two aspects are applied in practice by software, where illustrations and tables are used to clarify approximate solutions, errors and speed of convergence, and its relations to some of the numerical method parameters, such as step size and tolerance. MATLAB and Python are the most popular programming languages for mathematicians, scientists, and engineers. Both the two programming languages possess various libraries for numerical and symbolic computations and data representation and visualization. Proficiency with the computer programs contained in this book requires that the student have prior knowledge of the basics of the programming languages MATLAB and Python, such as branching, Loops, symbolic packages, and the graphical libraries. The MATLAB version used for this book is 2017b and the Python version is 3.7.4.

The book consists of 11 chapters divided into three parts: the first part is concerned with discussing numerical solutions for linear and nonlinear systems and numerical difficulties facing these types of problems with how to overcome these numerical difficulties. The second part deals with methods of completing functions, differential and numerical integration, and solutions of differential equations. The last part of the book discusses methods to solve linear and nonlinear programming and optimal control problems. It also contains some specialized software in Python language to solve some problems numerically. These software packages must be downloaded from a third party, such as Gekko which is used for the solutions of differential equations and linear and nonlinear programming in addition to the optimal control problems. Also, the Pulp package is used to solve linear programming problems and finally Pyomo a package is used for solving linear and nonlinear programming problems. How to install and run such a package is also presented in the book.

What distinguishes this book from many other numerical analysis books is that it contains some topics that are not usually found in other books, such as nonstandard finite difference methods for solving differential equations and solutions of optimal control problems. In addition, the book discusses implementations of methods with high convergence rates, such as Gauss integration methods discussed in the numerical differentiation and integration, exact finite difference schemes for solving differential equations discussed in the nonstandard finite differences Chapter. It also uses efficient python-based software for solving some kinds of mathematical problems numerically.

The parts of the book are separate from each other so that the student can study any part of it without having to read the previous parts of that part. The exception to this is the optimal control chapter in the third part, which requires studying numerical methods to solve the differential equations discussed in the second part.

After reading this book and implementing the programs contained on it, a student will be able to deal with and solve many kinds of mathematical problems such as differential equations, static, and dynamical optimization problems and apply the methods to real-life problems.

## Acknowledgment

I am very grateful to the African Institute of Mathematical Sciences (AIMS), Cape Town, South Africa, which hosted me on a research visit during which some parts of this book have been written. I would also like to thank the editorial team of this book under the leadership of publisher, Randi Cohen, for their continuous assistance in formatting, coordinating, editing, and directing the book throughout all stages. Special thanks go to all professors who taught me the courses of numerical analysis in the various stages of my under- and postgraduate studies, and, in particular, I thank Dr. Mohsin Hashim University of

Khartoum, Professor Arieh Iserles African Institute of Mathematical Sciences, Professor Kailash Patidar University of the Western Cape in Cape Town, and the spirit of my great teacher Professor David Mackey (Cambridge University and the African Institute for Mathematical Sciences) who passed away four years ago. Finally, I am also grateful to my family for continuous encouragement and patience while writing this book.

MATLAB® is a registered trademark of The MathWorks, Inc. For product information, please contact:

The MathWorks, Inc.
3 Apple Hill Drive
Natick, MA, 01760-2098 USA
Tel: 508-647-7000
Fax: 508-647-7001
E-mail: info@mathworks.com
Web: www.mathworks.com

# *Author*

**Eihab B. M. Bashier** obtained his PhD in 2009 from the University of the Western Cape in South Africa. His permanent job is in the Department of Applied Mathematics on the faculty of mathematical sciences and information technology, University of Khartoum. Currently, he is an Associate Professor of Applied Mathematics at the College of Arts and Applied Sciences at Dhofar University, Oman. His research interests are mainly in numerical methods for differential equations with applications to biology and in information and computer security with a focus in cryptography. In 2011, Dr. Bashier won the African Union and the Third World Academy of Science (AU-TWAS) Young Scientists National Award in Basic sciences, Technology and Innovation. Dr. Bashier is a reviewer for some international journals and a member of the IEEE and the EMS. (Email: eihab-bashier@gmail.com, eihabbash@aims.ac.za).

# Part I

# Solving Linear and Nonlinear Systems of Equations

# 1

## Solving Linear Systems Using Direct Methods

## Abstract

Linear systems of equations have many applications in mathematics and science. Many of the numerical methods used for solving mathematics problems such as differential or integral equations, polynomial approximations of transcendental functions and solving systems of nonlinear equations arrive at a stage of solving linear systems of equations. Hence, solving a linear system of equations is a fundamental problem in numerical computing.

This chapter discusses the direct methods for solving linear systems of equations, using Gauss and Gauss-Jordan elimination techniques and the matrix factorization approach. MATLAB® and Python implementations of such algorithms are provided.

## 1.1 Testing the Existence of the Solution

A linear system consisting of $m$ equations in $n$ unknowns, can be written in the matrix form:

$$A\boldsymbol{x} = \boldsymbol{b} \tag{1.1}$$

where,

$$A = \begin{pmatrix} a_{11} & a_{12} & \cdots & a_{1n} \\ a_{21} & a_{21} & \cdots & a_{2n} \\ \vdots & \vdots & \ddots & \vdots \\ a_{m1} & a_{m2} & \cdots & a_{mn} \end{pmatrix}, \boldsymbol{x} = \begin{pmatrix} x_1 \\ x_2 \\ \vdots \\ x_n \end{pmatrix} \text{ and } \boldsymbol{b} = \begin{pmatrix} b_1 \\ b_2 \\ \vdots \\ b_m \end{pmatrix}$$

Here, the coefficients $a_{ij}$ of matrix $A \in \mathbb{R}^{m \times n}$ are assumed to be real, $\boldsymbol{x} \in \mathbb{R}^n$ is the vector of unknowns and $\boldsymbol{b} \in \mathbb{R}^m$ is a known vector. Depending on the relationship between $m$ and $n$ three kinds of linear systems are defined [30, 53]:

1. **overdetermined linear systems:** there are more equations than unknown $(m > n)$.

2. **determined linear systems:** equal numbers of equations and unknowns $(m = n)$.

3. **underdetermined linear systems:** there are more unknowns than equations $(m < n)$.

Let $\tilde{A} = [A \mid b]$ be the augmented matrix of the linear system $Ax = b$. Then, the existence of a solution for the given linear system is subject to one of the two following cases:

1. $rank(\tilde{A}) = rank(A)$: in this case, there is at least one solution, and we have two possibilities:

   (a) $rank(\tilde{A}) = rank(A) = n$: in this case there is a unique solution.

   (b) $rank(\tilde{A}) = rank(A) < n$: in this case there is infinite number of solutions.

2. $rank(\tilde{A}) > rank(A)$: in this case there is no solution and we can look for a least squares solution.

If the linear system $Ax = b$ has a solution, it is called a `consistent` linear system, otherwise, it is an `inconsistent` linear system [30].

In MATLAB, the command rank can be used to test the rank of a given matrix $A$.

```
>> A = [1 2 3; 4 5 6; 7 8 9]
A =
1    2    3
4    5    6
7    8    9
>> b = [1; 1; 1]
b =
1
1
1
>> r1 = rank(A)
r1 =
2
>> r2 = rank([A b])
r2 =
2
```

In python, the function `matrix_rank` (located in `numpy.linalg`) is used to compute the rank of matrix $A$ and the augmented system $[Ab]$.

```
In [1]: import numpy as np
In [2]: A = np.array([[1, 2, 3], [4, 5, 6], [7, 8, 9]])
In [3]: b = np.array([1, 1, 1])
```

```
In [4]: r1, r2 = np.linalg.matrix_rank(A), np.linalg.matrix_rank
    (np.c_[A, b])
In [5]: r1
Out[5]: 2
In [6]: r2
Out[6]: 2
```

In the special case when $m = n$ ($A$ is a squared matrix) and there is a unique solution ($rank(\tilde{A}) = rank(A) = n$), this unique solution is given by:

$$x = A^{-1}b.$$

Hence, finding the solution of the linear system requires the inversion of matrix $A$.

## 1.2 Methods for Solving Linear Systems

This section considers three special types of linear systems which are linear systems with diagonal, upper triangular and lower triangular matrices.

### 1.2.1 Special Linear Systems

We consider the linear system:

$$Ax = b,$$

where $A \in \mathbb{R}^{n \times n}$, $x$ and $b \in \mathbb{R}^n$. We consider two cases.

1. **$A$ is a diagonal matrix:**
   In this case, matrix $A$ is of the form:

$$A = \begin{pmatrix} a_{11} & 0 & 0 & \cdots & 0 \\ 0 & a_{22} & 0 & \cdots & 0 \\ 0 & 0 & a_{33} & \cdots & 0 \\ \vdots & \vdots & \vdots & \ddots & \vdots \\ 0 & 0 & 0 & \cdots & a_{nn} \end{pmatrix}$$

   which leads to the linear system:

$$\begin{pmatrix} a_{11} & 0 & 0 & \cdots & 0 \\ 0 & a_{22} & 0 & \cdots & 0 \\ 0 & 0 & a_{33} & \cdots & 0 \\ \vdots & \vdots & \vdots & \ddots & \vdots \\ 0 & 0 & 0 & \cdots & a_{nn} \end{pmatrix} \begin{pmatrix} x_1 \\ x_2 \\ x_3 \\ \vdots \\ x_n \end{pmatrix} = \begin{pmatrix} b_1 \\ b_2 \\ b_3 \\ \vdots \\ b_n \end{pmatrix} \quad (1.2)$$

The solution of the linear system (1.2) is given by:

$$x_i = \frac{b_i}{a_{ii}}$$

The MATLAB code to compute this solution is given by:

```
1   function x = SolveDiagonalLinearSystem(A, b)
2       % This function solves the linear system Ax = b, where ...
            A is a diagonal matrix
3       % b is a known vector and n is the dimension of the ...
            problem.
4       n = length(b) ;
5       x = zeros(n, 1) ;
6       for j = 1: n
7           x(j) = b(j)/A(j, j) ;
8       end
```

We can apply this function to solve the diagonal system:

$$\begin{pmatrix} 2 & 0 & 0 \\ 0 & -1 & 0 \\ 0 & 0 & 3 \end{pmatrix} \begin{pmatrix} x_1 \\ x_2 \\ x_3 \end{pmatrix} = \begin{pmatrix} 4 \\ 1 \\ -3 \end{pmatrix}$$

by using the following MATLAB commands:

```
>> A = diag([2, -1, 3])
A =
2     0     0
0    -1     0
0     0     3
>> b = [4; 1; 3]
b =
4
1
3
>> x = SolveDiagonalLinearSystem(A, b)
x =
2
-1
1
```

The python code of the function SolveDiagonalLinearSystem is as follows.

```
1   import numpy as np
2   def SolveDiagonalLinearSystem(A, b):
3       n = len(b)
```

```
4        x = np.zeros((n, 1), 'float')
5        for i in range(n):
6            x[i] = b[i]/A[i, i]
7        return x
```

```
In [7]: A = np.diag([2, -1, 3])
In [8]: b = np.array([4, -1, 3])
In [9]: x = SolveDiagonalLinearSystem(A, b)
In [10]: print('x = \n', x)
x =
[[ 2.]
 [ 1.]
 [ 1.]]
```

2. **$A$ is an upper triangular matrix:**
In this case, matrix $A$ is of the form:

$$A = \begin{pmatrix} a_{11} & a_{12} & a_{13} & \cdots & a_{1n} \\ 0 & a_{22} & a_{23} & \cdots & a_{2n} \\ 0 & 0 & a_{33} & \cdots & a_{3n} \\ \vdots & \vdots & \vdots & \ddots & \vdots \\ 0 & 0 & 0 & \cdots & a_{nn} \end{pmatrix}$$

Therefore, we have the linear system:

$$\begin{pmatrix} a_{11} & a_{12} & a_{13} & \cdots & a_{1n} \\ 0 & a_{22} & a_{23} & \cdots & a_{2n} \\ 0 & 0 & a_{33} & \cdots & a_{3n} \\ \vdots & \vdots & \vdots & \ddots & \vdots \\ 0 & 0 & 0 & \cdots & a_{nn} \end{pmatrix} \begin{pmatrix} x_1 \\ x_2 \\ x_3 \\ \vdots \\ x_n \end{pmatrix} = \begin{pmatrix} b_1 \\ b_2 \\ b_3 \\ \vdots \\ b_n \end{pmatrix} \tag{1.3}$$

In this case we use the back substitution method for finding the solution of system 1.3. The MATLAB function SolveUpperSystem.m solves the linear system 1.3 using the back-substitution method.

```
1  function x = SolveUpperLinearSystem(A, b)
2      % This function uses the backward substitution method ...
         for solving
3      % the linear system Ax = b, where A is an upper ...
         triangular matrix
4      % b is a known vector and n is the dimension of the ...
         problem.
5      n = length(b) ;
6      x = zeros(n, 1) ;
7      x(n) = b(n)/A(n, n) ;
8      for j = n-1: -1 : 1
9          x(j) = b(j) ;
```

```
10          for k = j+1 : n
11              x(j) = x(j) - A(j, k)*x(k) ;
12          end
13          x(j) = x(j)/A(j, j) ;
14      end
```

The python code for the `SolveUpperSystem`, is as follows.

```python
1  import numpy as np
2  def SolveUpperLinearSystem(A, b):
3      n = len(b)
4      x = np.zeros((n, 1), 'float')
5      x[n-1] = b[n-1]/A[n-1, n-1]
6      for i in range(n-2, -1, -1):
7          x[i] = b[i]
8          for j in range(i+1, n):
9              x[i] -= A[i, j]*x[j]
10         x[i] /= A[i, i]
11     return x
```

3. **$A$ is a lower triangular system:**
In this case, matrix $A$ is of the form:

$$A = \begin{pmatrix} a_{11} & 0 & 0 & \dots & 0 \\ a_{21} & a_{22} & 0 & \dots & 0 \\ a_{31} & a_{32} & a_{33} & \dots & 0 \\ \vdots & \vdots & \vdots & \ddots & \vdots \\ a_{n1} & a_{n2} & a_{n3} & \dots & a_{nn} \end{pmatrix}$$

Therefore, we have the linear system:

$$\begin{pmatrix} a_{11} & 0 & 0 & \dots & 0 \\ a_{21} & a_{22} & 0 & \dots & 0 \\ a_{31} & a_{32} & a_{33} & \dots & 0 \\ \vdots & \vdots & \vdots & \ddots & \vdots \\ a_{n1} & a_{n2} & a_{n3} & \dots & a_{nn} \end{pmatrix} \begin{pmatrix} x_1 \\ x_2 \\ x_3 \\ \vdots \\ x_n \end{pmatrix} = \begin{pmatrix} b_1 \\ b_2 \\ b_3 \\ \vdots \\ b_n \end{pmatrix} \qquad (1.4)$$

The forward substitution method is used to find the solution of system 1.4.
The MATLAB function `SolveLowerSystem.m` solves the linear system 1.4
using the forward-substitution method.

```
1  function x = SolveLowerLinearSystem(A, b)
2      % This function uses the forward substitution method ...
           for solving
3      % the linear system Ax = b, where A is an lower ...
           triangular matrix
4      % b is a known vector and n is the dimension of the ...
           problem.
```

```
5      n = length(b) ;
6      x = zeros(n, 1) ;
7      x(1) = b(1)/A(1, 1) ;
8      for j = 2 : n
9          x(j) = b(j) ;
10         for k = 1 : j-1
11             x(j) = x(j) - A(j, k)*x(k) ;
12         end
13         x(j) = x(j)/A(j, j) ;
14     end
```

The python code of the function `SolveLowerSystem` is as follows.

```
1  def SolveLowerLinearSystem(A, b):
2      import numpy as np
3      n = len(b)
4      x = np.zeros((n, 1), 'float')
5      x[0] = b[0]/A[0, 0]
6      for i in range(1, n):
7          x[i] = b[i]
8          for j in range(i):
9              x[i] -= A[i, j]*x[j]
10         x[i] /= A[i, i]
11     return x
```

## 1.2.2 Gauss and Gauss-Jordan Elimination

Gauss and Gauss-Jordan elimination methods are related to each other. If given a matrix $A \in \mathbb{R}^{n \times n}$, then both Gauss and Gauss-Jordan apply elementary row operations through consequent steps over matrix $A$. The Gauss method stops after obtaining the row echelon form of matrix $A$ (If $A$ is non-singular, then its row echelon form is an upper triangular matrix), whereas Gauss-Jordan continuous until reaching the `reduced row echelon form` (If $A$ is nonsingular, then its reduced row echelon form is the identity matrix).

To illustrate the differences between the row echelon and the reduced row echelon forms, the two forms are computed for the matrix:

$$A = \begin{pmatrix} 4 & -1 & -1 \\ -1 & 4 & -1 \\ -1 & -1 & 4 \end{pmatrix}$$

Starting by finding the row echelon form for the given matrix.

$$A = \begin{pmatrix} 4 & -1 & -1 \\ -1 & 4 & -1 \\ -1 & -1 & 4 \end{pmatrix} \xrightarrow[R_3 \leftarrow 4R3+R_1]{R_2 \leftarrow 4R2+R_1} \begin{pmatrix} 4 & -1 & -1 \\ 0 & 15 & -5 \\ 0 & -5 & 15 \end{pmatrix} \xrightarrow{R_3 \leftarrow 3R3+R_2} \begin{pmatrix} 4 & -1 & -1 \\ 0 & 15 & -5 \\ 0 & 0 & 40 \end{pmatrix}$$

The upper triangular matrix

$$\begin{pmatrix} 4 & -1 & -1 \\ 0 & 15 & -5 \\ 0 & 0 & 40 \end{pmatrix}$$

is the row echelon form of matrix $A$.

Gauss-Jordan elimination continues above the pivot elements, to obtain the reduced row echelon form.

$$\begin{pmatrix} 4 & -1 & -1 \\ 0 & 15 & -5 \\ 0 & 0 & 40 \end{pmatrix} \xrightarrow{R_3 \leftarrow R3/40} \begin{pmatrix} 4 & -1 & -1 \\ 0 & 15 & -5 \\ 0 & 0 & 1 \end{pmatrix} \xrightarrow[R_2 \leftarrow R2+5R_3]{R_1 \leftarrow R_1+R_3} \begin{pmatrix} 4 & -1 & 0 \\ 0 & 15 & 0 \\ 0 & 0 & 1 \end{pmatrix}$$

$$\xrightarrow{R_2 \leftarrow R2/15} \begin{pmatrix} 4 & -1 & 0 \\ 0 & 1 & 0 \\ 0 & 0 & 1 \end{pmatrix} \xrightarrow{R_1 \leftarrow R_1+R_2} \begin{pmatrix} 4 & 0 & 0 \\ 0 & 1 & 0 \\ 0 & 0 & 1 \end{pmatrix} \xrightarrow{R_1 \leftarrow R_1/4} \begin{pmatrix} 1 & 0 & 0 \\ 0 & 1 & 0 \\ 0 & 0 & 1 \end{pmatrix}$$

### 1.2.3    Solving the System with the rref Function

The Gauss and Gauss-Jordan methods are two familiar approaches for solving linear systems. Both begin from the augmented matrix, obtain the row echelon form or the reduced row echelon form, respectively. Then, the Gauss method uses the back-substitution technique to obtain the solution of the linear system, whereas in Gauss-Jordan method the solution is located in the last column.

The MATLAB code below, reads a matrix $A$ and a vector $b$ from the user, then it applies the Gauss-Seidel elimination through applying the rref to the augmented system $[A \quad b]$

```
1   clear ; clc ;
2   A = input('Enter the matrix A: ') ; % Reading matrix A from ...
        the user
3   b = input('Enter the vector b: ') ; % Reading vector b from ...
        the user
4   [m, n] = size(A) ;                  % m and n are the matrix ...
        dimensions
5   r1 = rank(A) ;                      % the rank of matrix A is ...
        assigned to r1
6   r2 = rank([A b]) ;                  % the rank of the ...
        augmented system [A b] is assigned to r2
7   if r1 ≠ r2                          % testing whether rank(A) ...
        not equal rank([A b])
8       disp(['Rank(A) = ' num2str(r1) ' ≠ ' num2str(r2) ' = ...
            Rank([A b]).']) ;
9       fprintf('There is no solution.\n') ; % No solution in this ...
            case
10  end
11  if r1 == r2                        % testing whether rank(A) = ...
            rank([A b])
```

```
12      if r1 == n                      % if yes, testing whether the ...
            rank equals n
13          R = rref([A b]) ;       % the reduced row echelon form ...
                of [A b]
14          x = R(:, end) ;            % the solution is at the last ...
                column of the reduced
15          % row echelon form
16          disp(['Rank(A) = Rank([A b]) = ' num2str(r1) ' = ...
                #Col(A).']) ;
17          disp('There is a unique solution, given by: ') ; ...
                disp(x) ;
18          %displaying the solution of the linear system
19      else                            % rank(A) = rank([A b]) < n
20          disp(['Rank(A) = Rank([A b]) = ' num2str(r1) ' < ' ...
                num2str(n) ' = #Col(A).']) ;
21          fprintf('Infinite number of solutions.\n') ;
22      end
23  end
```

The result of executing the above MATLAB script is:

```
Enter the matrix A: [1 2 3; 4 5 6; 7 8 9]
Enter the vector b: [1;3;5]
Rank(A) = Rank([A b]) = 2 < 3 = #Col(A).
Infinite number of solutions.

Enter the matrix A: [1 3 5; 2 4 6; 7 8 9]
Enter the vector b: [1;1;1]
Rank(A) = 2 ~= 3 = Rank([A b]).
There is no solution.

Enter the matrix A: [2 2 -1; 1 2 1; -1 -1 2]
Enter the vector b: [2;4;1]
Rank(A) = Rank([A b]) = 3 = #Col(A).
There is a unique solution, given by:
0.6667
1.0000
1.3333
```

In Python, the built-in function `sympy.Matrix` is used to construct a matrix. The Matrix class has a method `rref` to compute the reduced row echelon from of the matrix.

```
1  import sympy as smp
2  A = smp.Matrix([[2, 2, -1], [1, 2, 1], [-1, -1, 2]])
3  b = smp.Matrix([[2], [4], [1]])
4  m, n = A.rows, A.cols
5  r1 = A.rank()
6  C = A.copy()
7  r2 = (C.row_join(b)).rank()
```

```
8   if r1 != r2:                            # testing whether rank(A) ...
        not equal rank([A b])
9       print('Rank(A) = ' +str(r1) +' != ' +str(r2) +' = Rank([A ...
            b]).')
10      print('There is no solution.\n') ; # No solution in this case
11  if r1 == r2:                            # testing whether rank(A) = ...
        rank([A b])
12      if r1 == n:                         # if yes, testing whether the ...
            rank equals n
13          R = (A.row_join(b)).rref()      # the reduced row ...
                echelon form of [A b]
14          x = R[0][:, -1]          # the solution is at the last ...
                column of the reduced
15                              # row echelon form
16          print('Rank(A) = Rank([A b]) = '+str(r1) +' = #Col(A).')
17          print('There is a unique solution, given by: ') ; ...
                print(x) ;
18          #displaying the solution of the linear system
19      else:                           # rank(A) = rank([A b]) < n
20          print('Rank(A) = Rank([A b]) = ' +str(r1) +' < ' ...
                +str(n) +' = #Col(A).')
21          print('Infinite number of solutions.\n')
```

By executing the code, the following results are shown:

```
Rank(A) = Rank([A b]) = 3 = #Col(A).
There is a unique solution, given by:
Matrix([[2/3], [1], [4/3]])
```

## 1.3   Matrix Factorization Techniques

Matrix factorization means to express a matrix $A$ as a multiplication of two or more matrices, each is called a factor [34, 21]. That is, to write:

$$A = A_1 \cdot A_2 \cdot \ldots \cdot A_n$$

In this section, three important matrix factorization techniques will be discussed; namely, the LU factorization, the QR factorization and the singular value decomposition (SVD). Then, the use of those factorization methods in solving linear systems of equations will be discussed.

Because cases of solving linear systems with upper or lower triangular matrices will be encountered, this section will start by writing MATLAB and Python codes for solving such a linear system.

### 1.3.1   The *LU* Factorization

In this factorization, the matrix $A$ is expressed as a multiplication of two matrices $L$ and $U$, where $L$ is an lower triangular matrix and $U$ is an upper triangular matrix. That is:

$$A = L \cdot U = \begin{pmatrix} l_{11} & 0 & 0 & \cdots & 0 \\ l_{21} & l_{22} & 0 & \cdots & 0 \\ l_{31} & l_{32} & l_{33} & \cdots & 0 \\ \vdots & \vdots & \vdots & \ddots & \vdots \\ l_{n1} & l_{n2} & l_{n3} & \cdots & l_{nn} \end{pmatrix} \cdot \begin{pmatrix} u_{11} & u_{12} & u_{13} & \cdots & u_{1n} \\ 0 & u_{22} & u_{23} & \cdots & u_{2n} \\ 0 & 0 & u_{33} & \cdots & u_{3n} \\ \vdots & \vdots & \vdots & \ddots & \vdots \\ 0 & 0 & 0 & \cdots & u_{nn} \end{pmatrix} \quad (1.5)$$

where $l_{jj} = 1$ for $j = 1, 2, \ldots, n$.

The function lu can be used for finding the $L$ and $U$ factors of matrix $S$. In MATLAB, this can be done as follows:

```
>> A = [4 -1 -1; -1 4 -1; -1 -1 4]
A =
 4     -1    -1
-1      4    -1
-1     -1     4

>> [L, U] = lu(A)
L =
 1.0000         0         0
-0.2500    1.0000         0
-0.2500   -0.3333    1.0000


U =
4.0000    -1.0000    -1.0000
0          3.7500    -1.2500
0          0          3.3333
```

In Python, the function lu is located in the scipy.linalg sub-package and can be used to find the LU factors of matrix $A$.

```
In [1]: import numpy as np, scipy.linalg as lg
In [2]: A = np.array([[4, -1, -1], [-1, 4, -1], [-1, -1, 4]])
In [3]: P, L, U = lg.lu(A)
In [4]: print('L = \n', L, '\nU = \n', U)
L =
[[ 1.          0.          0.        ]
 [-0.25        1.          0.        ]
 [-0.25       -0.33333333  1.        ]]
U =
[[ 4.         -1.         -1.        ]
 [ 0.          3.75       -1.25      ]
 [ 0.          0.          3.33333333]]
```

However, python can compact both the $L$ and $U$ factors of matrix $A$ using the function lu_factor.

```
In [5]: LU = lg.lu_factor(A)
In [6]: print('LU = \n', LU)
LU =
(array([[ 4.          , -1.          , -1.          ],
[-0.25        ,  3.75        , -1.25        ],
[-0.25        , -0.33333333,  3.33333333]]), array([0, 1, 2],
    dtype=int32))
```

Now, the linear system 1.1 becomes:

$$
\begin{pmatrix}
1 & 0 & 0 & \cdots & 0 \\
l_{21} & 1 & 0 & \cdots & 0 \\
l_{31} & l_{32} & 1 & \cdots & 0 \\
\vdots & \vdots & \vdots & \ddots & \vdots \\
l_{n1} & l_{n2} & l_{n3} & \cdots & 1
\end{pmatrix}
\cdot
\begin{pmatrix}
u_{11} & u_{12} & u_{13} & \cdots & u_{1n} \\
0 & u_{22} & u_{23} & \cdots & u_{2n} \\
0 & 0 & u_{33} & \cdots & u_{3n} \\
\vdots & \vdots & \vdots & \ddots & \vdots \\
0 & 0 & 0 & \cdots & u_{nn}
\end{pmatrix}
\begin{pmatrix}
x_1 \\ x_2 \\ x_3 \\ \vdots \\ x_n
\end{pmatrix}
=
\begin{pmatrix}
b_1 \\ b_2 \\ b_3 \\ \vdots \\ b_n
\end{pmatrix}
$$

$$(1.6)$$

The solution of the linear system 1.6 is found in three stages:

1. **First:** we let $y = Ux$, that is

$$
y =
\begin{pmatrix}
u_{11} & u_{12} & u_{13} & \cdots & u_{1n} \\
0 & u_{22} & u_{23} & \cdots & u_{2n} \\
0 & 0 & u_{33} & \cdots & u_{3n} \\
\vdots & \vdots & \vdots & \ddots & \vdots \\
0 & 0 & 0 & \cdots & u_{nn}
\end{pmatrix}
\begin{pmatrix}
x_1 \\ x_2 \\ x_3 \\ \vdots \\ x_n
\end{pmatrix}
$$

Then, solving system 1.6 is equivalent to solving the linear system

$$Ly = b$$

2. **Second:** we solve the system $Ly = b$ using the function SolveLower System.m to find $y$.

3. **Finally:** we solve the linear system $Ux = y$ using the back-substitution method, implemented by the MATLAB function SolveUpperSystem.

**Example 1.1** In this example, the LU-factors will be used to solve the linear system:

$$
\begin{pmatrix}
4 & -1 & -1 \\
-1 & 4 & -1 \\
-1 & -1 & 4
\end{pmatrix}
\begin{pmatrix}
x_1 \\ x_2 \\ x_3
\end{pmatrix}
=
\begin{pmatrix}
2 \\ 2 \\ 2
\end{pmatrix}
$$

In MATLAB, the following commands can be used:

```
>> A = [4 -1 -1; -1 4 -1; -1 -1 4] ;
>> b = [2; 2; 2] ;
>> [L, U] = lu(A)
```

```
L =
1.0000           0           0
-0.2500      1.0000           0
-0.2500     -0.3333      1.0000


U =
4.0000     -1.0000     -1.0000
0       3.7500     -1.2500
0            0       3.3333
>> y = SolveLowerLinearSystem(L, b, 3)
y =
2.0000
2.5000
3.3333
>> x = SolveUpperLinearSystem(U, y, 3)
x =
1.0000
1.0000
1.0000
```

In Python, similar steps can be followed to solve the linear system $Ax = b$ using the $LU$ factors of matrix $A$.

```
In [7]: y = lg.solve(L, b)
In [8]: x = lg.solve(U, y)
In [9]: print('x = \n', x)
x =
[[ 0.5]
 [ 0.5]
 [ 0.5]]
```

Python has the LU solver `lu_solve` located in `scipy.linalg` sub-package. It receives the matrix $LU$ obtained by applying the `lu_solve` function, to return the solution of the given linear system.

```
In [10]: x = lg.lu_solve(LU, b)
In [11]: print(x)
[[ 0.5]
 [ 0.5]
 [ 0.5]]
```

The Python's symbolic package `sympy` can also be used to find the LU factors of a matrix $A$. This can be done as follows:

```
In [10]: import sympy as smp
In [11]: A = smp.Matrix([[4., -1., -1.], [-1., 4., -1.],
    [-1., -1., 4.]])
In [12]: LU = B.LUdecomposition()
```

```
In [13]: LU
Out[13]:
(Matrix([
[    1,                     0, 0],
[-0.25,                     1, 0],
[-0.25, -0.333333333333333, 1]]), Matrix([
[4.0, -1.0,                  -1.0],
[  0, 3.75,                  -1.25],
[  0,    0, 3.33333333333333]]), [])
In [14]: LU[0]
Out[14]:
Matrix([
[    1,                     0, 0],
[-0.25,                     1, 0],
[-0.25, -0.333333333333333, 1]])
```

```
In [15]: LU[1]
Out[15]:
Matrix([
[4.0, -1.0,                  -1.0],
[  0, 3.75,                  -1.25],
[  0,    0, 3.33333333333333]])
```

The symbolic package **sympy** can be also used to solve a linear system, using the LU factors.

```
In [16]: b = [[2.0], [2.0], [2.0]]
In [17]: A.LUSolve(b)
Out[17]:
Matrix([
[1.0],
[1.0],
[1.0]])
```

## 1.3.2   The $QR$ Factorization

In this type of factorization, the matrix $A$ is expressed as a multiplication of two matrices $Q$ and $R$. The matrix $Q$ is orthogonal (its columns constitute an orthonormal set) and the matrix $R$ is an upper triangular.

From the elementary linear algebra, an orthogonal matrix satisfies the following two conditions:

1. $Q^{-1} = Q^T$, and

2. if $Q = [q_1 q_2 \ldots q_n]$, then,

$$(q_i, q_j) = q_i^T \cdot q_j = \begin{cases} 1 & i = j \\ 0 & i \neq j \end{cases}$$

In MATLAB, the function qr can be used for finding the QR factors of a matrix $A$. The command is as follows:

```
>> A = [4 -1 -1; -1 4 -1; -1 -1 4] ;
>> [Q, R] = qr(A)

Q =
-0.9428    -0.1421     0.3015
0.2357    -0.9239     0.3015
0.2357     0.3553     0.9045


R =
-4.2426     1.6499     1.6499
0    -3.9087     2.4873
0        0     3.0151
```

In Python, the function qr located in scipy.linalg can be used to find the QR factors of matrix $A$.

```
In [18]: Q, R = lg.qr(A)
In [19]: print('Q = \n', Q, '\nR =\n', R)
Q =
[[-0.94280904 -0.14213381  0.30151134]
 [ 0.23570226 -0.92386977  0.30151134]
 [ 0.23570226  0.35533453  0.90453403]]
R =
[[-4.24264069  1.64991582  1.64991582]
 [ 0.         -3.9086798   2.48734169]
 [ 0.          0.          3.01511345]]
```

The symbolic Python can also be used to find the QR-factors of matrix $A$

```
In [20]: QR = A.QRdecomposition()
In [21]: QR
Out[21]:
(Matrix([
[ 0.942809041582063,    0.14213381090374, 0.301511344577764],
[-0.235702260395516,   0.923869770874312, 0.301511344577764],
[-0.235702260395516,  -0.355334527259351, 0.904534033733291]]),
Matrix([[4.24264068711928, -1.64991582276861, -1.64991582276861],
[              0,  3.90867979985286, -2.48734169081546],
[              0,                 0,  3.01511344577764]]))
In [22]: QR[0]
Out[22]:
Matrix([
[ 0.942809041582063,    0.14213381090374, 0.301511344577764],
[-0.235702260395516,   0.923869770874312, 0.301511344577764],
```

```
[-0.235702260395516, -0.355334527259351, 0.904534033733291]])

In [23]: QR[1]
Out[23]:
Matrix([
[4.24264068711928, -1.64991582276861, -1.64991582276861],
[               0,  3.90867979985286, -2.48734169081546],
[               0,                 0,  3.01511344577764]])
```

To solve the linear system 1.1, using the QR factorization technique the following steps can be used:

1. Finding the $QR$ factors of matrix $A$ and rewrite the system $A\boldsymbol{x} = \boldsymbol{b}$ as $Q \cdot R\boldsymbol{x} = \boldsymbol{b}$.

2. multiplying the two sides of equation $Q \cdot R\boldsymbol{x} = \boldsymbol{b}$ by $Q^T$, giving:

$$Q^T Q R\boldsymbol{x} = Q^T \boldsymbol{b} \Rightarrow R\boldsymbol{x} = Q^T \boldsymbol{b}$$

3. solving the upper triangular system $R\boldsymbol{x} = \boldsymbol{y}$, where $\boldsymbol{y} = Q^T \boldsymbol{b}$ using the backward substitution method, implemented by the function SolveUpperLinearSystem.m.

**Example 1.2** In this example, the QR factors will be used to solve the linear system:

$$\begin{pmatrix} 4 & -1 & -1 \\ -1 & 4 & -1 \\ -1 & -1 & 4 \end{pmatrix} \begin{pmatrix} x_1 \\ x_2 \\ x_3 \end{pmatrix} = \begin{pmatrix} 2 \\ 2 \\ 2 \end{pmatrix}$$

```
>> A = [4, -1, -1; -1, 4, -1; -1, -1, 4] ;
>> b = [2; 2; 2] ;
>> [Q, R] = qr(A) ;
>> y = Q'*b
y =
-0.9428
-1.4213
3.0151
>> x = SolveUpperLinearSystem(R, y, 3)
x =
1.0000
1.0000
1.0000
```

Python can be used to solve the above linear system, using the QR factors as follows.

```
In [24]: import numpy as np, scipy.linalg as lg
In [25]: A = np.array([[4, -1, -1], [-1, 4, -1], [-1, -1, 4]])
```

```
In [26]: b = np.array([[2.0],[2.0],[2.0]])
In [27]: Q, R = lg.qr(A)
In [28]: y = np.matmul(Q.T, b)
In [29]: x = lg.solve(R, y)
In [30]: print('x = \n', x)
x =
[[ 1.0]
 [ 1.0]
 [ 1.0]]
```

*Another method is to use the symbolic package:*

```
In [31]: import sympy as smp
In [32]: A = smp.Matrix([[4.0, -1.0, -1.0], [-1.0, 4.0, -1.0],
         [-1.0, -1.0, 4.0]])
In [33]: b = smp.Matrix([[2.0], [2.0], [2.0]])
In [34]: x = A.QRsolve(b)
In [35]: x
Out[35]:
Matrix([
[1.0],
[1.0],
[1.0]])
```

### 1.3.3   The Singular Value Decomposition (SVD)

In the svd decomposition, the matrix $A$ is expressed as a multiplication of three matrices $U$, $S$ and $V^T$, that is:

$$A = U \cdot S \cdot V^T,$$

where both $U$ and $V$ are unitary matrices and $S$ is a diagonal matrix.

The columns of matrix $U$ are the eigenvectors of the matrix $AA^T$. The columns of matrix $V$ are the eigenvectors of the matrix $A^T A$. The diagonal elements of the matrix $S$ are the squares of the eigenvalues of matrix $A$.

If $\{u_1, u_n\}$ are the columns of $U$, and $\{v_1, v_2, \ldots, v_n\}$ the columns of $V$, then they satisfy the following conditions:

1. $U^{-1} = U^*$ and $V^{-1} = V^*$. If $U$ and $V$ have real entries, then $U^{-1} = U^T$ and $V^{-1} = V^T$.

2. the columns of $U$ and $V$ are orthonormal sets:

$$(u_i, u_j) = u_i^* \cdot u_j = \begin{cases} 1 & i = j \\ 0 & i \neq j \end{cases}$$

and

$$(v_i, v_j) = v_i^* \cdot v_j = \begin{cases} 1 & i = j \\ 0 & i \neq j \end{cases}$$

In MATLAB, the command svd is used to find the svd components of a matrix $A$.

```
>> A = [4 -1 -1; -1 4 -1; -1 -1 4] ;
>> [U, S, V] = svd(A)


U =
0.0000    -0.8165    -0.5774
-0.7071    0.4082    -0.5774
0.7071    0.4082    -0.5774


S =
5    0    0
0    5    0
0    0    2


V =
0    -0.8165    -0.5774
-0.7071    0.4082    -0.5774
0.7071    0.4082    -0.5774
```

In Python, the function svd is used to find the svd decomposition of matrix $A$.

```
In [36]: U, S, V = lg.svd(A)
In [37]: print('U = \n', U, '\nS = \n', S, '\nV = \n', V)
U =
[[ 2.69618916e-17  -8.16496581e-01  -5.77350269e-01]
 [ -7.07106781e-01   4.08248290e-01  -5.77350269e-01]
 [  7.07106781e-01   4.08248290e-01  -5.77350269e-01]]
S =
[ 5.    5.    2.]
V =
[[ 0.            -0.70710678   0.70710678]
 [-0.81649658    0.40824829   0.40824829]
 [-0.57735027   -0.57735027  -0.57735027]]
```

Now, solving system $Ax = b$ is equivalent to finding the solution of the linear system

$$USV^T x = b, \tag{1.7}$$

hence, multiplying the two sides of Equation (1.7) by $V \cdot S^{-1} U^T$, gives,

$$x = V \cdot S^{-1} U^T b$$

**Example 1.3** The svd will be used to solve the linear system:

$$\begin{pmatrix} 4 & -1 & -1 \\ -1 & 4 & -1 \\ -1 & -1 & 4 \end{pmatrix} \begin{pmatrix} x_1 \\ x_2 \\ x_3 \end{pmatrix} = \begin{pmatrix} 2 \\ 2 \\ 2 \end{pmatrix}$$

```
>> x = V*inv(S)*U'*b
x =
1.0000
1.0000
1.0000
```

In Python, the linear system is solved with the svd components as follows:

```
In [38]: x = lg.solve(V, lg.solve(np.diag(S), lg.solve(U,b)))
In [39]: print('x = \n', x)
x =
[[ 0.5]
 [ 0.5]
 [ 0.5]]
```

# 2

---

# Solving Linear Systems with Iterative and Least Squares Methods

---

## Abstract

The direct methods for solving a linear system $Ax = b$ are to try to find the exact solution of the linear system, by inverting the matrix $A$ directly or indirectly. The iterative methods aim at finding an approximate solution of the linear system by finding a sequence of vectors that is converging to the exact solution. In the case that the linear system does not have a solution, the problem turns into a least squares problem.

This chapter aims to find approximate and least squared solutions of linear systems. It is divided into three sections. In the first section, basic concepts such as error norm and convergence of vector sequences are introduced. Then, in the second section three iterative methods for finding approximate solutions of linear systems of equations are discussed and implemented in MATLAB® and Python. When a linear system does not have a solution, the problem turns into searching for a least squares solution that minimizes the error norm. Examples of least squares problems and best approximations of functions by polynomials are discussed and implemented in MATLAB and Python, in the third section.

---

## 2.1 Mathematical Backgrounds

This section present the concepts of vectors norms and the convergence of sequences in a vector space $\mathbb{R}^n$.

### 2.1.1 Convergent Sequences and Cauchi's Convergence

Let $a$ be any point in $\mathbb{R}$, and $\varepsilon \in \mathbb{R}^+$ be any positive real number. The $\varepsilon$-neighbourhood of $a$ (denoted by $\mathcal{N}(a, \varepsilon)$) is the set of all points $x \in \mathbb{R}$ that lie in the open interval $(a - \varepsilon, a + \varepsilon)$. That is, $\mathcal{N}(a, \varepsilon) = \{x \in \mathbb{R} : |x - a| < \varepsilon\}$ [46].

A sequence of real numbers $\{a_n\}_{n=0}^{\infty} = a_0, a_1, a_2, \ldots$ is said to converge to $a \in \mathbb{R}$, if for any choice of $\varepsilon > 0$, $\mathcal{N}(a, \varepsilon)$ contains an infinite sub-sequence

$\{a_n\}_{n=N}^{\infty} = a_N, a_{N+1}, a_{N+2}, \ldots$ of the sequence $\{a_n\}_{n=0}^{\infty}$. To express the convergence of the sequence $\{a_n\}_{n=0}^{\infty}$ to $a$ mathematically, we write

$$|a_n - a| < \varepsilon, \ \forall n \geq N$$

For example, the sequence

$$\left\{ \frac{n}{n+1} \right\}_{n=0}^{\infty} \to 1 \text{ as } n \to \infty$$

because, if we make any choice for $\varepsilon > 0$, then

$$\left| \frac{n}{n+1} - 1 \right| = \left| \frac{-1}{n+1} \right| = \frac{1}{n+1} < \varepsilon \Rightarrow n > \frac{1}{\varepsilon} - 1 = \frac{1 - \varepsilon}{\varepsilon},$$

and therefore, if we let

$$N = \left\lceil \frac{1-\varepsilon}{\varepsilon} \right\rceil,$$

then for any integer $n \geq N$ we find

$$\left| \frac{n}{n+1} - 1 \right| < \varepsilon.$$

**Definition 2.1 (Convergence of a sequence [46])** *Let $\{a_n\}$ be a real-valued sequence. We say that the sequence $\{a_n\}$ converges to a real number $a$, if for any $\varepsilon > 0$, there exists $N \in \mathbb{N}$ such that:*

$$|a_n - a| < \varepsilon, \ \forall n \geq N.$$

*In this case, we say that $a$ is the limit of sequence $\{a_n\}$, and we write*

$$\lim_{n \to \infty} a_n = a$$

It is worthy to notice that if $\{a_n\}$ is a convergent sequence (to some limit $a$), then,

$$|a_0 - a_1| \geq |a_1 - a_2| \geq \cdots \geq |a_n - a_{n+1}| \geq \ldots$$

and

$$\{|a_n - a_{n+1}|\} \to 0, \text{ as } n \to \infty$$

For example, in the sequence $\left\{ \frac{n}{n+1} \right\}$, the corresponding sequence $\{|a_n - a_{n+1}|\} = \frac{1}{(n+1)(n+2)}$, is

$$\frac{1}{2}, \frac{1}{6}, \frac{1}{12}, \frac{1}{20}, \ldots \to 0 \text{ as } n \to \infty.$$

**Definition 2.2** *A sequence $\{a_n\}$ is Cauchy if for any $\varepsilon > 0$, there exists $N \in \mathbb{N}$, such that*

$$|a_n - a_m| < \varepsilon, \text{for all } n, m \geq N.$$

Because $\mathbb{R}$ is complete, any convergent sequence is Cauchy's sequence and vice-versa.

## 2.1.2 Vector Norm

If $x$ is a vector with components $x_1, x_2, \ldots, x_n$, then its $p^{th}$-norm is defined by:

$$\|x\|_p = \left( \sum_{j=1}^{n} |x_j|^p \right)^{\frac{1}{p}} = (|x_1|^p + |x_2|^p + \ldots + |x_n|^p)^{\frac{1}{p}}$$

It satisfies $\|x\|_1 \geq \|x\|_2 \geq \ldots \geq \|x\|_\infty$. The norm $\|x\|_2$ gives the classical Euclidean distance:

$$\|x\|_2 = \sqrt{x_1^2 + x_2^2 + \ldots + x_n^2}$$

The Python function norm (located in `numpy.linalg` library) receives a vector $x$ and an integer $p$ or $np.inf$, and returns $\|x\|_p$.

```
In [27]: x = np.array([1, -1, 2, -2, 3, -3])
In [28]: from numpy.linalg import norm
In [29]: n1, n2, n3 = norm(x, 1), norm(x, 2), norm(x, np.inf)
In [30]: print(n1, n2, n3)
12.0
5.29150262213
3.0
```

## 2.1.3 Convergent Sequences of Vectors

The distance in a vector space can be defined by using any norm. If we have two vectors $x$ and $y$ both are in $\mathbb{R}^n$, then the distance between $x$ and $y$ can be defined by $\|x - y\|_1, \|x - y\|_2, \ldots$ or $\|x - y\|_\infty$. Usually, $\|x - y\|_2$ or $\|x - y\|_\infty$ are used as values for the distance between two vectors $x$ and $y$.

**Definition 2.3 (Convergence of sequences in vector spaces [4])** *Let* $x^{(0)}, x^{(1)}, \ldots$ *be a sequence of vectors in* $\mathbb{R}^n$. *We say that the sequence* $\{x^{(k)}\}_{k=0}^{\infty}$ *converges to a vector* $x \in \mathbb{R}^n$, *if for any* $\varepsilon > 0$, *there exists* $N \in \mathbb{N}$, *such that*

$$\|x^{(m)} - x\| < \varepsilon, \text{ for all } m \geq N.$$

**Definition 2.4 (Cauchy's Convergence of sequences in vector spaces [45])** *Let* $x^{(0)}, x^{(1)}, \ldots$ *be a sequence of vectors in* $\mathbb{R}^n$. *We say that the sequence* $\{x^{(k)}\}_{k=0}^{\infty}$ *is Cauchy convergent, if for any* $\varepsilon > 0$, *there exists* $N \in \mathbb{N}$, *such that*

$$\|x^{(m)} - x^{(s)}\| < \varepsilon, \text{ for all } m, s \geq N.$$

## 2.2 The Iterative Methods

The iterative methods for solving linear systems of equations, is to start from a given initial guess $x^{(0)}$ and generate a sequence of vectors $x^{(0)}, x^{(1)}, \ldots$ that converge to the solution of the linear system $x = x^*$. The generated sequence stops at some vector $x^{(s)}$ that satisfies:

$$\|x^{(s)} - x^{(s-1)}\| < \varepsilon$$

where $\| \ \|$ is some norm in $\mathbb{R}^n$ and $\varepsilon > 0$ is a given tolerance. Then, the solution of the linear system $x^*$ is approximated by $x^{(s)}$, that is $x^* \approx x^{(s)}$.

Generally, there are two kinds of iterative methods [47]:

1. **stationary iterative methods:** at an iteration $k$ the iterative method computes $x^{(k)}$ from $x^{(k-1)}$ without referring to the previous history. This class of methods includes the Jacobi, Gauss-Seidel and the relaxation methods.

2. **non-stationary iterative methods:** at iteration $k$, the iterative method refers to the whole history $x^{(0)}, x^{(1)}, \ldots, x^{(k-1)}$ for the computation of $x^{(k)}$. This class of methods includes the conjugate gradient and the GMRES subspace methods.

In this section, three stationary iterative methods for solving linear system will be discussed. The methods include the Jacobi, Gauss-Seidel and relaxation methods.

## 2.2.1    The General Idea

Given the linear system $Ax = b$, $\varepsilon > 0$ and an initial point $x^{(0)}$. The goal is to generate a sequence of vectors in $\mathbb{R}^n$

$$x^{(0)}, x^{(1)}, x^{(2)}, \ldots$$

that converges to the solution of the linear system $Ax = b$. That is

$$\lim_{k \to \infty} x^{(k)} = x^*$$

The generation of the sequence stops at some iteration $s$, with

$$\|x^{(s)} - x^{(s-1)}\| < \varepsilon$$

In a stationary iterative method, the matrix $A$ is expressed as a sum of two matrices $S$ and $T$, that is $A = S + T$, where $S$ is an invertible matrix. Then, the linear system $Ax = b$ is replaced by $(S + T)x = b$ or

$$Sx = b - Tx$$

Since $S$ is invertible, we multiply the two sides by $S^{-1}$ and obtain a linear system:

$$x = Bx + c$$

where $B = -S^{-1}T$ and $c = S^{-1}b$.

The solution of the above linear system is a fixed point $x^*$, where

$$x^* = Bx^* + c$$

The fixed point is approached iteratively, starting from the given initial point $x^{(0)}$, by using the iterative relationship:

$$x^{(k+1)} = Bx^{(k)} + c$$

Through the whole of this section, we assume that we write matrix $A$ is a sum of three matrices $L$, $D$ and $U$, where:

$$L = \begin{pmatrix} 0 & 0 & 0 & \cdots & 0 & 0 & 0 \\ a_{2,1} & 0 & 0 & \cdots & 0 & 0 & 0 \\ a_{3,1} & a_{3,2} & 0 & \cdots & 0 & 0 & 0 \\ \vdots & \vdots & \vdots & \ddots & \vdots & \vdots & \vdots \\ a_{n-2,1} & a_{n-2,2} & a_{n-2,3} & \cdots & 0 & 0 & 0 \\ a_{n-1,1} & a_{n-1,2} & a_{n-1,3} & \cdots & a_{n-1,n-2} & 0 & 0 \\ a_{n,1} & a_{n,2} & a_{n,3} & \cdots & a_{n,n-2} & a_{n,n-1} & 0 \end{pmatrix},$$

$$D = \begin{pmatrix} a_{1,1} & 0 & 0 & \cdots & 0 & 0 & 0 \\ 0 & a_{2,2} & 0 & \cdots & 0 & 0 & 0 \\ 0 & 0 & a_{3,3} & \cdots & 0 & 0 & 0 \\ \vdots & \vdots & \vdots & \ddots & \vdots & \vdots & \vdots \\ 0 & 0 & 0 & \cdots & a_{n-2,n-2} & 0 & 0 \\ 0 & 0 & 0 & \cdots & 0 & a_{n-1,n-1} & 0 \\ 0 & 0 & 0 & \cdots & 0 & 0 & a_{n,n} \end{pmatrix}$$

and

$$U = \begin{pmatrix} 0 & a_{1,2} & a_{1,3} & \cdots & a_{1,n-2} & a_{1,n-1} & a_{1,n} \\ 0 & 0 & a2,3 & \cdots & a_{2,n-2} & a_{2,n-1} & a_{2,n} \\ 0 & 0 & 0 & \cdots & a_{3,n-2} & a_{3,n-1} & a_{3,n} \\ \vdots & \vdots & \vdots & \ddots & \vdots & \vdots & \vdots \\ 0 & 0 & 0 & \cdots & 0 & a_{n-2,n-1} & a_{n-2,n} \\ 0 & 0 & 0 & \cdots & 0 & 0 & a_{n-1,n} \\ 0 & 0 & 0 & \cdots & 0 & 0 & 0 \end{pmatrix}$$

In MATLAB, matrices $L, D$ and $U$ can be obtained by using the commands:

```
>> D = diag(diag(A)) ;
>> L = tril(A) - D ;

>> U = triu(A) - D;
```

In Python, the `tril`, `triu` are implemented in the `scipy.linalg` package, and `diag` is implemented in both `scipy` and `numpy`. They can be obtained through the following commands:

```
In [1]: import scipy as sp, numpy as np
In [2]: A = np.array([[4, 1, -1], [-1, 5, 1], [0, -1, 4]])
In [3]: print('L = \n', sp.linalg.tril(A), '\nU = \n',
        sp.linalg.triu(A), ...
'\nD = \n', np.diag(np.diag(A)))
L =
[[ 4  0  0]
 [-1  5  0]
 [ 0 -1  4]]
U =
[[ 4  1 -1]
 [ 0  5  1]
 [ 0  0  4]]
D =
[[4 0 0]
 [0 5 0]
 [0 0 4]]
```

## 2.2.2   The Jacobi Method

Given the linear system of equations:

$$
\begin{aligned}
a_{11}x_1 + a_{12}x_2 + \ldots + a_{1n}x_n &= b_1 \\
a_{21}x_1 + a_{22}x_2 + \ldots + a_{2n}x_n &= b_2 \\
&\vdots \\
a_{i1}x_1 + a_{i2}x_2 + \ldots + a_{in}x_n &= b_i \\
&\vdots \\
a_{n1}x_1 + a_{n2}x_2 + \ldots + a_{nn}x_n &= b_b
\end{aligned}
\tag{2.1}
$$

From the above equation, follows that:

$$
\begin{aligned}
x_1 &= \frac{1}{a_{11}}(b_1 - a_{12}x_2 - \ldots - a_{n1}x_n) \\
x_2 &= \frac{1}{a_{22}}(b_2 - a_{21}x_1 - a_{23}x_3 - \ldots - a_{n1}x_n) \\
&\vdots \\
x_i &= \frac{1}{a_{ii}}(b_i - a_{i1}x_1 - \ldots - a_{ii-1}x_{i-1} - a_{ii+1}x_{i+1} - \ldots - a_{in}x_n) \quad (2.2) \\
&\vdots \\
x_n &= \frac{1}{a_{nn}}(b_n - a_{n1}x_1 - \ldots - a_{nn-1}x_{n-1})
\end{aligned}
$$

The Jacobi method is an iterative method, which starts from an initial guess for the solution $[x_1^{(0)}, x_2^{(0)}, \ldots, x_n^{(0)}]^T$. Then, the solution in iteration $k$ is used to find an approximation for the system solution in iteration $k+1$. This is done as follows:

$$x_1^{(k+1)} = \frac{1}{a_{11}} \left( b_1 - a_{12}x_2^{(k)} - \ldots - a_{n1}x_n^{(k)} \right)$$

$$x_2^{(k+1)} = \frac{1}{a_{22}} \left( b_2 - a_{21}x_1^{(k)} - a_{23}x_3^{(k)} - \ldots - a_{n1}x_n^{(k)} \right)$$

$$\vdots \quad \vdots$$

$$x_i^{(k+1)} = \frac{1}{a_{ii}} \left( b_i - a_{i1}x_1^{(k)} - \ldots - a_{ii-1}x_{i-1}^{(k)} - a_{ii+1}x_{i+1}^{(k)} - \ldots - a_{in}x_n^{(k)} \right) \quad (2.3)$$

$$\vdots \quad \vdots$$

$$x_n^{(k+1)} = \frac{1}{a_{nn}} \left( b_n - a_{n1}x_1^{(k)} - \ldots - a_{nn-1}x_{n-1}^{(k)} \right)$$

Generally, the solution in iteration $x_i^{(k+1)}$ can be written in the form:

$$x_i^{(k+1)} = \frac{1}{a_{ii}} \left( b_i - \sum_{j=1, j \neq i}^{n} a_{ij}x_j^{(k)} \right), \ i = 1, \ldots, n \quad (2.4)$$

The Jacobi iteration stops when,

$$\left\| x_i^{(k+1)} - x_i^{(k)} \right\| < \varepsilon,$$

for some arbitrary $\varepsilon > 0$.

**Example 2.1** Write the first three iterations of the Jacobi method, for the linear system:

$$\begin{pmatrix} 2 & -1 & 1 \\ -2 & 5 & -1 \\ 1 & -2 & 4 \end{pmatrix} \begin{pmatrix} x_1 \\ x_2 \\ x_3 \end{pmatrix} = \begin{pmatrix} -1 \\ 1 \\ 3 \end{pmatrix}$$

starting from the zeros vector

$$\boldsymbol{x}^{(0)} = \begin{pmatrix} 0 \\ 0 \\ 0 \end{pmatrix}$$

**Solution:**

We write:

$$x_1^{(k+1)} = \frac{1}{2} \left( -1 + x_2^{(k)} - x_3^{(k)} \right)$$

$$x_2^{(k+1)} = \frac{1}{5} \left( 1 + 2x_1^{(k)} + x_3^{(k)} \right)$$

$$x_3^{(k+1)} = \frac{1}{4} \left( 3 - x_1^{(k)} + 2x_2^{(k)} \right)$$

1. **First iteration** $k = 0$:

$$x_1^{(1)} = \frac{1}{2}\left(-1 + x_2^{(0)} - x_3^{(0)}\right) = \frac{1}{2}(-1 + 0 - 0) = -\frac{1}{2}$$

$$x_2^{(1)} = \frac{1}{5}\left(1 + 2x_1^{(0)} + x_3^{(0)}\right) = \frac{1}{5}(1 + 2(0) + 0) = \frac{1}{5}$$

$$x_3^{(1)} = \frac{1}{4}\left(3 - x_1^{(0)} + 2x_2^{(0)}\right) = \frac{1}{4}(3 - 0 + 2(0)) = \frac{3}{4}$$

2. **Second iteration** $k = 1$:

$$x_1^{(2)} = \frac{1}{2}\left(-1 + x_2^{(1)} - x_3^{(1)}\right) = \frac{1}{2}\left(-1 + \frac{1}{5} - \frac{3}{4}\right) = -\frac{31}{40}$$

$$x_2^{(2)} = \frac{1}{5}\left(1 + 2x_1^{(1)} + x_3^{(1)}\right) = \frac{1}{5}\left(1 + 2 \cdot \frac{-1}{2} + \frac{3}{4}\right) = \frac{3}{20}$$

$$x_3^{(2)} = \frac{1}{4}\left(3 - x_1^{(1)} + 2x_2^{(1)}\right) = \frac{1}{4}\left(3 - \frac{-1}{2} + 2\frac{1}{5}\right) = \frac{39}{40}$$

3. **Third iteration** $k = 2$:

$$x_1^{(3)} = \frac{1}{2}\left(-1 + x_2^{(2)} - x_3^{(2)}\right) = \frac{1}{2}\left(-1 + \frac{3}{20} - \frac{39}{40}\right) = -\frac{73}{80}$$

$$x_2^{(3)} = \frac{1}{5}\left(1 + 2x_1^{(2)} + x_3^{(2)}\right) = \frac{1}{5}\left(1 + 2 \cdot \frac{-31}{40} - \frac{-73}{80}\right) = \frac{17}{200}$$

$$x_3^{(3)} = \frac{1}{4}\left(3 - x_1^{(2)} + 2x_2^{(2)}\right) = \frac{1}{4}\left(3 - \frac{31}{40} + 2\frac{3}{20}\right) = \frac{163}{160}$$

**Example 2.2** The Jacobi method will be applied for solving the linear system:

$$\begin{pmatrix} -5 & 1 & -2 \\ 1 & 6 & 3 \\ 2 & -1 & -4 \end{pmatrix} \begin{pmatrix} x_1 \\ x_2 \\ x_3 \end{pmatrix} = \begin{pmatrix} 13 \\ 1 \\ -1 \end{pmatrix}$$

```
1   function x = JacobiSolve(A, b, Eps)
2       n = length(b) ;
3       x0 = zeros(3, 1) ;
4       x = ones(size(x0)) ;
5       while norm(x-x0, inf) ≥ Eps
6           x0 = x ;
7           for i = 1 : n
8               x(i) = b(i) ;
9               for j = 1 : n
10                  if j ≠ i
11                      x(i) = x(i) - A(i, j)*x0(j) ;
12                  end
13              end
14              x(i) = x(i) / A(i, i) ;
15      end
```

```
>> A = [-5 1 -2; 1 6 3; 2 -1 -4] ;
>> b = [13; 1; -1] ;
>> JacobiSolveLinSystem(A, b)
-2.0000
1.0000
-1.0000
```

Using Python, the function `JacobiSolve` has the following code:

```
1   def JacobiSolve(A, b, Eps):
2       import numpy as np
3       n = len(b)
4       x0, x = np.zeros((n, 1), 'float'), np.ones((n, 1), 'float')
5       while np.linalg.norm(x-x0, np.inf) >= Eps:
6           x0 = x.copy()
7           for i in range(n):
8               x[i] = b[i]
9               for j in range(n):
10                  if j != i:
11                      x[i] -= A[i][j]*x0[j]
12              x[i] /= A[i][i]
13      return x
```

By calling the function `JacobiSolve` to solve the given linear system, we obtain:

```
In [3]: A = np.array([[-5, 1, -2], [1, 6, 3], [2, -1, -4]])
In [4]: b = np.array([[13], [1], [-1]])
In [5]: x = JacobiSolve(A, b, Eps)
In [6]: print('x = \n', x)
Out[6]:
x =
[[-2.         ],
 [ 1.00000002],
 [-1.00000002]]
```

## 2.2.3 The Jacobi Method in the Matrix Form

Matrix $A$ can be expressed as:

$$A = L + D + U,$$

therefore, the linear system $A\boldsymbol{x} = \boldsymbol{b}$ can be written as:

$$(L + D + U)\boldsymbol{x} = \boldsymbol{b}$$

The Jacobi method chooses $S = D$ and $T = L + U$. It is worthy to notice that no diagonal element in $D$ can be 0. That is $d_{ii} \neq 0$, for all $i = 1, \ldots, n$. The Jacobi method is of the form:

$$\boldsymbol{x}^{(n+1)} = B\boldsymbol{x}^{(n)} + \boldsymbol{c},$$

where, $B = -D^{-1}(L + U)$ and $\boldsymbol{c} = D^{-1}\boldsymbol{b}$.

The following MATLAB code implements the Jacobi method in the matrix form:

```
1   function [x, Iters] = JacobiIter_VectorForm(A, b, Eps, x0)
2       D = diag(diag(A)) ;
3       B = -D\(A-D) ;
4       c = D\b ;
5       Iters = 1 ;
6       x1 = B*x0 + c ;
7       while norm(x1-x0, inf) >= Eps
8           x0 = x1 ;
9           x1 = B*x0 + c ;
10          Iters = Iters + 1 ;
11      end
12      x = x1 ;
```

Applying the above code to the linear system in Example 2.2:

```
>> A = [-5 1 -2; 1 6 3; 2 -1 -4] ;
>> b = [13; 1; -1] ;
>> [x, Iters] = JacobiIter_VectorForm(A, b, Eps, x0)
x =
-2.0000
1.0000
-1.0000
Iters =
26
```

The Python code for the function `JacobiIter_VectorForm` is:

```
1   def JacobiIter_VectorForm(A, b, x0, Eps):
2       import numpy as np, scipy as sp
3       D = np.diag(np.diag(A))
4       B = -sp.linalg.solve(D, A-D)
5       c =  sp.linalg.solve(D, b)
6       Iters = 1
7       x = np.matmul(B, x0)+c
8       while np.linalg.norm(x-x0) >= Eps:
9           x0 = x.copy()
10          x = np.matmul(B, x0)+c
11          Iters += 1
12      return x, Iters
```

From the IPython console, the following commands can be used to solve the linear system:

```
In [7]: x, Iterations = JacobiSolve_VectorForm(A, b, x0, Eps)
In [8]: print('Iterations = ', Iterations, '\nx = \n', x)
Iterations =  24
x =
[[ -2.        ]
```

```
[ 1.00000002]
[-1.00000002]]
```

### 2.2.3.1  The Gauss-Seidel Iterative Method

The Gauss-Seidel method is close to the Jacobi method, but it benefits from the recently updated variables $x_1^{(k+1)}, \ldots, x_{i-1}^{(k+1)}$ in iteration $k+1$ to update $x_i^{(k+1)}$, whereas the other variables $x_{i+1}^{(k)}, \ldots, x_n^{(k)}$ are taken from the previous iteration $k$. The Gauss-Seidel is of the form:

$$
\begin{aligned}
a_{1,1}x_1^{(k+1)} &= b_1 - a_{12}x_2^{(k)} - \ldots - a_{n1}x_n^{(k)} \\
a_{2,1}x_1^{(k+1)} + a_{2,2}x_2^{(k+1)} &= b_2 - a_{23}x_3^{(k)} - \ldots - a_{n1}x_n^{(k)} \\
&\vdots \\
a_{i,1}x_1^{(k+1)} + \ldots + a_{i,i}x_i^{(k+1)} &= b_i - a_{i,i+1}x_{i+1}^{(k)} - \ldots - a_{i,n}x_n^{(k)} \quad (2.5) \\
&\vdots \\
a_{n,1}x_1^{(k+1)} + \ldots + a_{i,i}x_i^{(k+1)} &= b_n
\end{aligned}
$$

From equation (2.5),

$$
\begin{aligned}
x_1^{(k+1)} &= \frac{1}{a_{11}}\left(b_1 - a_{12}x_2^{(k)} - \ldots - a_{n1}x_n^{(k)}\right) \\
x_2^{(k+1)} &= \frac{1}{a_{22}}\left(b_2 - a_{21}x_1^{(k+1)} - a_{23}x_3^{(k)} - \ldots - a_{n1}x_n^{(k)}\right) \\
&\vdots \\
x_i^{(k+1)} &= \frac{1}{a_{ii}}\left(b_i - a_{i1}x_1^{(k+1)} - \ldots - a_{ii-1}x_{i-1}^{(k+1)} - a_{ii+1}x_{i+1}^{(k)} \right. \quad (2.6) \\
&\qquad\qquad \left. - \ldots - a_{in}x_n^{(k)}\right) \\
&\vdots \\
x_n^{(k+1)} &= \frac{1}{a_{nn}}\left(b_n - a_{n1}x_1^{(k+1)} - \ldots - a_{nn-1}x_{n-1}^{(k+1)}\right)
\end{aligned}
$$

The following MATLAB function implements the Gauss-Seidel method in the equation form:

```
1  function [x, Iters] = GaussSeidel(A, b, Eps, x0)
2      Iters = 1 ;
3      x1 = zeros(size(x0)) ; n = length(b) ;
4      for i = 1 : n
5          x1(i) = b(i) ;
6          for j = 1 : i-1
7              x1(i) = x1(i) -A(i,j)*x1(j) ;
8          end
9          for j = i+1 : n
```

```
10              x1(i) = x1(i) - A(i, j)*x0(j) ;
11          end
12          x1(i) = x1(i)/A(i,i) ;
13      end
14      while norm(x1-x0, inf) ≥ Eps
15          x0 = x1 ;
16          for i = 1 : n
17              x1(i) = b(i) ;
18              for j = 1 : i-1
19                  x1(i) = x1(i)-A(i,j)*x1(j) ;
20              end
21              for j = i+1 : n
22                  x1(i) = x1(i) - A(i, j)*x0(j) ;
23              end
24              x1(i) = x1(i)/A(i,i) ;
25          end
26          Iters = Iters + 1 ;
27      end
28      x = x1 ;
```

Applying the above code to the linear systems in examples 2.1 and 2.2:

```
>> Eps = 1e-8 ; x0 = [0;0;0] ;
>> A = [2 -1 1; -2 5 -1; 1 -2 4] ; b = [-1;1;3] ;
>> [x, Iters] = GaussSeidel(A, b, Eps, x0)
x =
-1.0000
0
1.0000
Iters =
11
>> A = [-5 1 -2; 1 6 3; 2 -1 -4] ; b = [13; 1; -1] ;
>> [x, Iters] = GaussSeidel(A, b, Eps, x0)
x =
-2.0000
1.0000
-1.0000
Iters =
15
```

The python code for the function GaussSeidel is:

```
1  import numpy as np
2  def GaussSeidelSolve(A, b, x0, Eps):
3      n = len(b)
4      x = np.ones((n, 1), 'float')
5      Iterations = 1
6      while np.linalg.norm(x-x0, np.inf) ≥ Eps:
7          x0 = x.copy()
8          for i in range(n):
9              x[i] = b[i]
```

```
10              for j in range(i):
11                  x[i] -= A[i][j]*x[j]
12                  for j in range(i+1, n):
13                      x[i] -= A[i][j]*x0[j]
14                  x[i] /= A[i][i]
15          Iterations += 1
16      return x, Iterations
17
18  A = np.array([[-5, 1, -2], [1, 6, 3], [2, -1, -4]])
19  b = np.array([[13], [1], [-1]])
20  x0 = np.zeros((3, 1), 'float')
21  Eps = 1e-8
22  x, Iterations = GaussSeidelSolve(A, b, x0, Eps)
23  print('Iterations = ', Iterations, '\nx = \n', x)
```

By executing the above code we obtain:

```
Iterations =  16
x =
[[-2.]
 [ 1.]
 [-1.]]
```

## 2.2.4  The Gauss-Seidel Method in the Vector Form

The Gauss-Seidel method chooses $S = L + D$ and $T = U$. The Jacobi method is of the form:

$$x^{(n+1)} = Bx^{(n)} + c,$$

where, $B = -(L+D)^{-1}U$ and $c = (L+D)^{-1}b$.

The following MATLAB code implements the Gauss-Seidel method in the matrix form:

```
1  function [x, Iters] = GaussSeidelIter(A, b, Eps, x0)
2  LD = tril(A) ; U = A - LD ;
3  B = -LD\U ; c = LD\b ;
4  Iters = 1 ;
5  x1 = B*x0 + c ;
6  while norm(x1-x0, inf) ≥ Eps
7      x0 = x1 ;
8      x1 = B * x0 + c ;
9      Iters = Iters + 1 ;
10 end
11 x = x1 ;
```

Applying the above code to examples 2.1 and 2.2:

```
>> Eps = 1e-8 ; x0 = [0;0;0] ;
>> A = [2 -1 1; -2 5 -1; 1 -2 4] ; b = [-1;1;3] ;
>> [x, Iters] = GaussSeidelIter(B, b, Eps, x0)
x =
```

```
-1.0000
0
1.0000
Iters =
11
>> A = [-5 1 -2; 1 6 3; 2 -1 -4] ; b = [13; 1; -1] ;
>> [x, Iters] = GaussSeidelIter(A, b, Eps, x0)
x =
-2.0000
1.0000
-1.0000
Iters =
15
```

The function `GaussSeidelIter` can be implemented in python as:

```
 1  def GaussSeidelIter(A, b, Eps, x0):
 2      import numpy as np, scipy as sp
 3      LD = sp.linalg.tril(A)
 4      U = A - LD
 5      B = -sp.linalg.solve(LD, U)
 6      c =  sp.linalg.solve(LD, b)
 7      Iters = 1
 8      x = np.matmul(B, x0) + c
 9      while np.linalg.norm(x-x0, np.inf) >= Eps:
10          x0 = x.copy()
11          x = np.matmul(B, x0) + c
12          Iters += 1
13      return x, Iters
14
15  A = np.array([[-5, 1, -2], [1, 6, 3], [2, -1, -4]])
16  b = np.array([[13], [1], [-1]])
17  x0 = np.zeros((3, 1), 'float')
18  Eps = 1e-8
19  x, Iterations = GaussSeidelIter(A, b, x0, Eps)
20  print('Iterations = ', Iterations, '\n x = \n', x)
```

The result after executing the above code is:

```
Iterations =   15
x =
[[-2.]
 [ 1.]
 [-1.]]
```

## 2.2.5    The Relaxation Methods

A relaxation method is an iterative method for solving linear systems based on Gauss-Seidel iteration and involves a parameter $0 \leq \omega \leq 2$ with the purpose of accelerating the convergence rate of the iterative method to the solution of the linear system.

Multiplying the linear system $\boldsymbol{Ax} = \boldsymbol{b}$ by $w$ and replacing $\boldsymbol{A}$ by $\boldsymbol{U} + \boldsymbol{D} + \boldsymbol{L}$ gives

$$(\omega\boldsymbol{U} + \omega\boldsymbol{D} + \omega\boldsymbol{L})\boldsymbol{x} = \omega\boldsymbol{b} \Rightarrow (\omega\boldsymbol{U} + (1 - (1 - \omega))\boldsymbol{D} + \omega\boldsymbol{L})\boldsymbol{x} = \omega\boldsymbol{b}$$

from which,

$$(\boldsymbol{D} + \omega\boldsymbol{L})\boldsymbol{x} = ((1 - \omega)\boldsymbol{D} - \omega\boldsymbol{U})\boldsymbol{x} + \omega\boldsymbol{b}$$

The iterative method is obtained by replacing $\boldsymbol{x}$ at the left-hand side by $\boldsymbol{x}^{(k+1)}$ and by $\boldsymbol{x}^{(k)}$ at the right-hand side, giving the formula

$$(\boldsymbol{D} - \omega\boldsymbol{L})\boldsymbol{x}^{(k+1)} = ((1 - \omega)\boldsymbol{D} - \omega\boldsymbol{U})\boldsymbol{x}^{(k)} + \omega\boldsymbol{b} \qquad (2.7)$$

When $\omega < 1$, the iterative method in (2.7) is `under relaxation method`. When $\omega = 1$ the method is identical to Gauss-Seidel iteration, and when $\omega > 1$ (2.7) is `over-relaxation method`. In the latest case ($\omega > 1$), the resulting method is called the `successive over-relaxation method` and is abbreviated as `SOR`.

By writting $\boldsymbol{S}_\omega = \boldsymbol{D} - \omega\boldsymbol{L}$, $\boldsymbol{T}_\omega = (1 - \omega)\boldsymbol{D} + \omega\boldsymbol{U}$, $\boldsymbol{B}_\omega = \boldsymbol{S}_\omega^{-1}\boldsymbol{T}_\omega$ and $\boldsymbol{c}_\omega = \boldsymbol{S}_\omega^{-1}\omega\boldsymbol{b}$, the SOR iterative method is:

$$\boldsymbol{x}^{(k+1)} = \boldsymbol{B}_\omega\boldsymbol{x}^{(k)} + \boldsymbol{c}_\omega \qquad (2.8)$$

The MATLAB function `SOR.m` receives a matrix $\boldsymbol{A}$, a vector $\boldsymbol{b}$, an initial starting vector $\boldsymbol{x}_0$, a real value $\omega$ and a tolerance $\varepsilon$, and returns an approximate solution of the system $\boldsymbol{Ax} = \boldsymbol{b}$ within the given tolerance together with the number of iterations.

```
1  function [x, Iters] = SOR(A, b, w, x0, Eps)
2      D = diag(diag(A)) ;
3      L = tril(A) - D ; U = A - (L+D) ;
4      B = -(D+w*L)\((w-1)*D+w*U) ; c = (D+w*L)\(w*b) ;
5      Iters = 1 ;
6      x = B*x0 + c ;
7      while norm(x-x0, inf) >= Eps
8          x0 = x ;
9          x = B * x0 + c ;
10         Iters = Iters + 1 ;
11     end
12 end
```

Applying the above code to examples 2.1 and 2.2:

```
>> A = [2 -1 1; -2 5 -1; 1 -2 4] ; b = [-1;1;3] ;
>> x0 = [1.; 1.; 1.]; w = 1.25; Eps = 1e-8 ;
>> [x, Iters] = SOR(A, b, w, x0, Eps)

x =
-1.0000
```

```
0.0000
1.0000
>> A = [-5 1 -2; 1 6 3; 2 -1 -4] ; b = [13; 1; -1] ;
>> [x, Iters] = SOR(A, b, w, x0, Eps)


x =
-2.0000
1.0000
-1.0000


Iters =
33
```

The Python code `SORIter.py` applies the SOR to solve the linear systems in examples 2.1 and 2.2.

```
1   import numpy as np
2   from scipy.linalg import tril, solve
3   def SOR(A, b, w, x0, Eps):
4       D = np.diag(np.diag(A))
5       L = tril(A) - D
6       U = A - (L+D)
7       B = solve(-(D+w*L), (w-1)*D+w*U)
8       c = solve((D+w*L), w*b)
9       Iters = 1
10      x = np.matmul(B, x0) + c
11      while np.linalg.norm(x-x0, np.inf) >= Eps:
12          x0 = x.copy()
13          x = np.matmul(B, x0) + c
14          Iters += 1
15      return x, Iters
16
17  print('Solving the first linear system:')
18  A = np.array([[2, -1, 1], [-2, 5, -1], [1, -2, 4]])
19  b = np.array([[-1], [1], [3]])
20  w, x0 = 1.25, np.zeros((3, 1), 'float')
21  Eps = 1e-8
22  x, Iterations = SOR(A, b, w, x0, Eps)
23  print( 'x = \n', x, '\nIterations = ', Iterations)
24
25  print('Solving the second linear system:')
26  A = np.array([[-5, 1, -2], [1, 6, 3], [2, -1, -4]])
27  b = np.array([[13], [1], [-1]])
28  x, Iterations = SOR(A, b, w, x0, Eps)
29  print( 'x = \n', x, '\nIterations = ', Iterations)
```

Executing the above code gives the following results:

```
runfile('D:/PyFiles/SORIter.py', wdir='D:/PyFiles')
Solving the first linear system:
x =
[[-1.00000000e+00]
```

```
[-1.94732286e-10]
[ 1.00000000e+00]]
Iterations =  16
Solving the second linear system:
x =
[[-2.]
[ 1.]
[-1.]]
Iterations =  32
```

The Gauss Seidel method solved Example 2.2 in 16 iterations, and the SOR method with $\omega = 1.25$ solved the example in 33 iterations. From this example, it is clear that the SOR is not guaranteed to converge faster than Gauss-Seidel method if the parameter $\omega$ is not selected carefully. The selection of the relaxation parameter $\omega$ plays a key role in the convergence rate of the SOR. Hence when the optimal value of the parameter is selected, the SOR method achieves the best convergence rate.

In the following example, 15 values in $[0.1, 1.5]$ are selected for parameter $w$ and the corresponding numbers of iterations are computed. Then the values of $w$ are plotted against the corresponding numbers of iterations.

```
1  W = np.arange(0.1, 1.6, 0.1)
2  Iterations = np.zeros_like(W)
3  for j in range(len(W)):
4      x, Iters = SOR(A, b, W[j], x0, Eps)
5      Iterations[j] = Iters
6  import matplotlib.pyplot as plt
7  plt.figure(1)
8  plt.plot(W, Iterations, marker='s', color='purple', lw=3)
9  plt.xticks(np.arange(0.1, 1.6, 0.2), fontweight = 'bold')
10 plt.yticks(np.arange(25, 225, 25), fontweight = 'bold')
11 plt.grid(True, ls=':')
12 plt.xlabel('w', fontweight='bold')
13 plt.ylabel('Iterations', fontweight='bold')
```

Executing this code gives Figure 2.1. From figure 2.1 the optimal value of parameter $\omega$ lies some where between 0.8 and 1.1. Hence, zooming more in this interval, the optimal value of the parameter $\omega$ lies somewhere between 0.92 and 0.98 as shown in Figure 2.2.

## 2.3  The Least Squares Solutions

If $A \in \mathbb{R}^{m \times n}$ and $b \in \mathbb{R}^m$, the linear system $Ax = b$ will have a solution $x \in \mathbb{R}^n$ if and only if $b$ is in the column space of matrix $A$ ($b \in col(A)$) [30]. In the

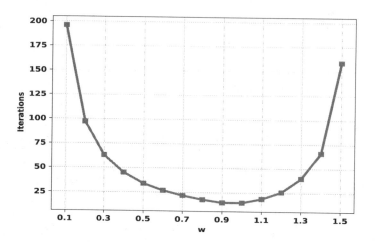

FIGURE 2.1: Different values of $\omega$ in $[0.1, 1.5]$ VS the corresponding numbers of iterations.

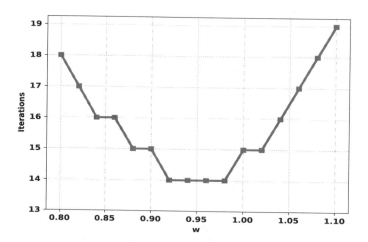

FIGURE 2.2: Plot of the data (w, Iters) for $\omega$ in $[0.8, 1.1]$ VS the corresponding numbers of iterations.

case that $b$ does not lie in $col(A)$, the problem of solving $Ax = b$ becomes an approximation problem, in which we look for some $\hat{x} \in \mathbb{R}^n$ such that

$$\|b - A\hat{x}\| = \min_{x \in \mathbb{R}^n} \{\|b - Ax\|\},$$

or in other words:

$$\text{find } \hat{\boldsymbol{x}} \; : \; \|\boldsymbol{b} - A\hat{\boldsymbol{x}}\| = \min_{\boldsymbol{x} \in \mathbb{R}^n} \{\|\boldsymbol{b} - A\boldsymbol{x}\|\}. \tag{2.9}$$

There are three interesting cases of 2.9 [47]:

1. the **minimax** problem, where the minimization is over the infinity-norm:

$$\text{find } \hat{\boldsymbol{x}} \; : \; \|\boldsymbol{b} - A\hat{\boldsymbol{x}}\|_\infty = \min_{\boldsymbol{x} \in \mathbb{R}^n} \{\|\boldsymbol{b} - A\boldsymbol{x}\|_\infty\} = \min_{\boldsymbol{x} \in \mathbb{R}^n} \{\max_{i=1,\dots,m} \{|b_i - (A\boldsymbol{x})_i|\}\},$$

2. the **absolute deviation** problem, where the minimization is over $\|\cdot\|_1$:

$$\text{find } \hat{\boldsymbol{x}} : \|\boldsymbol{b} - A\hat{\boldsymbol{x}}\|_1 = \min_{\boldsymbol{x} \in \mathbb{R}^n} \sum_{j=1}^m |b_j - (A\boldsymbol{x})_j|,$$

3. the **least squares** problem, where the minimization is under the classical Euclidean distance in $\mathbb{R}^m$:

$$\text{find } \hat{\boldsymbol{x}} : \|\boldsymbol{b} - A\hat{\boldsymbol{x}}\|_2 = \min_{\boldsymbol{x} \in \mathbb{R}^n} \left( \sum_{j=1}^m |b_j - (A\boldsymbol{x})_j|^2 \right)^{\frac{1}{2}}.$$

Given an over-determined linear system:

$$A\boldsymbol{x} = \boldsymbol{b},$$

where $A \in \mathbb{R}^{m \times n}$, $\boldsymbol{x} \in \mathbb{R}^n$, $\boldsymbol{b} \in \mathbb{R}^m$ and $m > n$, the aim is to find a least squares solution of the least squares problem:

$$find \; \hat{\boldsymbol{x}} : \|\boldsymbol{b} - A\hat{\boldsymbol{x}}\|_2 = \min_{\boldsymbol{x} \in \mathbb{R}^n} \left( \sum_{j=1}^m |b_j - (A\boldsymbol{x})_j|^2 \right)^{\frac{1}{2}}. \tag{2.10}$$

If $\hat{\boldsymbol{x}} \in \mathbb{R}^n$ is a least squares solution of the linear system $A\boldsymbol{x} = \boldsymbol{b}$, then it satisfies:

$$\|\boldsymbol{b} - A\hat{\boldsymbol{x}}\| \le \|\boldsymbol{b} - A\boldsymbol{x}\|, \; \forall \boldsymbol{x} \in \mathbb{R}^n \tag{2.11}$$

From the elementary linear algebra, $\hat{b}b = A\hat{\boldsymbol{x}}$ is the orthogonal projection of $\boldsymbol{b}$ in $Col(A)$ ($\boldsymbol{b} - \hat{b}b = \boldsymbol{b} - A\hat{\boldsymbol{x}} \perp Col(A)$). That is $\boldsymbol{b} - A\hat{\boldsymbol{x}}$ is orthogonal to each column of $A$, hence, if $\boldsymbol{a}_j$ is the $j^{th}$-column of matrix $A$, then, $\boldsymbol{a}_j \cdot (\boldsymbol{b} - A\hat{x}b) = 0$, $j = 1, \dots, n$. As a result, we get:

$$A^T(\boldsymbol{b} - A\hat{\boldsymbol{x}}) = \boldsymbol{0},$$

or

$$A^T A\hat{\boldsymbol{x}} = A^T \boldsymbol{b} \tag{2.12}$$

Equation (2.12) is called the *normal equations*, and the least squares solution $\hat{\boldsymbol{x}}$ is equivalent to solve the normal equations.

**Example 2.3** Find the least squares solution of the inconsistent linear system $Ax = b$, where:

$$A = \begin{bmatrix} 1 & 2 & 4 \\ 3 & 1 & 5 \\ 1 & 1 & 1 \\ 2 & 2 & 1 \\ 3 & 1 & 3 \end{bmatrix} \text{ and } b = \begin{bmatrix} 5 \\ 3 \\ 3 \\ 1 \\ 4 \end{bmatrix}$$

**Solution:** The normal equations are given by $A^T A \hat{x} = A^T b$, where,

$$A^T A = \begin{bmatrix} 24 & 13 & 31 \\ 13 & 11 & 19 \\ 31 & 19 & 52 \end{bmatrix} \text{ and } A^T b = \begin{bmatrix} 31 \\ 22 \\ 51 \end{bmatrix}$$

The least squares solution $\hat{x}$ is obtained by solving the normal equations. In MATLAB:

```
>> A = [1 2 4; 3 1 5; 1 1 1; 2 2 1; 3 1 3]
A =
1    2    4
3    1    5
1    1    1
2    2    1
3    1    3
>> b = [5;3;3;1;4]
b =
5
3
3
1
4
>> xh = A'*A\(A'*b)
xh =
-0.1639
0.8970
0.7507
```

In Python, the solution of the least squares problem is obtained by using the following Python commands:

```
In  [9]: import numpy as np
In [10]: A = np.array([[1, 2, 4], [3, 1, 5], [1, 1, 1],
         [2, 2, 1], [3, 1, 3]])
In [11]: b = np.array([[5],[3],[3],[1],[4]])
In [12]: xh = np.linalg.solve(A.T@A, A.T@b)
In [13]: print(xh)
[[-0.16388616]
```

```
[ 0.8969578 ]
[ 0.75073602]]
```

**Remark 2.1** The inconsistent linear system $Ax = b$ has a unique least-squares solution, if $A^T A$ is a full-rank matrix. That is $A^T A$ is nonsingular [30].

**Remark 2.2** If $\hat{x}$ is a least-squares solution of the inconsistent linear system $Ax = b$, then $\|b - A\hat{x}\|_2$ defines the *least squares error* [53].

For example, the least squares error in the above example is obtained by using the following MATLAB command:

```
>> LSError = norm(b-A*xh, 2)
LSError =   2.6570
```

In Python, the least squares error is obtained by using the command:

```
In [14]: LSError = np.linalg.norm(b-A@xh, 2)
In [15]: print(LSError)
2.6570401973629143
```

### 2.3.1   Some Applications of Least Squares Solutions

In this section, Three applications of the least squares solutions will be discussed, namely, fitting linear models to data, polynomials to data and approximating a function by a polynomial.

i **Fitting a linear model to data:** The problem is described as follows: Given a set of data points

| $x_1$ | $x_2$ | $\cdots$ | $x_n$ |
|-------|-------|----------|-------|
| $y_1$ | $y_2$ | $\cdots$ | $y_n$ |

find $\hat{\alpha}$ and $\hat{\beta}$ in $\mathbb{R}$, such that the linear model $y = \hat{\alpha} + \hat{\beta}x$ gives the best fit of the linear model to the data $(x_i, y_i)$, $i = 1, \ldots, n$. This problem is equivalent to the problem:

$$\text{Find } \hat{\alpha}, \hat{\beta} \in \mathbb{R} : \sum_{j=1}^{n} \left( y_j - \left( \hat{\alpha} + \hat{\beta}x_j \right) \right)^2 = \min_{\alpha, \beta \in \mathbb{R}} \left\{ \sum_{j=1}^{n} (y_j - (\alpha + \beta x_j))^2 \right\}.$$

The parameters $\hat{\alpha}$ and $\hat{\beta}$ are called the `regression coefficients`. To find them, we substitute the data points $(x_j, y_j)$, $j = 1 \ldots, n$; in the linear model $y = \alpha + \beta x$. Then, we get:

$$
\begin{aligned}
y_1 &= \alpha + \beta x_1 \\
y_2 &= \alpha + \beta x_2 \\
&\vdots \qquad \vdots \\
y_n &= \alpha + \beta x_n
\end{aligned}
\tag{2.13}
$$

Equations (2.13) can be written in the matrix form:

$$A\boldsymbol{x} = \boldsymbol{y} \tag{2.14}$$

where,

$$A = \begin{bmatrix} 1 & x_1 \\ 1 & x_2 \\ \vdots & \vdots \\ 1 & x_n \end{bmatrix}, \; \boldsymbol{x} = \begin{bmatrix} \alpha \\ \beta \end{bmatrix} \text{ and } \boldsymbol{y} = \begin{bmatrix} y_1 \\ y_2 \\ \vdots \\ y_n \end{bmatrix}$$

The regression coefficients $\hat{\alpha}$ and $\hat{\beta}$ can be obtained from the normal equations

$$A^T A \hat{\boldsymbol{x}} = A^T \hat{\boldsymbol{y}},$$

where,

$$A^T A = \begin{bmatrix} 1 & 1 & \cdots & 1 \\ x_1 & x_2 & \cdots & x_n \end{bmatrix} \begin{bmatrix} 1 & x_1 \\ 1 & x_2 \\ \vdots & \vdots \\ 1 & x_n \end{bmatrix} = \begin{bmatrix} n & \sum_{j=1}^{n} x_j \\ \sum_{j=1}^{n} x_j & \sum_{j=1}^{n} x_j^2 \end{bmatrix}$$

and

$$A^T \boldsymbol{y} = \begin{bmatrix} 1 & 1 & \cdots & 1 \\ x_1 & x_2 & \cdots & x_n \end{bmatrix} \begin{bmatrix} y_1 \\ y_2 \\ \vdots \\ y_n \end{bmatrix} = \begin{bmatrix} \sum_{j=1}^{n} y_j \\ \sum_{j=1}^{n} x_j y_j \end{bmatrix}$$

Now,

$$\left(A^T A\right)^{-1} = \frac{1}{det(A^T A)} adj(A^T A) = \frac{1}{n \sum_{j=1}^{n} x_j^2 - \left(\sum_{j=1}^{n} x_j\right)^2}$$

$$\times \begin{bmatrix} \sum_{j=1}^{n} x_j^2 & -\sum_{j=1}^{n} x_j \\ -\sum_{j=1}^{n} x_j & n \end{bmatrix}$$

The regression coefficients $\hat{\alpha}$ and $\hat{\beta}$ can be obtained from the solution:

$$\hat{\boldsymbol{x}} = \left(A^T A\right)^{-1} A^T \boldsymbol{y},$$

as follows:

$$\hat{\alpha} = \frac{\sum_{j=1}^{n} x_j^2 \sum_{j=1}^{n} y_j - \sum_{j=1}^{n} x_j \sum_{j=1}^{n} x_j y_j}{n \sum_{j=1}^{n} x_j^2 - \left(\sum_{j=1}^{n} x_j\right)^2} \tag{2.15}$$

$$\hat{\beta} = \frac{n \sum_{j=1}^{n} x_j y_j - \sum_{j=1}^{n} x_j \sum_{j=1}^{n} y_j}{n \sum_{j=1}^{n} x_j^2 - \left(\sum_{j=1}^{n} x_j\right)^2} \tag{2.16}$$

We write a MATLAB function `LinearRegCoefs` that receives two vectors x and y and returns the regression coefficients `ahat` and `bhat`:

```
1    function [ahat, bhat] = LinearRegCoefs(x, y)
2    n = length(x) ;
3    ahat = (sum(x.^2)*sum(y)-sum(x)*sum(x.*y))/(n*sum(x.^2)
4        -sum(x)^2) ;
5    bhat = ...
         (n*sum(x.*y)-sum(x)*sum(y))/(n*sum(x.^2)-sum(x)^2) ;
```

Now, we test the function `LinearRegCoefs`, using data points of heights (in meters) and weights (in kilograms) taken for nine students as in Table 2.1.

TABLE 2.1: Heights (in centimeters) vs. weights (in kilograms) for nine male students in secondary school

| Height | 1.65 | 1.67 | 1.68 | 1.72 | 1.77 | 1.82 | 1.86 | 1.89 | 1.90 |
|--------|------|------|------|------|------|------|------|------|------|
| Weight | 57.0 | 61.0 | 64.0 | 69.0 | 75.0 | 83.0 | 90.0 | 97.0 | 100.0 |

```
1    H = [1.65, 1.67, 1.68, 1.72, 1.77, 1.82, 1.86, 1.89, 1.90] ;
2    W = [57.0, 61.0, 64.0, 69.0, 75.0, 83.0, 90.0, 97.0, 100.0] ;
3    [ahat, bhat] = LinearRegCoefs(H, W) ;
4    WW = ahat + bhat * H ;
5    plot(H, W, 'mo', H, WW, 'b-', 'LineWidth', 2, ...
         'MarkerFaceColor', 'm')
6    axis([1.6, 1.95, 50, 110])
7    xlabel('Height (in meters)', 'fontweight','bold')
8    ylabel('Weight (in kilograms)', 'fontweight','bold')
9    grid on
10   set(gca,'GridLineStyle', '--') ;
11   set(gca, 'fontweight', 'bold') ;
```

In Python, the code to compute the regression coefficients, plotting the linear model and scattering the data is as follows:

```
1    import numpy as np
2    import matplotlib.pylab as plt
3    def LinearRegCoefs(x, y):
4        n = len(x)
5        ahat = ...
             (sum(x*x)*sum(y)-sum(x)*sum(x*y))/(n*sum(x*x)-sum(x)**2)
6        bhat = (n*sum(x*y)-sum(x)*sum(y))/(n*sum(x*x)-sum(x)**2)
7        return ahat, bhat
8
9    H = np.array([1.65, 1.67, 1.68, 1.72, 1.77, 1.82, 1.86, ...
         1.89, 1.90])
10   W = np.array([57.0, 61.0, 64.0, 69.0, 75.0, 83.0, 90.0, ...
         97.0, 100.0])
11   ahat, bhat = LinearRegCoefs(H, W) ;
12   WW = ahat + bhat * H ;
```

```
13  plt.plot(H, W, 'mo', lw=2)
14  plt.plot(H, WW, 'b-', lw=2)
15  plt.axis([1.6, 1.95, 50, 110])
16  plt.xlabel('Height (in meters)', fontweight='bold')
17  plt.ylabel('Weight (in kilograms)', fontweight='bold')
18  plt.grid(True, ls='--')
```

Executing the above Python code, will give the a similar graph as in Figure 2.3.

ii **Fitting a polynomial to data:** Given a table of data points:

| $x_1$ | $x_2$ | ... | $x_n$ |
|-------|-------|-----|-------|
| $y_1$ | $y_2$ | ... | $y_n$ |

it is possible to fit a polynomial model with degree not exceeding $n-1$ to the data. Supposing that the relationship between the variable $y$ and the variable $x$ is of the form:

$$y = \alpha_0 + \alpha_1 x + \alpha_2 x^2 + \cdots + \alpha_k x^k, \ 1 \leq k \leq n-1, \qquad (2.17)$$

where the regression coefficients $\hat{\alpha}_0, \ldots, \hat{\alpha}_k$ are to be computed, such that model 2.17 is a least-square solution. We substitute the data points $(x_j, y_j)$, $j = 1, \ldots, n$ in 2.17 to give $n$ equations in the $k+1$ unknowns $\alpha_0, \ldots, \alpha_k$:

$$
\begin{aligned}
y_1 &= \alpha_0 + \alpha_1 x_1 + \alpha_2 x_1^2 + \cdots + \alpha_k x_1^k \\
y_2 &= \alpha_0 + \alpha_1 x_2 + \alpha_2 x_2^2 + \cdots + \alpha_k x_2^k \\
&\vdots \\
y_n &= \alpha_0 + \alpha_1 x_n + \alpha_2 x_n^2 + \cdots + \alpha_k x_n^k
\end{aligned}
$$

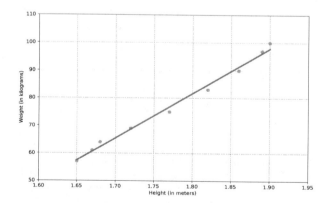

FIGURE 2.3: The linear model vs. the scattered data of heights and weights.

Equations (2.18) can be written in the matrix form:

$$
\begin{bmatrix} y_1 \\ y_2 \\ \vdots \\ y_n \end{bmatrix} = \begin{bmatrix} 1 & x_1 & x_1^2 & \cdots & x_1^k \\ 1 & x_2 & x_2^2 & \cdots & x_2^k \\ \vdots & \vdots & \vdots & \ddots & \vdots \\ 1 & x_n & x_n^2 & \cdots & x_n^k \end{bmatrix} \begin{bmatrix} \alpha_0 \\ \alpha_1 \\ \vdots \\ \alpha_k \end{bmatrix} \tag{2.18}
$$

The regression coefficients $\hat{\alpha}_0, \ldots, \hat{\alpha}_k$ can be found by solving the normal equations:

$$
\begin{bmatrix} 1 & 1 & \cdots & 1 \\ x_1 & x_1^2 & \cdots & x_1^k \\ \vdots & \vdots & \ddots & \vdots \\ x_n & x_n^2 & \cdots & x_n^k \end{bmatrix} \begin{bmatrix} 1 & x_1 & x_1^2 & \cdots & x_1^k \\ 1 & x_2 & x_2^2 & \cdots & x_2^k \\ \vdots & \vdots & \vdots & \ddots & \vdots \\ 1 & x_n & x_n^2 & \cdots & x_n^k \end{bmatrix} \begin{bmatrix} \alpha_0 \\ \alpha_1 \\ \vdots \\ \alpha_k \end{bmatrix} = \begin{bmatrix} 1 & 1 & \cdots & 1 \\ x_1 & x_1^2 & \cdots & x_1^k \\ \vdots & \vdots & \ddots & \vdots \\ x_n & x_n^2 & \cdots & x_n^k \end{bmatrix} \begin{bmatrix} y_1 \\ y_2 \\ \vdots \\ y_n \end{bmatrix}
$$
$$\tag{2.19}$$

**Example 2.4** In this example, we write MATLAB and Python codes to fit a model $y = \alpha_0 + \alpha_1 x + \cdots + \alpha_k x^k$, for $k = 2, 4, 7$ and $8$, to random data from the normal distribution, as in Table 2.2:

TABLE 2.2: Virtual data of class centers vs. frequencies

| x | 5 | 15 | 25 | 35 | 45 | 55 | 65 | 75 | 85 | 95 |
|---|---|----|----|----|----|----|----|----|----|----|
| y | 2 | 6 | 10 | 16 | 25 | 21 | 10 | 5 | 3 | 2 |

The MATLAB code is:

```
1  clear ; clc ; clf ;
2  x = [5    15    25    35    45    55    65    75    85    95] ;
3  y = [2     6    10    16    25    21    10     5     3     2] ;
4  xx = 1 : 100 ;
5  deg = [2 4 7 9] ;
6  Model = zeros(4, length(x)) ;
7  Mdl = zeros(4, length(xx)) ;
8  for l = 1 : 4
9      k = deg(l) ;
10     A = zeros(length(x), k+1) ;
11     for j = 1 : k+1
12         A(:, j) = x(:).^(j-1) ;
13     end
14     Q = (A'*A)\(A'*y(:)) ;% polyfit(x, y, k) ;
15     for j = 1 : k+1
16         Model(l, :) = Model(l, :) + Q(j).*x.^(j-1) ;
17     end
18     Mdl(l, :) = pchip(x, Model(l, :), xx) ;
19     FittingError(l) = norm(y-Model(l, :), 2) ;
20 end
21 plot(x, y, 'ro', xx, Mdl(1, :), ':b', xx, Mdl(2, :), '--r', ...
       xx, Mdl(3, :), '-.m', xx, Mdl(4, :), '-k', ...
       'MarkerFaceColor', 'y', 'MarkerSize', 8, 'LineWidth', ...
       2) ;
```

```
22  legend('Data', ['k = ' num2str(deg(1))], ['k = ' ...
        num2str(deg(2))], ['k = ' num2str(deg(3))], ['k = ' ...
        num2str(deg(4))], 'fontweight','bold') ;
23  xlabel('x', 'fontweight','bold') ;
24  ylabel('y', 'fontweight','bold') ;
25  grid on ;
26  ax = gca ;
27  ax.FontWeight = 'bold' ;
28  ax.FontSize = 14 ;
29  set(gca, 'GridLineStyle','--') ;
30  axis([0, 100, 0, 30]) ;
```

The above code uses the piece-wise cubic Hermite interpolating polynomials (**pchip**) to approximate the solution by a smooth curve, instead of the linear interpolation in which the solution curve between two points is approximated by a straight line. By executing the above code, we obtain the Figure 2.4:

The Python code to fit polynomials of degrees 2, 4, 7 and 9 to the tabular data and plot the least-square curves is:

```
1  import numpy as np
2  from scipy.interpolate import pchip
3  import matplotlib.pyplot as plt
4  x = np.array([5.,    15.,    25.,    35.,    45.,    55.,    65., ...
        75.,    85.,    95.])
5  y = np.array([2.,     6.,    10.,    16.,    25.,    21.,    10., ...
        5.,     3.,     2.])
6  xx = np.arange(1., 101.)
7  deg = np.array([2, 4, 7, 9])
8  Model = np.zeros((4, len(x)), float) ;
```

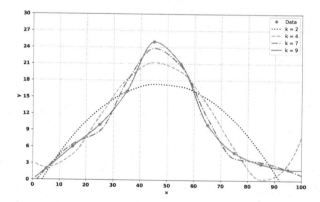

FIGURE 2.4: Fitting polynomials of degrees 2, 4, 7 and 9 to the data in Table 2.2.

```
 9  Mdl = np.zeros((4, len(xx)), float) ;
10  for l in range(4):
11      k = deg[l]
12      A = np.zeros((len(x), k+1), float) ;
13      for j in range(k+1):
14          A[:, j] = x[:]**(j) ;
15      Q = np.linalg.solve(A.T@A,A.T@y[:]) ;# polyfit(x, y, k) ;
16      for j in range(k+1):
17          Model[l, :] = Model[l, :] + Q[j]*x**(j) ;
18      Mdl[l, :] = pchip(x, Model[l, :])(xx) ;
19  plt.plot(x, y, 'ro', mfc = 'y', lw = 3, label = 'Data')
20  plt.plot(xx, Mdl[0, :], ':b', label = 'k = 2', lw = 2)
21  plt.plot(xx, Mdl[1, :], '--r', label = 'k = 4', lw = 2)
22  plt.plot(xx, Mdl[2, :], '-.m', label = 'k = 7', lw = 2)
23  plt.plot(xx, Mdl[3, :], '-k', label = 'k = 9', lw = 2)
24  plt.legend() ;
25  plt.xlabel('x', fontweight = 'bold') ;
26  plt.ylabel('y', fontweight = 'bold') ;
27  plt.grid(True, ls = '--')
28  plt.axis([0, 100, 0, 30]) ;
29  plt.show()
```

The variable `FittingError` contains the least-squares errors in fitting the
polynomials of degrees 2, 4, 7 and 9 to the tabulated data in Table 2.2.
Those least-squares errors are listed in Table 2.3

TABLE 2.3: Least squares errors corresponding to fitting with polynomials of
degrees 2, 4, 7 and 9

| Degree | 2 | 4 | 7 | 9 |
|--------|---|---|---|---|
| LS Error | $1.29467 \times 10^1$ | 6.89477 | 3.01662 | $1.50627 \times 10^{-3}$ |

Both MATLAB and Python have a function `polyfit` that receives two vectors $x = (x_1,\ldots,x_n), y = (y_1,\ldots,y_n)$ and a positive integer $k$ and returns
the optimal parameters $\hat{\alpha}_1,\ldots,\hat{\alpha}_{k+1}$. In MATLAB, the code lines:

```
1  A = zeros(length(x), k+1) ;
2  for j = 1 : k+1
3  A(:, j) = x(:).^(j-1) ;
4  end
5  Q = (A'*A)\(A'*y(:)) ;
```

can be replaced by:

```
1  Q = polyfit(x, y, k) ;
```

In Python, the code lines:

```
1  A = np.zeros((len(x), k+1), float) ;
2  for j in range(k+1):
3  A[:, j] = x[:]**(j) ;
4  Q = np.linalg.solve(A.T@A,A.T@y[:]) ;
```

can be replaced by:

```
1  Q = np.polyfit(x, y, k)
```

Therefore, we have the following remarks:

(a) As the degree of the fitting polynomial increases, the least-squares error decreases.

(b) if given $n$ data points, the degree of the fitting polynomial cannot exceed $n-1$, otherwise, the matrix of the normal equations will be singular.

(c) if given $n$ data points, and the fitting polynomial is of degree $n-1$, the matrix of coefficients is called the `Vandermonde matrix`.

(d) The Vandermonde matrix is nonsingular, but as the number of data points $n$ increases, it gets closer to be singular.

iii **Least-squares Approximations of functions:**
Given a function $f(x)$ that is continuous in an interval $[a,b] \subset \mathbb{R}$. If $P_n(x) = \alpha_n x^n + \alpha_{n-1} x^{n-1} + \cdots + \alpha_0$, $\alpha_0, \ldots, \alpha_n \in \mathbb{R}$ is a polynomial of degree $n$ that approximates $f(x)$ in $[a,b]$, then the approximation error $E(\alpha_0, \alpha_1, \ldots, \alpha_n)$ is defined by:

$$E(\alpha_0, \alpha_1, \ldots, \alpha_n) = \|f(x) - P_n(x)\|_2^2 = \int_a^b \left( f(x) - \sum_{j=0}^n \alpha_j x^j \right)^2 dx \quad (2.20)$$

The problem is to find the optimal coefficients $\hat{\alpha}_0, \hat{\alpha}_1, \ldots, \hat{\alpha}_n$ such that the $n^{th}$-degree polynomial $P_n(x) = \sum_{j=0}^n \hat{\alpha}_j x^j$ is a least squares solution of the approximation problem *i.e.*:

$$E(\hat{\alpha}_0, \hat{\alpha}_1, \ldots, \hat{\alpha}_n) = \int_a^b \left( f(x) - \sum_{j=0}^n \hat{\alpha}_j x^j \right)^2 dx$$

$$= \min_{\alpha_0, \ldots, \alpha_n \in \mathbb{R}} \{ E(\alpha_0, \alpha_1, \ldots, \alpha_n) \} \quad (2.21)$$

At the optimal parameters $\hat{\alpha}_0, \ldots, \hat{\alpha}_n$, we find that

$$\left[ \frac{\partial E}{\partial \alpha_k} \right]_{\alpha_k = \hat{\alpha}_k} = 0, \ k = 0, \ldots, n \quad (2.22)$$

From Equation (2.20),

$$
\left[\frac{\partial E}{\partial \alpha_k}\right]_{\alpha_k = \hat{\alpha}_k} = \left[\frac{\partial}{\partial \alpha_k}\right]_{\alpha_k = \hat{\alpha}_k} \int_a^b \left(f(x) - \sum_{j=0}^n \alpha_j x^j\right)^2 dx
$$

$$
= \int_a^b \left[2\left(f(x) - \sum_{j=0}^n \hat{\alpha}_j x^j\right) x^k\right] dx
$$

$$
= 2\int_a^b x^k f(x)dx - 2\sum_{j=0}^n \hat{\alpha}_j \int_a^b x^{j+k}\, dx
$$

$$
= 2\int_a^b x^k f(x)dx - 2\sum_{j=0}^n \hat{\alpha}_j \frac{b^{j+k+1} - a^{j+k+1}}{j+k+1} = 0 \quad (2.23)
$$

From Equation (2.23) we obtain the normal equations:

$$
\begin{bmatrix}
b-a & \frac{b^2-a^2}{2} & \frac{b^3-a^3}{3} & \cdots & \frac{b^{n+1}-a^{n+1}}{n+1} \\
\frac{b^2-a^2}{2} & \frac{b^3-a^3}{3} & \frac{b^4-a^4}{4} & \cdots & \frac{b^{n+2}-a^{n+2}}{n+2} \\
\vdots & \vdots & \vdots & \ddots & \vdots \\
\frac{b^{n+1}-a^{n+1}}{n+1} & \frac{b^{n+2}-a^{n+2}}{n+2} & \frac{b^{n+3}-a^{n+3}}{n+3} & \cdots & \frac{b^{2n+1}-a^{2n+1}}{2n+1}
\end{bmatrix}
\begin{bmatrix}
\hat{\alpha}_0 \\ \hat{\alpha}_1 \\ \hat{\alpha}_2 \\ \vdots \\ \hat{\alpha}_n
\end{bmatrix}
=
\begin{bmatrix}
\int_a^b f(x)dx \\ \int_a^b x f(x)dx \\ \int_a^b x^2 f(x)dx \\ \vdots \\ \int_a^b x^n f(x)dx
\end{bmatrix}
$$

$$(2.24)$$

The optimal coefficients $\hat{\alpha}_0, \ldots, \hat{\alpha}_n$ are obtained by solving the normal equations (2.24).

We write a function **FunApproxCoef** that receives a function **f**, limits of interval **a** and **b** and the degree of the least-squares approximating polynomial **n**. The function returns a vector **alph** whose components are the optimal coefficients $\hat{\alpha}_0, \ldots, \hat{\alpha}_n$.

The MATLAB code of the function **FunApproxCoef**:

```
1   function alph = FunApproxCoef(f, a, b, n)
2       A = zeros(n+1) ;
3       y = zeros(n+1, 1) ;
4       j = 0 ;
5       while j ≤ n
6           k = 0 ;
7           while k ≤ j
8               A(j+1, k+1) = (b^(j+k+1)-a^(j+k+1))/(j+k+1) ;
9               A(k+1, j+1) = A(j+1, k+1) ;
10              k = k + 1 ;
11          end
12          j = j + 1 ;
13      end
```

```
14      for k = 1 : n+1
15          y(k) = integral(@(x) x.^(k-1).*f(x), a, b) ;
16      end
17      alp = A\y ;
18      alph = zeros(1, length(alp)) ;
19      for j = 1 : length(alp)
20          alph(j) = alp(n-j+2) ;
21      end
```

The Python code of the function `FunApproxCoef`:

```
1   import numpy as np
2   from scipy import integrate
3   import matplotlib.pyplot as plt
4
5   def FunApproxCoef(f, a, b, n):
6       A = np.zeros((n+1, n+1), float)
7       y = np.zeros((n+1, 1), float)
8       j = 0
9       while j <= n:
10          k = 0 ;
11          while k <= j:
12              A[j, k] = (b**(j+k+1)-a**(j+k+1))/(j+k+1)
13              A[k, j] = A[j, k]
14              k += 1
15          j += 1
16      for k in range(n+1):
17          y[k] = integrate.quad(lambda x: x**k*f(x), a, b)[0]
18      alp = np.linalg.solve(A,y)
19      alph = list(alp)
20      alph.reverse()
21      alph = np.array(alph)
22      return alph
```

To test the function `FunApproxCoef`, we find approximations to the exponential function $f(x) = e^x$ in the interval $[0,3]$, using polynomials of degrees 1,2 and 4. The MATLAB code to do this is:

```
1   a = 0; b = 1 ;
2   x = linspace(0, 3) ;
3   f = @(x) exp(x) ;
4   P = zeros(4, length(x)) ;
5   for l = 1 : 4
6       n = l ;
7       alph = FunApproxCoef(f, a, b, n) ;
8       P(l, :) = polyval(alph, x) ;
9   end
10  plot(t, f(t), '-b', t, P(1,:), '--r', t, P(2, :), '-.m', t, ...
        P(4, :), ':k', 'LineWidth', 2) ;
11  xlabel('x', 'fontweight','bold') ;
12  ylabel('y', 'fontweight','bold') ;
13  legend('y = exp(x)', 'n = 1', 'n = 2', 'n = 4') ;
14  grid on ;
```

```
15   ax = gca ;
16   ax.FontWeight = 'bold' ;
17   ax.FontSize = 14 ;
```

The Python code is:

```
1    import matplotlib.pyplot as plt
2    a, b = 0, 1
3    t = np.linspace(0, 3, 301)
4    f = lambda t: np.exp(t)
5    P = np.zeros((4, len(t)), float)
6    for l in range(4):
7        n = l+1
8        alph = FunApproxCoef(f, a, b, n)
9        P[l, :] = np.polyval(alph, t)
10   plt.plot(t, f(t), '-b', lw = 2, label='y=exp(x)')
11   plt.plot(t, P[0, :], '--r', lw = 2, label = 'n = 1')
12   plt.plot(t, P[1, :], '-.m', lw = 2, label= 'n= 2')
13   plt.plot(t, P[3, :], ':k', lw = 2, label = 'n = 4')
14   plt.xlabel('x', fontweight = 'bold')
15   plt.ylabel('y', fontweight= 'bold')
16   plt.legend()
17   plt.grid(True, ls = '--')
```

The exponential function $f(x) = e^x$ is approximated by polynomials of degrees $1, 2$ and $4$, and the approximating polynomials are shown in Figure 2.5.

**Example 2.5** In this example, we find approximations of the function $f(x) = \cos(\pi x)$ in the interval $[0,1]$, using polynomials of degrees $n = 1, 3, 5, \ldots, 15$. We show the errors $\| \cos(\pi x) - P_n(x)$ for $x \in [0,1]$ and plot the approximate functions $P_1(x)$ and $P_3(x)$.

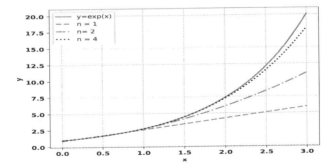

FIGURE 2.5: Approximation of the exponential function $f(x) = e^x$ with polynomials of degrees 1, 2 and 4.

The Python code is:

```
1   a, b = 0, 1
2   t = np.linspace(a, b, 101)
3   f = lambda t: np.cos(np.pi*t)
4   P = np.zeros((7, len(t)), float)
5   print('n \t ||f(t)-P_n(t)||_2)\n')
6   print('----------------------------------\n')
7   for l in range(7):
8       n = 2*l+1
9       alph = FunApproxCoef(f, a, b, n)
10      P[l, :] = np.polyval(alph, t)
11      print(n, '\t', ...
               '{:7.4e}'.format(np.linalg.norm(f(t)-P2[l, :]), '\n'))
12  plt.plot(t, f(t), '-m', lw=2, label='cos(pi*t)')
13  plt.plot(t, P[0, :], '--r', lw = 2, label = 'n = 1')
14  plt.plot(t, P[2, :], ':b', lw = 4, label = 'n = 3')
15  plt.xlabel('x', fontweight = 'bold')
16  plt.ylabel('y', fontweight= 'bold')
17  plt.xticks(np.arange(0.0, 1.1, 0.1))
18  plt.legend()
19  plt.grid(True, ls = '--')
```

The result of executing the code is:

| n  | \|\|f(t)-P_n(t)\|\|_2 |
|----|------------------------|
| 1  | 1.4890e+00             |
| 3  | 4.8910e-02             |
| 5  | 7.4470e-04             |
| 7  | 6.5437e-06             |
| 9  | 3.7348e-08             |
| 11 | 7.2770e-08             |
| 13 | 1.8264e-07             |
| 15 | 3.8105e-08             |

The approximate polynomials $P_1(x)$ and $P_3(x)$ are explained in Figure 2.6, where $P_1(x)$ is plotted with a dashed line and $P_3(x)$ is plotted with a dotted line. The function $\cos(\pi x)$ is plotted with a solid line.

The MATLAB code for approximating $f(x) = \cos(\pi x)$ with polynomial of degrees $n = 1, \ldots, 13$ is:

```
1   clear ; clc ; clf ;
2   a = 0; b = 1 ;
3   t = linspace(a, b) ;
4   f = @(t)cos(pi*t) ;
5   P = zeros(8, length(t)) ;
6   l = 1 ;
7   fprintf('n \t||f(t)-P_n(t)||_2\n') ; ...
        fprintf('--------------------\n') ;
```

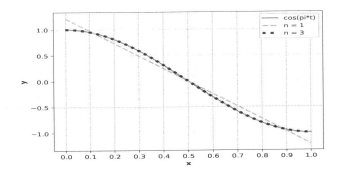

FIGURE 2.6: Approximate functions of $f(x) = \cos(\pi x)$, $x \in [0,1]$ with polynomials of degrees 1 and 3

```
8   for k = 1 : 8
9       n = 2*k-1 ;
10      alph = FunApproxCoef(f, a, b, n) ;
11      P(k, :) = polyval(alph, t) ;
12      fprintf('%i\t\t%7.4e\n', n, norm(f(t)-P(k, :), 2))
13  end
14  plot(t, f(t), '-m', t, P(1, :), '--r', t, P(2, :), ':b', ...
        'LineWidth', 2) ;
15  xlabel('x', 'fontweight','bold') ;
16  ylabel('y', 'fontweight','bold') ;
17  legend('y = cos(pi*x)', 'n = 1', 'n = 3') ;
18  grid on ;
19  ax = gca ;
20  ax.FontWeight = 'bold' ;
21  ax.FontSize = 14 ;
```

We notice that the approximate solution improves as we increase the degree of the polynomial from $n = 1$ to $n = 9$, After that the approximate solution does not improve. In fact, when we run the MATLAB code, we see warning messages connected to $n = 11, 13$ and $15$.

```
n          ||f(t)-P_n(t)||_2
---------------------------
1               8.7431e-01
3               2.9178e-02
5               4.5233e-04
7               4.0581e-06
9               2.3696e-08
11              3.0460e-08
> In FunApproxCoef (line 17)
In PlotApproxcos (line 10)
Warning: Matrix is close to singular or badly scaled.
Results may be inaccurate. RCOND =  3.190019e-19.
```

```
13                      1.6090e-07
> In FunApproxCoef (line 17)
In PlotApproxcos (line 10)
Warning: Matrix is close to singular or badly scaled.
Results may be inaccurate. RCOND =  5.742337e-20.
15                      3.1356e-07
```

The reason behind these error messages will be discussed in the next chapter.


**Remark 2.3** When approximating a function $f(x)$ by a polynomial $P_n(x)$ of degree $n$ in the interval $[0,1]$, the resulting coefficient matrix is called Hilbert matrix. The Hilbert matrix of type $n \times n$ $(H_n)$ is given by:

$$H_n = \begin{bmatrix} 1 & \frac{1}{2} & \frac{1}{3} & \cdots & \frac{1}{n} \\ \frac{1}{2} & \frac{1}{3} & \frac{1}{4} & \cdots & \frac{1}{n+1} \\ \frac{1}{3} & \frac{1}{4} & \frac{1}{5} & \cdots & \frac{1}{n+2} \\ \vdots & \vdots & \vdots & \ddots & \vdots \\ \frac{1}{n} & \frac{1}{n+1} & \frac{1}{n+2} & \cdots & \frac{1}{2n+1} \end{bmatrix} \tag{2.25}$$

MATLAB has a function hilb, which recieves an integer $n$ and return the corresponding Hiblbert matrix $H_n$.

```
>> hilb(3)
ans =
1.0000    0.5000    0.3333
0.5000    0.3333    0.2500
0.3333    0.2500    0.2000

>> hilb(5)
ans =
1.0000    0.5000    0.3333    0.2500    0.2000
0.5000    0.3333    0.2500    0.2000    0.1667
0.3333    0.2500    0.2000    0.1667    0.1429
0.2500    0.2000    0.1667    0.1429    0.1250
0.2000    0.1667    0.1429    0.1250    0.1111
```

# 3

## Ill-Conditioning and Regularization Techniques in Solutions of Linear Systems

### Abstract

If a small perturbation is introduced to either the coefficient matrix $A$ or vector $b$, it might lead to a big change in the solution vector $x$. Hence, both the direct and iterative methods are not guaranteed to give accurate solution of the given linear system.

This chapter is divided into two sections. The first section presents the concept of ill-conditioning in linear systems and how to use MATLAB® and Python to measure the condition numbers of matrices. In the second section, some regularization techniques are presented to stabilize the solutions of ill-conditioned systems.

## 3.1 Ill-Conditioning in Solutions of Linear Systems

An ill-conditioned linear system is a linear system which responds to a small perturbation on the coefficient matrix or the vector at the right-hand side with a large change in the system solution [28, 42]. To see this, two kinds of little perturbations will be presented to two examples that will be considered. In the first example a slight change on one component of the coefficient matrix will be introduced and in the second example the perturbation will be made on the vector at the right-hand side.

**Example 3.1** In this example we consider the linear system $Fx = d$, with

$$A = \begin{pmatrix} 5.0 & 1.0 & 1.0 \\ 1.0 & 5.0 & 1.0 \\ 1.0 & 1.0 & 5.0 \end{pmatrix} \text{ and } b = \begin{pmatrix} 5.0 \\ -3.0 \\ 5.0 \end{pmatrix}$$

To solve this linear system in MATLAB, we use the following commands:

```
>> A = [5., 1., 1.; 1., 5., 1.; 1., 1., 5.]
A =
```

```
5      1      1
1      5      1
1      1      5

>> b = [5.; -3.; 5.]
b =
     5
    -3
     5

>> x = A\b
x =
    1.0000
   -1.0000
    1.0000
```

In Python:

```
In [1]: import numpy as np
In [2]: A = np.array([[5., 1., 1.], [1., 5., 1.], [1., 1., 5.]])
In [3]: b = np.array([[5.], [-3.], [5.]])
In [4]: x = np.linalg.solve(A, b)
In [5]: print('x = \n', x)
x =
[[ 1.]
[-1.]
[ 1.]]
```

Now, the first perturbation we consider will be on the component $A_{11}$, where we set $B = A$ and $B_{11} = A_{11} + 10^{-4} = A11 + 0.0001$

```
>> B = A
B =
5      1      1
1      5      1
1      1      5
>> B(1, 1) = B(1, 1) + 0.0001
B =
5.0001    1.0000    1.0000
1.0000    5.0000    1.0000
1.0000    1.0000    5.0000
>> y = B\b
y =
1.0000
-1.0000
1.0000
>> disp('x - y = '), disp(num2str(x-y))
```

```
x - y =
2.1428e-05
-3.5714e-06
-3.5714e-06
>> disp(['|| x-y||_2 = ' num2str(norm(x-y, 2))])
|| x-y||_2 = 2.2015e-05
```

In Python,

```
In [7]: B = A.copy()
In [8]: B[0, 0] += 1e-4
In [9]: y = np.linalg.solve(B, b)
In [10]: print('y = \n', y)
y =
[[ 0.99997857]
[-0.99999643]
[ 1.00000357]]
In [11]: print('x - y = \n', x-y)
x - y =
[[ 2.14281123e-05]
[-3.57135204e-06]
[-3.57135204e-06]]
In [12]: print('||x-y||_2 = \n', np.linalg.norm(x-y, 2))
||x-y||_2 =
2.201529253996403e-05
```

The second perturbation under consideration will be on the right-hand side, where $b_3$ will be replaced by $b_3 + 10^{-4}$ giving a vector $c$ in the right-hand side. The original coefficient matrix $A$ will remain without a change and the linear system $Az = c$ will be solved.

```
>> c = b ;
>> c(3) = c(3) + 1e-4

c =

5.0000
-3.0000
5.0001

>> z = A\c

z =

1.0000
-1.0000
1.0000
```

```
>> disp('x - z = '), disp(num2str(x-z))
x - z =
3.5714e-06
3.5714e-06
-2.1429e-05
>> disp(['|| x-z||_2 = ' num2str(norm(x-z, 2))])
|| x-z||_2 = 2.2016e-05
```

In Python,

```
In [12]: c = b.copy()
In [13]: c[-1] += 1e-4
In [14]: z = np.linalg.solve(A, c)
In [15]: print('z = \n', z)
z =
[[ 0.99999643]
[-1.00000357]
[ 1.00002143]]
In [16]: print('x - z = \n', x-z)
x - z =
[[ 3.57142857e-06]
[ 3.57142857e-06]
[-2.14285714e-05]]
In [17]: print('||x-z||_2 = \n', np.linalg.norm(x-z, 2))
||x-z||_2 =
2.2015764296112095e-05
```

From Example 3.1 it can be noticed that small changes in some components of the coefficient matrix or the vector at the right-hand side lead to small changes in the solution. Hence, the given linear system is not sensitive to small perturbations. A system that is not sensitive to small perturbations is called `well-posed system`.

**Example 3.2** In this example a linear system $F\boldsymbol{x} = \boldsymbol{d}$ is considered, where

$$F = \begin{pmatrix} 1001 & -999 & 999 \\ 1 & 1 & -1 \\ 1000 & -1000 & 1000 \end{pmatrix} \text{ and } \boldsymbol{d} = \begin{pmatrix} 1001 \\ 1 \\ 1000 \end{pmatrix}$$

The purpose from this example is to show how a small change in one entry of the coefficient matrix $F$ or vector $\boldsymbol{d}$ can cause a drastic change in the solution of the linear system $F\boldsymbol{x} = \boldsymbol{d}$.

MATLAB is used to solve the linear system $A\boldsymbol{x} = \boldsymbol{b}$, with the commands:

```
>> F = [1001, -999, 999; 1, 1, -1; 1000, -1000, 1000] ;
>> d = [1001; 1; 1000] ;
>> x = F\d
```

```
Warning: Matrix is close to singular or badly scaled. Results
   may be inaccurate.
RCOND =  1.067985e-16.


x =
1
0.4147
0.4147
```

The warning indicates that the matrix is close to being singular, hence the results might be inaccurate. The MATLAB solution of the linear system is far from the exact solution $[1.0, 1.0, 1.0]^T$ as noticed.

In Python, the above linear system can be solved by using the Python commands:

```
In [1]: import numpy as np
In [2]: A = np.array([
[1.001e+03, -9.990e+02,  9.990e+02],
[1.000e+00,  1.000e+00, -1.000e+00],
[1.000e+03, -1.000e+03,  1.000e+03]])
In [3]: b = np.array([[1001], [1], [1000]])
In [4]: x = np.linalg.solve(A, b)
In [5]: print('x =\n', x)
x =
[[1.        ]
[0.93969727]
[0.93969727]]
```

Again, despite Python having better accuracy in finding a solution of the linear system than MATLAB, still the error in solving the linear system is not small.

Again, small perturbations will be introduced to the coefficient matrix $F$ and vector $d$. First, a perturbation $10^{-5}$ will be added to $F_{11}$ giving a matrix $G$. Then, computing the solution of the linear system $Gy = d$ and $\|y - x\|_2$:

```
>> G = F ;
>> format long g
>> G(1, 1) = G(1, 1) + 1e-5
G =
1001.00001                     -999                    999
0.9999999999999     1.0000000000001    -0.9999999999999
1000                  -1000                    1000
>> y = G\d
Warning: Matrix is close to singular or badly scaled. Results
   may be inaccurate.
RCOND =  1.092847e-16.
```

```
y =
0.999997714598233
22853063.4012923
22853063.4012946
>> disp(norm(y-x, 2))
32319111.6174026
```

In Python, the commands will be used:

```
In [6]: G = F.copy()
In [7]: G[0, 0] += 1e-5
In [8]: y = np.linalg.solv(G, d)
In [9]: print('y = ', y)
y =
[[9.99997654e-01]
[2.34012849e+07]
[2.34012849e+07]]
In [10]: print('y-x = \n', y-x)
y-x =
[[-2.34606334e-06]
[ 2.34012840e+07]
[ 2.34012840e+07]]
In [11]: print("%20.16e"% np.linalg.norm(x-y, 2))
3.3094413214227043e+07
```

Second, a small value $10^{-5}$ is added to the third component of vector $d$ to get a vector $g$ and solve the linear system $Fz = g$. In MATLAB, the following commands are used:

```
>> g = d ;
>> g(3) = g(3) + 1e-5
g =
1001
1
1000.00001
>> z = F\g
Warning: Matrix is close to singular or badly scaled. Results
    may be inaccurate.
RCOND =  1.067985e-16.
z =
0.999997659959457
23385189.5233867
23385189.523389
>> format long e
>> disp('||z - x||_2 = '), disp(norm(z-x, 2))
||z - x||_2 =
3.307165159616157e+07
```

In Python:

```
In [12]: g = d.copy()
In [13]: g[-1] += 1e-5
In [14]: z = np.linalg.solve(F, g)
In [15]: print('z = \n', z)
z =
[[ 9.99977257e-01]
[-2.94647268e+11]
[-2.94647268e+11]]
In [16]: print('||x-z||_2 = \n', "%20.16e"%
    np.linalg.norm(x-z, 2))
||x-z||_2 =
4.1669416392824341e+11
```

Python does not release a message to complain about the closeness of the matrix to being singular, but it is true that the matrix is close to being singular.

The results obtained in Example 3.2 show that either a small change in the coefficient matrix or the vector at the right-hand side lead to huge changes in the solution of the linear system. Sensitivity to small changes indicate that such a linear system is `ill-conditioned` [42].

### 3.1.1 More Examples of Ill-Posed System

In this section two least squares approximation problems are considered. In the first problem, a Vandermonde matrix will be constructed to find the regression coefficients that give best fit of tabular data. In the second example, a Hilbert matrix will be constructed to find the coefficient of a polynomial of specific degree to approximate a given function.

**Example 3.3** In this example, the data given in Table 2.1 is considered. It consists of nine data pairs of heights and weights for nine persons. The purpose is to find regression coefficients $\alpha_8, \ldots, \alpha_0$ such that $W = \sum_{j=0}^{9} \alpha_j H^j$ is a least squares solution. In MATLAB we compute the regression coefficients as follows:

```
>> H = [1.65, 1.67, 1.68, 1.72, 1.77, 1.82, 1.86, 1.89, 1.9]' ;
>> W = [57, 61, 64, 69, 75, 83, 90, 97, 100]' ;
>> V = vander(H) ;
>> A = V'*V ;
>> b = V'*W ;
>> x = A\b
Warning: Matrix is close to singular or badly scaled. Results
    may be inaccurate.
RCOND =  1.040749e-20.
```

```
x =
3918.1
-21520
35375
-46361
2.1855e+05
-4.3689e+05
52491
5.8691e+05
-4.1416e+05
```

In Python:

```
In [17]: import numpy as np
In [18]: H = np.array([1.65, 1.67, 1.68, 1.72, 1.77, 1.82, 1.86,
    1.89, 1.9])
In [19]: W = np.array([57, 61, 64, 69, 75, 83, 90, 97, 100])
In [20]: V = np.fliplr(np.vander(H))
In [21]: A = V.T@V
In [22]: b = V.T@W
In [23]: x = np.linalg.solve(A, b)
In [24]: print('x = \n', x)
x =
[-1.26559701e+06  3.39786878e+06 -3.76621053e+06  2.29662532e+06
-8.88505759e+05  2.23976721e+05 -1.87100975e+04 -9.10775041e+03
2.24200821e+03]
```

Now we present a small change on the third component of the height vector and see how this change will affect the resulting regression coefficients.

```
>> H1 = H  ;
>> H1(3) = H1(3) + 0.01 ;
>> V1 = vander(H1) ;
>> B = V1'*V1 ;
>> b1 = V1'*W ;
>> y = B\b1
Warning: Matrix is close to singular or badly scaled. Results
    may be inaccurate. RCOND =  1.439393e-20.
y =
2214
-13856
28739
-31202
90976
-2.4644e+05
2.4675e+05
-20840
-63193
```

```
>> disp(norm(x-y))
7.6365e+05
```

In Python:

```
In [25]: H1 = H.copy()
In [26]: H1[2] += 0.01
In [27]: V1 = np.fliplr(np.vander(H1))
In [28]: B = V1.T@V1
In [29]: b1 = V1.T@W
In [30]: y = np.linalg.solve(B, b1)
In [31]: print('y = \n', y)
y =
[-1.67279152e+05  2.06762159e+05  1.09086145e+05 -2.55301791e+05
 8.12931633e+04  3.93681517e+04 -2.63131683e+04  3.30116939e+03
 2.38312295e+02]
In [32]: print('||x - y||_2 = ', '{0:1.6e}'.format(np.linalg
    .norm(x-y)))
||x - y||_2 =  5.821901e+06
```

**Example 3.4** In this example, we approximate the function $y(x) = 5xe^{-2x^2}$ by a polynomial $P(x)$ of degree 12 in an interval $[\alpha, \beta]$. The coefficients of the polynomial are computed by the function `FunApproxCoef` developed in the previous chapter.

The coefficient matrix $A$ is of type $13 \times 13$ defined by

$$A_{ij} = \frac{\beta^{i+j-1} - \alpha^{i+j-1}}{i+j-1}, i, j = 1 \ldots, 13$$

and the right-hand side is a vector $b$ of type $13 \times 1$ defined by

$$b_j = \int_{\alpha}^{\beta} x^{j-1} y(x) dx, j = 1, \ldots, 13.$$

At the beginning the problem is solved without any change in either matrix $A$ or vector $b$. Then we make a small change in the last component in $b$ such that $b_{13} = b_{13} + 10^{-4}$.

The MATLAB code to compute the polynomials is:

```
 1  f = @(x) 2 + 5 * x.* exp(-2 * x.^2) ;
 2  A = zeros(13, 13) ; b = zeros(13, 1) ;
 3  a = 0.0 ; b = 2.0 ;
 4  al = 0.0 ; bt = 2.0 ;
 5  for i = 1 : 13
 6      for j = 1 : i
 7          k = i + j - 1 ;
 8          A(i, j) = (bt^k - al^k)/k ;
 9          A(j, i) = A(i, j) ;
10      end
11      b(i) = integral(@(x) x.^(i-1).*f(x), al, bt) ;
12  end
```

```
13  C = A\b(:) ; C = wrev(C) ;
14  b1 = b ; b1(end) = b1(end) + 1e-4 ;
15  C1 = A\b1(:) ; C1 = wrev(C1) ;
16  t = linspace(al, bt, 401) ; F = f(t) ;
17  Pe = polyval(C, t) ; Pp = polyval(C1, t) ;
18
19  subplot(1, 2, 1) ; plot(t, F, '--b', 'LineWidth', 3) ;
20  xlabel('x', 'fontweight','bold') ; ylabel('y', ...
        'fontweight','bold') ;
21  legend('y(x) = 5xe^{-2x^2}') ; grid on ; ax = gca ;
22  ax.FontWeight = 'bold' ; ax.FontSize = 12 ;
23  set(gca, 'XTick', linspace(al, bt, 9)) ; set(gca, 'YTick', ...
        linspace(1, 5, 9)) ;
24  axis([al, bt, 1, 5]) ;
25
26  subplot(1, 2, 2) ; plot(t, Pe, '--r', t, Pp, ':k', ...
        'LineWidth', 3) ;
27  xlabel('x', 'fontweight','bold') ; ylabel('y', ...
        'fontweight','bold') ;
28  legend('Without perturbation', 'With perturbation') ; grid on ;
29  ax = gca ; ax.FontWeight = 'bold' ; ax.FontSize = 12 ;
30  set(gca, 'XTick', linspace(al, bt, 9)) ; set(gca, 'YTick', ...
        linspace(1, 5, 9)) ;
31  axis([al, bt, 1, 5]) ;
```

The Python code is:

```
1   import numpy as np
2   import matplotlib.pyplot as plt
3   from scipy import integrate
4   f = lambda x: 2.+5*x*np.exp(-2*x**2)
5   A, b = np.zeros((13, 13), 'float'), np.zeros((13, 1), 'float')
6   al, bt = 0., 2.
7   for i in range(13):
8       for j in range(i+1):
9           k = i+j+1
10          A[i, j] = (bt**k-al**k)/k
11          A[j, i] = A[i, j]
12      b[i] = integrate.quad(lambda x: x**i*f(x), al, bt)[0]
13  C = np.linalg.solve(A, b)
14  c = list(C.T[0])
15  c.reverse()
16  C = np.array(c)
17  t = np.linspace(al, bt+(bt-al)/400, 401)
18  F = f(t)
19  Pe = np.polyval(C, t)
20  b[-1] += 1e-4
21  C1 = np.linalg.solve(A, b)
22  c = list(C1.T[0])
23  c.reverse()
24  C1 = np.array(c)
25  Pp = np.polyval(C1, t)
26
27  plt.figure(1, figsize=(20, 8))
28  plt.subplot(1, 2, 1)
29  plt.plot(t, F, '--b', lw = 3, label = 'y(x) = 5 x e^{-2x^2}')
30  plt.xlabel('x', fontweight='bold')
```

```
31  plt.ylabel('y', fontweight='bold')
32  plt.legend()
33  plt.xticks(np.arange(al, bt+(bt-al)/8, (bt-al)/8), ...
        fontweight='bold')
34  plt.yticks(np.arange(1., 5., 0.5), fontweight='bold')
35  plt.grid(True, ls=':')
36  plt.axis([al, bt, 1., 5.])
37
38  plt.subplot(1, 2, 2)
39  plt.plot(t, Pe, '--b', lw = 3, label = 'Without perturbation')
40  plt.plot(t, Pp, ':k', lw = 3, label = 'With perturbation')
41  plt.xlabel('x', fontweight='bold')
42  plt.ylabel('y', fontweight='bold')
43  plt.legend(loc="upper center")
44  plt.xticks(np.arange(al, bt+(bt-al)/8, (bt-al)/8), ...
        fontweight='bold')
45  plt.yticks(np.arange(1., 5., 0.5), fontweight='bold')
46  plt.grid(True, ls=':')
47  plt.axis([al, bt, 1., 5.])
```

Figure 3.1 contains two subgraphs. In the first subgraph (at the left side) the original function $y(x) = 5xe^{-2x}$ is plotted. In the second subgraph (at the right side) the coefficients resulting from the linear systems without and with changing the right-hand side are used to graph the approximating polynomials.

## 3.1.2 Condition Numbers and Ill-Conditioned Matrices

In this section we give an interpretation as to why some linear systems are more sensitive to small changes than other linear systems. Before we do so, we discuss the following concepts:

(I) **Matrix norm:** If $A$ is a matrix of dimensions $m \times n$, with elements $a_{i,j}, i = 1, \ldots, m; j = 1, \ldots, n$, then:

   (a) $\|A\|_1 = \max \left\{ \sum_{i=1}^{m} |a_{i,1}|, \sum_{i=1}^{m} |a_{i,2}|, \ldots, \sum_{i=1}^{m} |a_{i,n}| \right\}$

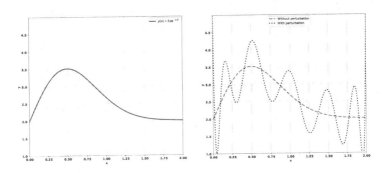

FIGURE 3.1: Left: the exact function. Right: the approximating polynomials of degree 12 without and with changes in the last component of vector $b$.

(b) $\|A\|_\infty = \max \left\{ \sum_{j=1}^n |a_{1,j}|, \sum_{j=1}^n |a_{2,j}|, \ldots, \sum_{j=1}^n |a_{m,j}| \right\}$

(c) $\|A\|_2 = \max \left\{ \sqrt{\sigma_1}, \ldots, \sqrt{\sigma_m} \right\}$, where $\sigma_j$ is the $j^{th}$ eigenvalue of $A \cdot A^T$.

We can use the Python command `norm` to compute $\|A\|_p$ of matrix $A$.

```
In [31]: from numpy.linalg import norm
In [32]: A = np.array([[1, -2], [2, -3], [3, -4]])
In [33]: n1, n2, n3 = norm(A, 1), norm(A, 2), norm(A, np.inf)
In [34]: print(n1, '\n', n2, '\n', n3)
9.0
6.54675563644
7.0
```

In MATLAB, also a function `norm` is used to compute $\|A\|_p$ of matrix $A$.

```
>> A = [1, -2; 2, -3; 3, -4] ;
>> n1 = norm(A, 1), n2 = norm(A, 2), n3 = norm(A, inf)
n1 =
    9
n2 =
    6.5468
n3 =
    7
```

Let $A$ and $B$ be two matrices of type $m \times n$, $\alpha$ be scaler and $\boldsymbol{x}$ be a vector in $\mathbb{R}$. The matrix norm is a function $\|\cdot\| : \mathbb{R}^{m \times n} \to \mathbb{R}$ that satisfies the following properties:

(a) $\|A\| \geq 0$.

(b) $\|\alpha A\| = |\alpha| \cdot \|A\|$.

(c) $\|A + B\| \leq \|A\| + \|B\|$.

(d) $\|A\boldsymbol{x}\| \leq \|A\| \|\boldsymbol{x}\|$.

(e) $\|AB\| \leq \|A\| \|B\|$.

(II) **Machine precession:** A digital computer has limited storage capacity to represent a number with floating point. Subject to the International Electrical and Electronic Engineering (IEEE) standards, a double-precession number is represented by 8 bytes (64 bits). These 64 bits are divided into three segments: the most significant bit (the most left bit) is used for representing the sign of the number. The next 11 bits are used to represent the exponent part and the last 52 bits (called the `mantissa`) are used to represent the binary fraction bits. Under this setup, the minimum floating number that can be represented by 64 bits is approximately $-1.7977 \times 10^{308}$ and the largest floating number is approximately $1.7977 \times 10^{308}$.

Both MATLAB and Python adopt the IEEE double precision numbers as default type of their numerical variables with floating points. The smallest binary fraction that can be represented by 52 bits is $2^{-52} \approx 2.2204 \times 10^{-16}$. This number is called the `machine precision` and is denoted by $\varepsilon$. In MATLAB the machine precision can be seen by typing *eps* in the command window:

```
>> disp(eps)
2.2204e-16
```

In Python it can be seen by typing:

```
In [35]: print(np.finfo(float).eps)
2.220446049250313e-16
```

If the fractional part of some floating number requires more than 52 bits to be represented, then the computer can represent only 52 bits of the number and ignores the rest. The difference between the true fractional part of the number and the computer representation is called the `round off error`. For example it is well known that $\sin n\pi = 0, n = 0, \pm 1, \ldots$. But Python and MATLAB do not give 0 exactly if we choose $n \neq 0$:

```
In [36] np.sin(0.0)
Out[36]: 0.0
In [37]: np.sin(np.pi)
Out[37]: 1.2246467991473532e-16
```

```
>> disp(sin(0.0))
   0
>> disp(sin(2*pi))
  -2.4493e-16
```

Now to understand how a small change in either vector $\boldsymbol{b}$ at the right-hand side of the linear system $A\boldsymbol{x} = \boldsymbol{b}$, or in matrix $A$, can change the solution of the linear system, the properties of matrix norm are used to explore such a phenomena.

First, it is assumed that a change in the right-hand side from $\boldsymbol{b}$ to $\boldsymbol{b} + \Delta\boldsymbol{b}$ can lead to a change in the solution of the system from $\boldsymbol{x}$ to $\boldsymbol{x} + \Delta\boldsymbol{x}$. Then,

$$A(\boldsymbol{x} + \Delta\boldsymbol{x}) = \boldsymbol{b} + \Delta\boldsymbol{b} \Rightarrow A\boldsymbol{x} + A\Delta\boldsymbol{x} = \boldsymbol{b} + \Delta\boldsymbol{b} \Rightarrow A\Delta\boldsymbol{x} = \Delta\boldsymbol{b} \Rightarrow \Delta\boldsymbol{x} = A^{-1}\Delta\boldsymbol{b}$$

because $A\boldsymbol{x} = \boldsymbol{b}$. Then,

$$\|\boldsymbol{b}\| = \|A\boldsymbol{x}\| \leq \|A\|\|\boldsymbol{x}\| \Rightarrow \|A\|\|\boldsymbol{x}\| \geq \|\boldsymbol{b}\|, \tag{3.1}$$

and

$$\|\Delta\boldsymbol{x}\| = \|A^{-1}\Delta\boldsymbol{b}\| \leq \|A^{-1}\|\|\Delta\boldsymbol{b}\|. \tag{3.2}$$

By dividing Equation (3.2) by Equation (3.1) gives:

$$\frac{\|x\|}{\|A\|\|x\|} \leq \frac{\|A^{-1}\| \cdot \|\Delta b\|}{\|b\|}$$

from which,

$$\frac{\|\Delta x\|}{\|x\|} \leq \|A\|\|A^{-1}\|\frac{\|\Delta b\|}{\|b\|} \tag{3.3}$$

The number $\kappa(A) = \|A\|\|A^{-1}\|$ is called the `condition number` of matrix $A$. The bigger the condition number, the higher sensitivity of the linear system to perturbation on the right-hand side and vice versa. Equation (3.3) shows the relationship between the relative change in solution of the linear system compared to the relative change in the right-hand side of the linear system.

Second, if it is assumed that matrix $A$ is changed to $A + \Delta A$ leads to change in the solution from $x$ to $x + \Delta x$, then

$$(A + \Delta A)(x + \Delta x) = b \Rightarrow Ax + A\Delta x + \Delta A(x + \Delta x)$$
$$= b \Rightarrow A\Delta x + \Delta A(x + \Delta x) = 0$$

Multiplying the two sides by $A^{-1}$ gives

$$\Delta x + A^{-1}\Delta A(x + \Delta x) = 0 \Rightarrow -\Delta x = A^{-1}\Delta A(x + \Delta x)$$

By taking the norms for the two sides, we get:

$$\|\Delta x\| = \|A^{-1}\Delta A(x + \Delta x)\| \leq \|A^{-1}\| \cdot \|\Delta A\| \cdot \|x + \Delta x\|$$

Dividing the two sides by $\|x + \Delta x\|$ and multiplying the right-hand side by $\frac{\|A\|}{\|A\|}$ gives:

$$\frac{\|\Delta x\|}{\|x + \Delta x\|} \leq \kappa(A)\frac{\|\Delta A\|}{\|A\|} \tag{3.4}$$

Equation (3.4) shows the relationship between the relative change in the solution vector $x$ compared to the relative change in matrix $A$, in terms of the condition number of matrix $A$.

It can be seen that the smaller the condition number of a matrix $A$, the less the linear system will be sensitive to perturbations, and the resulting system is well-posed. Also, the larger the condition number, the more sensitive the linear system is to perturbations, and the resulting system is ill-posed. The condition number of the $3 \times 3$ zero matrix is $\infty$ and for the $3 \times 3$ identity matrix is 1. The MATLAB command `cond` can be used for finding the condition number of a matrix $A$.

```
>> A = [100 1 0; 0 1 2; 0 2 4.003]
A =
100.0000     1.0000          0
0            1.0000     2.0000
0            2.0000     4.0000
```

```
0               2.0000          4.0030
>> cond(A)
ans =
1.6676e+05
>> cond(zeros(3))
ans =
Inf
>> cond(eye(3))
ans =
1
```

The Python command `numpy.linalg.cond` can be used for computing the condition number of a matrix $A$.

```
In [38]: A = np.array([[100, 1, 0], [0, 1, 2], [0, 2, 4.003]])
In [39]: print('{0:1.6e}'.format(np.linalg.cond(A)))
1.667617e+05
In [40]: print('{0:1.6e}'.format(np.linalg.cond
    (np.zeros((3, 3)))))
inf
In [41]: print('{0:1.6e}'.format(np.linalg.cond(np.eye(3))))
1.000000e+00
```

A matrix with a large condition number is called ill-conditioned, otherwise, it is well-conditioned [7, 6]. The condition number for any singular matrix is $\infty$. For the identity matrix, it is 1. For any nonsingular matrix, the condition number lies in $[1, \infty)$.

An equivalent measure to the condition number $\kappa(A)$ is the `reciprocal condition number` (rcond), where

$$rcond(A) = \frac{1}{\|A\|_1 \cdot \|A^{-1}\|_1}$$

Under this measure, the reciprocal condition number of any singular matrix is 0, for the identity matrix is 1 and for any other nonsingular matrix is in the interval $(0.0, 1.0)$. A matrix $A$ is closer to singular if $0 \neq \kappa(A) \approx 0$ [42].

### 3.1.3 Linking the Condition Numbers to Matrix Related Eigenvalues

If $A$ is a squared $n$ by $n$ matrix with eigenvalues $\{\lambda_1, \dots, \lambda_n\}$, then the condition number of matrix $A$ is the ratio between the largest (in magnitude) and smallest (in magnitude) eigenvalues. That is:

$$\kappa(A) = \frac{\lambda_{max}}{\lambda_{min}},$$

where $\lambda_{max} = \max\{|\lambda_j|, j = 1, \dots, n\}$ and $\lambda_{min} = \min\{|\lambda_j|, j = 1, \dots, n\}$.

If $A$ is an $m$ by $n$ matrix, where $m \neq n$, and $\{\lambda_1, \ldots, \lambda_m\}$ are the eigenvalues of $A \cdot A^T$, then the condition number of matrix $A$ is given by:

$$\kappa(A) = \sqrt{\frac{\lambda_{max}}{\lambda_{min}}},$$

where $\lambda_{max} = \max\{|\lambda_j|, j = 1, \ldots, m\}$ and $\lambda_{min} = \min\{\lambda_j, j = 1, \ldots, m\}$

In Example 2.5 the function $f(x) = \cos \pi x$ is approximated with polynomials $P_n(x); n = 1, 3, \ldots, 15$. It was expected that as the degree of the polynomial increases, the accuracy of the solution increases. But, what was noticed was that the accuracy of the approximating polynomial increases up to degree 9, after that the accuracy drops. That happened because of the increase in the condition number as the polynomial degree increases.

The following MATLAB commands show the condition numbers of the Hilbert matrices $H_n$ for $n = 2, \ldots, 13$

```
>> fprintf('n\t\t ||H_n||\n'); for n = 2 : 13
     fprintf('%i\t\t%10.6e\n', n, cond(hilb(n))) ; end
n         ||H_n||
2                  1.928147e+01
3                  5.240568e+02
4                  1.551374e+04
5                  4.766073e+05
6                  1.495106e+07
7                  4.753674e+08
8                  1.525758e+10
9                  4.931541e+11
10                 1.602520e+13
11                 5.227499e+14
12                 1.629550e+16
13                 1.682118e+18
14                 2.715269e+17
15                 2.777738e+17
```

In Python, the code is:

```
In [41]: from numpy.linalg import cond
In [42]: from scipy.linalg import hilbert as hilb
In [43]: for n in range(2, 16): print(n, '\t\t',
     '{0:1.6e}'.format(cond(hilb(n))))
2                  1.928147e+01
3                  5.240568e+02
4                  1.551374e+04
5                  4.766073e+05
6                  1.495106e+07
7                  4.753674e+08
8                  1.525758e+10
9                  4.931534e+11
```

| 10 | 1.602503e+13 |
|----|--------------|
| 11 | 5.220207e+14 |
| 12 | 1.621164e+16 |
| 13 | 4.786392e+17 |
| 14 | 2.551499e+17 |
| 15 | 2.495952e+17 |

This shows that as the dimensions of Hilbert matrices increase, they become more ill-conditioned. It is also noticed that MATLAB and Python do agree on the values of condition numbers as long as $rcond(H_n) \geq eps$. When $rcond(H_n) < eps$ they could have different roundoff errors, causing them to produce different condition numbers.

Another example is the Vandermonde matrix, used with least-squares approximations. As the number of data points increases, the condition number of the corresponding Vandermone matrix increases, so it becomes more ill-conditioned. The following MATLAB commands show the condition numbers of Vandermone matrix for different numbers of data points:

```
>> fprintf('n\t\t ||V_n||\n'); for n = 2 : 13
   fprintf('%i\t\t%10.6e\n', n, cond(vander(H(1:n)))) ; end
n        ||V_n||
2               3.755673e+02
3               1.627586e+05
4               3.051107e+07
5               3.481581e+09
6               2.480023e+19
7               2.970484e+21
8               6.929557e+24
9               7.174378e+27
10              8.795704e+31
11              2.868767e+35
12              1.380512e+39
13              1.532255e+42
```

In Python:

```
In [44]: for n in range(2, len(H)): print(n, '\t\t',
   '{0:1.6e}'.format(cond(vander(H[:n]))))
2               3.755673e+02
3               1.627586e+05
4               3.051107e+07
5               3.481581e+09
6               2.479949e+19
7               2.970484e+21
8               6.929557e+24
9               7.174382e+27
10              8.796918e+31
11              2.862293e+35
```

```
12                      1.375228e+39
13                      1.740841e+42
```

If the condition number of some matrix $A$ is of order $10^\ell$, then when solving a linear system $Ax = b$ up to $\ell$ decimal places from the right can be inaccurate (with the notice that 16 decimal places are truly represented for double-precision numbers). To see this, we consider a Vandermonde matrix generated by a random vector $v$:

```
>> v = rand(10, 1)
v =
4.4559e-01
6.4631e-01
7.0936e-01
7.5469e-01
2.7603e-01
6.7970e-01
6.5510e-01
1.6261e-01
1.1900e-01
4.9836e-01
>> V = fliplr(vander(v)) ;
```

In Python, the vector $v$ and the Vandermonde matrix $V$ of $v$ can be generated by using the Python commands:

```
In [45]: v = np.random.rand(10)
In [46]: v
Out[46]:
array([0.47585697, 0.31429675, 0.73920316, 0.45044728, 0.16221156,
0.8241245, 0.9038605, 0.28001448, 0.85937663, 0.07834397])
In [47]: V = np.fliplr(np.vander(v))
```

We can measure the condition number of matrix $V$ in MATLAB:

```
>> Cv = cond(V)
Cv =
1.730811304916736e+10
```

In Python:

```
In [48]: cV = np.linalg.cond(V)
In [49]: print('{0:1.6e}'.format(cV))
8.671331e+07
```

Let $x$ be a column vector of ones of dimension 10 and $b = Vx$.

```
>> x = ones(10, 1) ;
>> b = V * x ;
```

Each component of $b$ is a sum of row elements of $V$.

Now let us pretend that we don't know $x$, and we want to retrieve it by solving the linear system:

$$V y = b$$

The vector $y$ is given by:

```
y = V\b ;
```

To find out how close $y$ to $x$, we measure the infinity norm of the difference between them.

```
>> Error = norm(x-y, inf)
Error =
6.900445270741074e-07
```

which tell that the Error is of $\mathcal{O}(10^{-7})$. That means the accuracy has been lost in 9 decimal places.

The above steps can be executed in Python, by using the commands:

```
In [50]: x = np.ones((10, 1))
In [51]: b = V @ x
In [52]: y = np.linalg.solve(V, b)
In [53]: Error = np.linalg.norm(y-x, np.inf)
In [54]: print('{0:1.6e}'.format(Error))
3.949043e-09
```

### 3.1.4 Further Analysis on Ill-Posed Systems

Given a linear system of equations

$$\boldsymbol{F x} = \boldsymbol{y} \tag{3.5}$$

where $\boldsymbol{F} \in \mathbb{R}^{m \times n}$, $\boldsymbol{x} \in \mathbb{R}^n$ is a vector of unknowns, $\boldsymbol{y} \in \mathbb{R}^m$ is a vector of exact data and $n \leq m$.

We want to compute a solution $\boldsymbol{x}^* \in \mathbb{R}^n$ to the linear sytem (3.5) such that,

(1) $\boldsymbol{x}^*$ is a least-squares solution to problem (3.5). The existence of a least-squares solution $\boldsymbol{x}_{LS}$ is characterized by

$$\boldsymbol{x}^* = \boldsymbol{x}_{LS} = \arg \min_{\boldsymbol{x} \in \mathbb{R}^n} \| F \boldsymbol{x} - \boldsymbol{y} \|^2 \tag{3.6}$$

$\boldsymbol{x}_{LS}$ is a *least squares solution*.

(2) $\boldsymbol{x}^*$ is *unique*. The uniqueness of a least-squares solution $\boldsymbol{x}_{LSMN}$ is characterized by

$$\boldsymbol{x}^* = \boldsymbol{x}_{LSMN} = \arg \min_{\boldsymbol{x}_{LS} \in \mathbb{R}^n} \{ \| \boldsymbol{x}_{LS} \|^2 \} = \boldsymbol{F}^{-1} \boldsymbol{y} \tag{3.7}$$

$\boldsymbol{x}_{LSMN}$ is referred to as a *least squares minimum norm* solution.

(3) the computed solution $x^*$ is stable. This is characterized by the existence of $F^{-1}$.

Problem (3.5) is a *well-posed problem* if conditions (1)-(3) are satisfied. It is an *ill-posed problem* if it is *not a well-posed problem*.

In real life, the data $y$ is usually obtained from measurements which are contaminated by small errors [42]. Solving the inverse problem using the noisy data in an ill-posed system will result in a catastrophic erroneous solution which is irrelevant to the true solution, as had seen in the previous section.

To find an interpretation to what is going behind, one should consider the singular value decomposition of the matrix $F$. Let $(U, \Sigma, V^T)$ be the singular value decomposition of the matrix $F$, where $U \in \mathbb{R}^{m \times m}$ and $V \in \mathbb{R}^{n \times n}$ are two unitary matrices and $\Sigma \in \mathbb{R}^{m \times n}$ is a diagonal matrix. The inverse of the matrix $F$ is given by $F^{-1} = (U\Sigma V^T)^{-1} = V\Sigma^{-1}U^T$. Now the solution $x$ of the linear system $Fx = y$ is given by

$$x = F^{-1}y = \sum_{i=1}^{n} \frac{(U_i^T y)}{\sigma_i} V_i \qquad (3.8)$$

where $U_i^T$ and $V_i$ are the $i^{th}$ columns of the matrices $U$ and $V$, and the $\sigma_i$'s are the singular values of $F$, $i = 1, \ldots, m$, with $\sigma_1 \geq \sigma_2 \geq \ldots \geq \sigma_n \geq 0$.

Therefore, the solution $x$ is a linear combination of $\{V_1, \ldots, V_n\}$, the columns of $V$, with coefficients $\frac{U_i^T y}{\sigma_i}, i = 1, \ldots, n$.

When the noisy data $y^\epsilon$ is used instead of the exact data $y$, we get

$$F^{-1}y^\epsilon = \sum_{i=1}^{m} \frac{(U_i^T y^\epsilon)}{\sigma_i} V_i = \sum_{i=1}^{n} \frac{(U_i^T y)}{\sigma_i} V_i + \sum_{i=1}^{n} \frac{(U_i^T \epsilon)}{\sigma_i} V_i = x + \sum_{i=1}^{n} \frac{(U_i^T \epsilon)}{\sigma_i} V_i$$
$$(3.9)$$

From equation (3.9), we find that, $\|y - x\|_2^2 = \sum_{i=1}^{n} \left(\frac{U_i^T \epsilon}{\sigma_i}\right)^2$. As the matrix $F$ tends to be singular, some of its singular values tend to be zeros, and hence, some of the coefficients $\frac{(U_i^T \epsilon)}{\sigma_i}$ gets very large. This tells us that, the residual norm is not effected by only how small the noise is, but also by how small a singular value of the matrix $F$ is.

As an explanation, let us look at the singular values of the $20 \times 20$ Hilbert matrix:

```
>> H = hilb(20) ;
>> [U, S, V] = svd(H) ;
>> format short e ;
>> D = diag(S)
D =
1.9071e+00
```

```
4.8704e-01
7.5596e-02
8.9611e-03
8.6767e-04
7.0334e-05
4.8305e-06
2.8277e-07
1.4140e-08
6.0361e-10
2.1929e-11
6.7408e-13
1.7384e-14
3.7318e-16
1.5057e-17
1.2511e-17
7.3767e-18
5.4371e-18
1.7279e-18
5.8796e-19
```

We see that the last 6 singular values of $H(=USV^T)$ are below $\varepsilon(=2.2204 \times 10^{-16})$, which are not discriminated from 0 in the $64-bit$ systems. Dividing by such a singular value when computing $H^{-1} = VS^{-1}U^T$ causes huge errors, since the diagonal elements of $S^{-1}$ are the reciprocals of the diagonal elements of $S$.

According to the distribution of the singular values of $\boldsymbol{F}$, ill-posed problems (whose coefficient matrices are ill-conditioned) are divided into two classes. Namely, the `rank deficient problems` and the `discrete ill posed problems`. These two classes of problems can be distinguished by using the following properties [36]

1. in the rank deficient problems, there is a small cluster of small singular values, while in discrete ill-posed problems there is a large cluster of small singular values.

2. in the rank deficient problems there is a clear gap between the large and small singular values, while in discrete ill-posed problems the singular values decay gradually without gaps between the large and small singular values.

3. for the rank deficient problems, there could be a formulation which can eliminate the ill-conditioning but no such formulation exists for the ill-posed problems.

## 3.2    Regularization of Solutions in Linear Systems

Regularization is the process of stabilizing the solution of an ill-posed problem [60, 59, 58]

$$Fx = y, F \in \mathbb{R}^{m \times n}, x \in \mathbb{R}^n \text{ and } y \in \mathbb{R}^m.$$

The regularization is done through introducing a parametric family of approximate inverse operators

$$\{\Gamma_\alpha : \mathbb{R}^n \to \mathbb{R}^m : \Gamma_\alpha \approx F^{-1}, \alpha \in (0, \infty)\}$$

such that, for each $y^\epsilon$ satisfying $\|y^\epsilon - y\| \le \epsilon$ there exists $\alpha \in (0, \infty)$ such that, $x^\alpha \overset{def}{=} \Gamma_\alpha y^\epsilon \to x$ as $\epsilon \to 0$.

Some regularization techniques include the truncated SVD method, the Tikhonov regularization method, the L-curve method and the Morosov discrepancy principle [8, 9].

In this section we discuss these methods.

### 3.2.1    The Truncated SVD (TSVD) Method

Considering the linear system

$$Fx = b, \ F \in \mathbb{R}^{n \times n}, \ x \in \mathbb{R}^n \text{ and } y \in \mathbb{R}^n,$$

and supposing that $U, \Sigma$ and $V$ are the svd factors of matrix $F$, such that $F = U \Sigma V^T$. Hence, $F^{-1} = V \Sigma^{-1} U^T$. The unique solution of the linear system $fx = b$ is given by:

$$x = V\Sigma^{-1}U^Tb = \sum_{j=1}^{n} \frac{u_j^T b}{\sigma_j} v_j = \frac{u_1^T b}{\sigma_1} v_1 + \cdots + \frac{u_n^T b}{\sigma_n} v_n$$

where $u_j$ and $v_j$ are the $j^{th}$ columns of $U$ and $V$ respectively, and $\sigma_j$ is the $j^{th}$ diagonal element of $\Sigma$, with the notice that $\sigma_1 \ge \sigma_2 \ge \ldots \ge \sigma_n$.

The idea behind the regularization method based on the truncated SVD is to truncate terms that contain very small singular values from the summation in equation (3.9). That is if $\alpha \ge \varepsilon$ ($\varepsilon$ is the machine precision) and $\sigma_1 \ge \sigma_2 \ge \ldots \ge \sigma_\ell \ge \alpha \ge \sigma_{\ell+1} \ge \ldots \ge \sigma_n$, then

$$x \approx x^\alpha = \sum_{j=1}^{\ell} \frac{u_j^T b}{\sigma_j} v_j = \frac{u_1^T b}{\sigma_1} v_1 + \cdots + \frac{u_\ell^T b}{\sigma_\ell} v_\ell \qquad (3.10)$$

Up to some optimal regularization level $\alpha^*$, the TSVD method performs well (for $\alpha \le \alpha^*$), after that (when $\alpha$ exceeds $\alpha^*$) the solution is affected by truncating more terms and as $\alpha \ge \alpha^*$ increases, $x^\alpha$ goes away from $x$.

The MATLAB function `SolveWithTSVD` receives a matrix $F$, vector $b$ and regularization parameter $\alpha$. It applies the truncated SVD regularization to return a solution $x$.

```
1  function x = SolveWithTSVD(F, b, Alpha)
2      [U, S, V] = svd(A) ;
3      D = diag(S) ;
4      [m, n] = size(A) ;
5      x = zeros(m, 1) ;
6      for i = 1 : length(S)
7          if D(i) > Alpha
8              x = x + U(:, i)'*b/D(i)*V(:, i) ;
9      end
10 end
```

The function `SolveWithTSVD` is tested for $20 \times 20$ Hilbert matrix, the vector $b$ at the RHS components are the summations of $H$ rows. The exact solution $y$ is a vector of ones of type $20 \times 1$.

Using a script `SolveForHilb.m` an unregularized solution $z$ and a regularized solution $w$ at level $\alpha = \varepsilon$ are computed and shown in Figure 3.2. Also, the error norms for different values of the regularization parameters $\alpha_j = 10^{j-1}\varepsilon$ are shown.

```
1  H = hilb(20) ;
2  y = ones(20, 1) ;
3  b = H * y ;
4  z = H\b ;
5  subplot(1, 2, 1) ;
6  plot(1:20, y, '-b', 1:20, z, 'r:', 'LineWidth', 3) ;
7  legend('Exact Solution', 'Unregularized Solution') ;
8  xlabel('Solution Component') ;
9  ylabel('Component Value') ;
10 grid on ;
11 format short e
12 disp('Error norm of unregularized solution ||y-z||_2'), ...
       disp(norm(y-z))
```

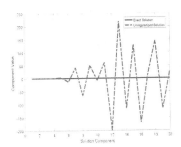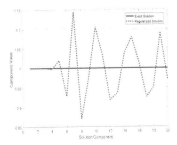

FIGURE 3.2: Unregularized and regularized solutions of $Hx = b$.

```
13  w = SolveWithTSVD(H, b, eps) ;
14  subplot(1, 2, 2) ;
15  plot(1:20, y, '-b', 1:20, w, 'r:', 'LineWidth', 3) ;
16  legend('Exact Solution', 'Regularized Solution') ;
17  xlabel('Solution Component') ;
18  ylabel('Component Value') ;
19  grid on
20  disp('Error norm of regularized solution ||y-w||_2'), ...
          disp(norm(y-w))
21
22  x = zeros(20, 15) ;
23  Err = zeros(15, 2) ;
24  disp('   Alpha      ||x-y||_2') ;
25  disp('-----------------------') ;
26  for n = 1 : 15
27  Alpha = 10^(n-1)*eps ;
28  x(:, n) = SolveWithTSVD(H, b, Alpha) ;
29  Err(n, 1) = Alpha ;
30  Err(n, 2) = norm(y-x(:, n), inf) ;
31  end
32  disp(Err) ;
```

The script `SolveForHilb` is executed by using the command:

```
>> SolveForHilb
> In SolveForHilb (line 4)
Warning: Matrix is close to singular or badly scaled. Results
    may be inaccurate. RCOND =  5.231543e-20.
Error norm of unregularized solution ||y-z||_2
4.4218e+02


Error norm of regularized solution ||y-w||_2
3.0301e-01


    Alpha     ||x-y||_2
-----------------------
2.2204e-16   1.4427e-01
2.2204e-15   2.6146e-02
2.2204e-14   3.3896e-04
2.2204e-13   3.3896e-04
2.2204e-12   2.4002e-06
2.2204e-11   7.8511e-06
2.2204e-10   7.8511e-06
2.2204e-09   4.5607e-05
2.2204e-08   2.1812e-04
2.2204e-07   2.1812e-04
2.2204e-06   1.0205e-03
2.2204e-05   4.2420e-03
2.2204e-04   1.5665e-02
```

```
2.2204e-03    5.1211e-02
2.2204e-02    1.6937e-01
```

The Python code is:

```
 1  import numpy as np
 2  from scipy.linalg import hilbert as hilb
 3  from numpy.linalg import norm, svd
 4  import matplotlib.pyplot as plt
 5  def SolveWithTSVD(A, b, Alpha):
 6      U, S, V = svd(A)
 7      x = np.zeros_like(b)
 8      n = len(S)
 9      for j in range(n):
10          if S[j] >= Alpha:
11              x += np.dot(U[:,j], b)/S[j]*V.T[:,j]
12          else:
13              continue
14      return x
15
16  H = hilb(20)
17  y = np.ones((20,), 'float')
18  b = H@y
19  z = np.linalg.solve(H, b)
20  Eps = np.spacing(1.0)
21  Alpha=Eps
22  w = SolveWithTSVD(H, b, Alpha)
23  print('Error norm for Unregularized solution = ', norm(y-z))
24  print('Error norm for Regularized solution = ', norm(y-w))
25  plt.figure(1)
26  plt.subplot(1, 2, 1)
27  t = np.arange(1, len(b)+1)
28  plt.plot(t, y, '-b', lw=3, label='Exact solution')
29  plt.plot(t, z, '-.r', lw=3, label='Unregularized solution')
30  plt.xlabel('Solution component', fontweight='bold')
31  plt.ylabel('Component value', fontweight='bold')
32  plt.grid(True, ls=':')
33  plt.legend()
34
35  plt.subplot(1, 2, 2)
36  t = np.arange(1, len(b)+1)
37  plt.plot(t, y, '-b', lw=3, label='Exact solution')
38  plt.plot(t, w, ':m', lw=3, label='Regularized solution')
39  plt.xlabel('Solution component', fontweight='bold')
40  plt.ylabel('Component value', fontweight='bold')
41  plt.grid(True, ls=':')
42
43  Err = np.zeros((16, 2), 'float')
44  for j in range(16):
45      Alpha = 10**j*np.spacing(1.)
46      w = SolveWithTSVD(H, b, Alpha)
47      Err[j, 0] = Alpha
48      Err[j, 1] = norm(y-w)
49  print(Err)
```

Executing the code gives:

```
runfile('D:/PyFiles/regWithTSVD.py', wdir='D:/PyFiles')
Error norm for Unregularized solution =   136.97028071675066
Error norm for Regularized solution =   0.5943628935591088
[[2.22044605e-16 5.94362894e-01]
 [2.22044605e-15 4.08083331e-04]
 [2.22044605e-14 4.08083331e-04]
 [2.22044605e-13 4.08083331e-04]
 [2.22044605e-12 1.19446030e-05]
 [2.22044605e-11 2.17705662e-05]
 [2.22044605e-10 2.17705662e-05]
 [2.22044605e-09 1.12283185e-04]
 [2.22044605e-08 5.33801614e-04]
 [2.22044605e-07 5.33801614e-04]
 [2.22044605e-06 2.34255106e-03]
 [2.22044605e-05 9.49123719e-03]
 [2.22044605e-04 3.54621361e-02]
 [2.22044605e-03 1.21845461e-01]
 [2.22044605e-02 3.83219676e-01]
 [2.22044605e-01 1.09386370e+00]]
```

## 3.2.2   Tikhonov Regularizaton Method

*Tikhonov regularization methods* are the most well known methods for the regularization of an ill-conditioned system. In a Tikhonov regularization method a penalty is added to a solution with a large norm. If $\varphi(x) = \|Fx - y\|_2^2$ is the square of the length of the residual vector $Fx - y$, the problem is to minimize the functional $\varphi_\alpha = \varphi_\alpha(x) = \|Fx - y\|_2^2 + \alpha\|x\|_2^2$, where the penalty $\alpha\|x\|_2^2$ is added to penalize the components of the vector $x$ on the rapid variations from positive to negative values and vice-versa.

We write

$$\varphi_\alpha(x) = <Fx - y, Fx - y> + \alpha < x, x > = (Fx - y)^T(Fx - y) + \alpha x^T x$$
$$= x^T(F^T F + \alpha I)x - x^T F^T y - y^T Fx - y^T y$$

If the functional $\varphi_\alpha(x)$ takes its minimum value at $x = x^\alpha$, then $\left.\frac{\partial \varphi_\alpha(x)}{\partial x}\right\|_{x=x^\alpha} = 0 = 2(F^T F + \alpha I)x^\alpha - 2F^T y$ and $(F^T F + \alpha I)x^\alpha = F^T y$.

Now, $F^T F + \alpha I = V(\Sigma^2 + \alpha I)V^T$, then

$$x^\alpha = \sum_{i=1}^{n}(U_i^T y)\frac{\sigma_i}{\alpha + \sigma_i^2}V_i \qquad (3.11)$$

We consider the term $\frac{\sigma_i}{\alpha + \sigma_i^2}$. For large eigenvalues $\sigma_i$'s, the term $\frac{\sigma_i}{\alpha + \sigma_i^2} \approx \frac{1}{\sigma_i}$, and for very small values for $\sigma_i$, $\frac{\sigma_i}{\alpha + \sigma_i^2} \approx \frac{\sigma_i}{\alpha}$. Suitable choices for $\alpha$ do reduce the high frequencies resulted by the small eigenvalues.

A python code to solve the linear system $H\boldsymbol{x} = \boldsymbol{b}$, with $H$ the Hilbert matrix of type $20 \times 20$ and $\boldsymbol{b}$ the $20 \times 1$ vector whose components are the sums of $H$ rows.

```python
import numpy as np
from scipy.linalg import hilbert as hilb
from numpy.linalg import norm, svd
import matplotlib.pyplot as plt
def SolveWithTikhonov(A, b, Alpha):
    U, S, V = svd(A)
    x = np.zeros_like(b)
    n = len(S)
    for j in range(n):
        x += np.dot(U[:,j], b)*S[j]/(S[j]**2+Alpha)*V.T[:,j]
    return x

H = hilb(20)
y = np.ones((20,), 'float')
b = H@y
z = np.linalg.solve(H, b)
Eps = np.spacing(1.0)
Alpha=Eps
w = SolveWithTikhonov(H, b, Alpha)
print('Error norm for Unregularized solution = ', ...
    '{0:1.8e}'.format(norm(y-z)))
print('Error norm for Regularized solution = ', ...
    '{0:1.8e}'.format(norm(y-w)))
plt.figure(1)
plt.subplot(1, 2, 1)
t = np.arange(1, len(b)+1)
plt.plot(t, y, '-b', marker='s', lw=3, label='Exact solution')
plt.plot(t, z, '-.r', marker='o', lw=3, label='Unregularized ...
    solution')
plt.xlabel('Solution component', fontweight='bold')
plt.ylabel('Component value', fontweight='bold')
plt.grid(True, ls=':')
plt.legend()

plt.subplot(1, 2, 2)
t = np.arange(1, len(b)+1)
plt.plot(t, y, '-.b', marker='s', lw=3, label='Exact solution')
plt.plot(t, w, ':m', marker='o', lw=3, label='Regularized ...
    solution')
plt.xlabel('Solution component', fontweight='bold')
plt.ylabel('Component value', fontweight='bold')
plt.grid(True, ls=':')
plt.legend()

Err = np.zeros((16, 2), 'float')
for j in range(16):
    Alpha = 10**j*np.spacing(1.)
    w = SolveWithTikhonov(H, b, Alpha)
    Err[j, 0] = Alpha
    Err[j, 1] = norm(y-w)
    print('{0:1.6e}'.format(Err[j, 0]), '\t', ...
        '{0:1.8e}'.format(Err[j, 1]))
```

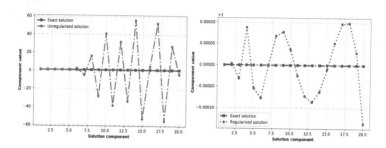

FIGURE 3.3: Unregularized solution and regularized solution with reg. param $\alpha = \varepsilon$ of $H\boldsymbol{x} = \boldsymbol{b}$.

In Figure 3.3 the unregularized solution and the regularized solution $\boldsymbol{x}^\alpha$ for $\alpha = \varepsilon$ are shown.

The norm error of regularized solutions $\boldsymbol{x}_j^\alpha$ where $\alpha_j = 10^{j-1}\varepsilon, j = 1, \ldots, 16$ are computed through the above Python code, and the outputs are as follows:

```
Error norm for Unregularized solution =  1.36970281e+02
Error norm for Regularized solution =  2.96674549e-04
2.220446e-16       2.96674549e-04
2.220446e-15       4.95587019e-04
2.220446e-14       7.25227169e-04
2.220446e-13       1.76193763e-03
2.220446e-12       2.40230468e-03
2.220446e-11       5.05832063e-03
2.220446e-10       8.76991238e-03
2.220446e-09       1.41598120e-02
2.220446e-08       2.96964698e-02
2.220446e-07       4.37394775e-02
2.220446e-06       9.44807071e-02
2.220446e-05       1.42053519e-01
2.220446e-04       2.95683135e-01
2.220446e-03       4.69592530e-01
2.220446e-02       9.24797653e-01
2.220446e-01       1.61864335e+00
```

The MATLAB code is:

```matlab
1   H = hilb(20) ;
2   y = ones(20, 1) ;
3   b = H * y ;
4   z = H\b ;
5   subplot(1, 2, 1) ;
6   plot(1:20, y, '-b', 1:20, z, '-.m', 'LineWidth', 3) ;
7   legend('Exact Solution', 'Unregularized Solution') ;
```

```
 8   xlabel('Solution Component') ;
 9   ylabel('Component Value') ;
10   grid on ;
11   format short e
12   disp('Error norm of unregularized solution ||y-z||_2'), ...
         disp(norm(y-z))
13   w = SolveWithTikhonov(H, b, eps) ;
14   subplot(1, 2, 2) ;
15   plot(1:20, y, '-b', 1:20, w, 'r:', 'LineWidth', 3) ;
16   legend('Exact Solution', 'Regularized Solution') ;
17   xlabel('Solution Component') ;
18   ylabel('Component Value') ;
19   grid on
20   disp('Error norm of regularized solution ||y-w||_2'), ...
         disp(norm(y-w))
21
22   x = zeros(20, 15) ;
23   Err = zeros(15, 2) ;
24   disp('-----------------------') ;
25   disp('      Alpha         ||x-y||_2') ;
26   disp('-----------------------') ;
27   for n = 1 : 15
28       Alpha = 10^(n-1)*eps ;
29       x(:, n) = SolveWithTikhonov(H, b, Alpha) ;
30       Err(n, 1) = Alpha ;
31       Err(n, 2) = norm(y-x(:, n), inf) ;
32   end
33   disp(Err) ;
34
35   function x = SolveWithTikhonov(A, b, Alpha)
36       [U, S, V] = svd(A) ;
37       D = diag(S) ;
38       [m, n] = size(A) ;
39       x = zeros(m, 1) ;
40       for i = 1 : n
41           x = x + U(:, i)'*b*D(i)/((D(i))^2+Alpha)*V(:, i) ;
42       end
43   end
```

In Tikhonov regularization further smoothing can be imposed on the vector of solution $x$, by imposing the smoothness of its $k^{th}$ derivative, using the $k^{th}$-order differential operator, $k = 1, 2, \ldots$. Then we consider a functional of the form

$$\varphi_\alpha(x) = \|Fx - y\|_2^2 + \alpha\|Lx\|_2^2 \tag{3.12}$$

where $L = D_k$ is the $k^{th}$ order differential operator.

For $k = 1$, $L = D_1$ is the $(n-1) \times n$ differential operator defined by

$$D_1 = \begin{bmatrix} -1 & 1 & 0 & \ldots & 0 & 0 \\ 0 & -1 & 1 & \ldots & 0 & 0 \\ \vdots & \vdots & \vdots & \ddots & \vdots & \vdots \\ 0 & 0 & 0 & \ldots & -1 & 1 \end{bmatrix}$$

and for $k = 2$, $L = D_2$ is the $(n-2) \times n$ differential operator defined by

$$D_2 = \begin{bmatrix} 1 & -2 & 1 & 0 & \cdots & 0 & 0 & 0 \\ 0 & 1 & -2 & 1 & \cdots & 0 & 0 & 0 \\ \vdots & \vdots & \vdots & \vdots & \ddots & \vdots & \vdots & \vdots \\ 0 & 0 & 0 & 0 & \cdots & 1 & -2 & 1 \end{bmatrix}$$

Using the first-order differential operator $L = D_1$, equation (3.12) takes the form

$$\varphi_\alpha(\boldsymbol{x}) = \|\boldsymbol{Fx} - \boldsymbol{y}\|_2^2 + \alpha \|D_1 \boldsymbol{x}\|_2^2 = \|\boldsymbol{Fx} - \boldsymbol{y}\|_2^2 + \alpha \sum_{i=1}^{m-1} (\boldsymbol{x}_{i+1} - \boldsymbol{x}_i)^2 \quad (3.13)$$

Using the second-order differential operator $L = D_2$, equation (3.12) takes the form

$$\varphi_\alpha(\boldsymbol{x}) = \|\boldsymbol{Fx} - \boldsymbol{y}\|_2^2 + \alpha \|D_2 \boldsymbol{x}\|_2^2 = \|\boldsymbol{Fx} - \boldsymbol{y}\|_2^2 + \alpha \sum_{i=1}^{m-2} (\boldsymbol{x}_{i+2} - 2\boldsymbol{x}_{i+1} - \boldsymbol{x}_i)^2$$

$$(3.14)$$

Minimizing $\varphi_\alpha(\boldsymbol{x})$ which appears in equation (3.12) is equivalent to solving the normal equations

$$(\boldsymbol{F}^T \boldsymbol{F} + \alpha L^T L)\boldsymbol{x}^\alpha = \boldsymbol{F}^T \boldsymbol{y} \qquad (3.15)$$

Hansen [22] considered the use of the generalized singular value decomposition (GSVD) to obtain the solution of problem (3.15). The generalized singular value decomposition for a pair $(\boldsymbol{F}, L)$, where $\boldsymbol{F} \in \mathbb{R}^{m \times n}$ and $L \in \mathbb{R}p \times n$, with $p \le n \le m$ is given by the form

$$\boldsymbol{F} = \boldsymbol{U} \begin{pmatrix} \Sigma & 0 \\ 0 & I_{n-p} \end{pmatrix} \boldsymbol{X}^{-1} \qquad L = \boldsymbol{V}(M \quad 0)\boldsymbol{X}^{-1} \qquad (3.16)$$

where

$$\Sigma = \begin{pmatrix} \sigma_1 & 0 & \cdots & 0 \\ 0 & \sigma_2 & \cdots & 0 \\ \vdots & \vdots & \ddots & \vdots \\ 0 & 0 & \cdots & \sigma_p \end{pmatrix}, \qquad M = \begin{pmatrix} \mu_1 & 0 & \cdots & 0 \\ 0 & \mu_2 & \cdots & 0 \\ \vdots & \vdots & \ddots & \vdots \\ 0 & 0 & \cdots & \mu_p \end{pmatrix}$$

with $0 \le \sigma_1 \le \sigma_2 \le \ldots \le \sigma_p \le 1$ and $0 < \mu_p \le \mu_{p-1} \le \ldots \le \mu_1 \le 1$
The sets $\{\sigma_i\}$ and $\{\mu_i\}$ are normalized such that $\sigma_i + \mu_i = 1$.
The solution $\boldsymbol{x}^\alpha$ of equation (3.15) is given by

$$\boldsymbol{x}^\alpha = \sum_{i=1}^{p} (\boldsymbol{U}_i^T \boldsymbol{y}) \frac{\sigma_i}{\sigma_i^2 + \alpha \mu_i} \boldsymbol{X}_i + \sum_{i=p+1}^{n} (\boldsymbol{U}_i^T \boldsymbol{y}) \boldsymbol{X}_i \qquad (3.17)$$

### 3.2.3 The L-curve Method

Using small values for the regularization parameter $\alpha$ on equations (3.11) and (3.15) may not inhibit the quantities $\frac{\sigma_i}{\sigma_i^2+\alpha}$ and $\frac{\sigma_i}{\sigma_i^2+\alpha\mu_i}$ effectively. Hence, not much improvement can be obtain for the smoothness of the unregularized solution, but it causes the regularized solution to stay close to the unregularized solution. On the other hand, using large values for the regularization parameter $\alpha$ do impose more smoothness on the regularized solution, but causes it to go away from the minimum residual value, obtained for the unregularized solution. Between these two extremes, optimal regularization parameters do lay.

When we plot the points $(\|\boldsymbol{F}\boldsymbol{x}^\alpha - \boldsymbol{y}\|, \|\boldsymbol{L}\boldsymbol{x}^\alpha\|)$ for different values of $\alpha$ on the loglog scale, it takes the shape of an L-curve. The L-curve does clearly display the compromise between minimization of the two quantities $\|\boldsymbol{F}\boldsymbol{x}^\alpha - \boldsymbol{y}\|$ and $\|\boldsymbol{L}\boldsymbol{x}^\alpha\|$) [22]. At the corner of this L-curve, optimal values for the regularization parameter $\alpha$ are located.

### 3.2.4 The Discrepancy Principle

Linz and Wang [33] considered the problem of solving a linear system $F\boldsymbol{x} = \boldsymbol{b}$, where $F$ is an $m \times n$ matrix with $m \geq n$, $\boldsymbol{x} \in \mathbb{R}^n$ and $\boldsymbol{b} \in \mathbb{R}^m$. He stated that, a solution $\boldsymbol{x}$ for this linear system is referred to as *acceptable* if it satisfies the condition

$$\|F\boldsymbol{x} - \boldsymbol{b}\| \leq \varepsilon$$

where $\varepsilon$ is the total error of the data and computations.

According to Linz and Wang [33] $\boldsymbol{x}$ is a `plausible` solution only if

$$\|L\boldsymbol{x}\| \leq M$$

where $M$ is a chosen positive number. Therefore, the task becomes to select $\alpha$, so that the regularized solution $\boldsymbol{x}^\alpha$ is both acceptable and plausible.

The solution $\boldsymbol{x}$ is the *most plausible acceptable* solution, if it solves the problem

$$\underset{\boldsymbol{x}\in\mathbb{R}^n}{\text{minimize}} = \|L\boldsymbol{x}\| \tag{3.18}$$

subject to the constraint

$$\|F\boldsymbol{x} - \boldsymbol{b}\| \leq \varepsilon \tag{3.19}$$

Problem (3.18)-(3.19) is equivalent to solve the problem

$$(F^T F + \alpha B^T B)\boldsymbol{x}_\alpha = F^T \boldsymbol{b} \tag{3.20}$$

subject to

$$\|F\boldsymbol{x}_\alpha - \boldsymbol{b}\| = \varepsilon \tag{3.21}$$

and the *Morozov discrepancy principle* is the problem of selecting $\alpha$ such that, (3.20-3.21) is satisfied.

# 4

## Solving a System of Nonlinear Equations

## Abstract

This chapter discusses the solutions of nonlinear systems using MATLAB®
and Python. It is divided into two sections. The first section presents four
numerical methods for solving single nonlinear equation. The second section
discusses numerical methods for solving a system of nonlinear equations.

## 4.1   Solving a Single Nonlinear Equation

Consider a nonlinear equation of the form:

$$f(x) = 0$$

where $f : \mathbb{R} \to \mathbb{R}$ is a nonlinear function. The problem is to find a point $x = x^*$
such that $f(x^*) = 0$. Then, $x = x^*$ is called a **root** of the function $f$. For
example, if $f(x) = e^{-x} - \cos x + 0.5$, then the root of $f(x) = e^{-x} - \cos x + 0.5 = 0$
lies at the intersection points of the curves $e^{-x} + 0.5$ and $\cos(x)$ as can be seen
in Figure 4.1.

In this section four methods for solving the given nonlinear equation,
namely, the bisection method, the Newton-Raphson method, the secant and the
iterative method will be discussed and implemented in MATLAB and Python.

### 4.1.1   The Bisection Method

Assume that the function $f(x)$ is continuous in an interval $[a, b]$ and is changing
its sign from positive to negative or vice versa as it moves from $a$ to $b$. A
direct result from the intermediate value theorem is that a continuous function
cannot change its sign in an interval, without passing through zero [52].

The bisection method is an iterative method, which in each iteration divides
the interval that contains the root into two equally subintervals, drops the half
which does not contain the root and looks for the root in the other half.

If the interval $[a, b]$ is divided into two equal subintervals $[a, c]$ and $[c, b]$,
where $c = (a + b)/2$ is the midpoint of the interval $[a, b]$. The root of the

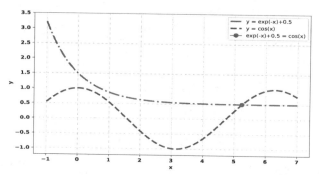

FIGURE 4.1: Graphs of the functions $e^{-x}+0.5$ (dashed curve) and $\cos(x)$ (dash-dotted curve). The x-coordinate of circle at the intersection of the two curves is the root of $e^{-x}-\cos x+0.5=0$ in $[4,6]$.

function $f$ either lies in $[a,c]$ or $[c,b]$. If it lies in $[a,c]$, then $f(a)\cdot f(c)<0$. In this case, we know that the root does not lie in the interval $[c,b]$ and we look for the root in $[a,c]$. If it lies in $[c,b]$, then $f(c)\cdot f(b)<0$ and in this case, we know that the root does not lie in $[a,c]$, so we look for it in $[c,b]$. The bisection method continues dividing the interval, which contains the root into two sub-intervals, such that one sub-interval contains the root whereas the other does not; therefore, the method considers only the interval which contains the root and drops the other half.

The MATLAB code that implements the bisection method is as follows:

```
1   function x = Bisection(f, a, b, Epsilon)
2       while b-a ≥ Epsilon
3           c  = (a+b)/2 ;
4           if f(a)*f(c) < 0
5               b = c ;
6           elseif f(b)*f(c) < 0
7               a = c ;
8           else
9               x = c ;
10          end
11      end
12      x = c ;
```

Now, we can call the function `Bisection` from the command prompt:

```
>> format long
>> Epsilon = 1e-8 ;
>> f = @(x) x^2 - 3 ;
>> r = Bisection(f, 1, 2, Epsilon)
r =
1.732050813734531

>> s = Bisection(f, -2, -1, Epsilon)
s =
-1.732050813734531
```

The Python code to implement the `Bisection` function is:

```
 1  def Bisection(f, a, b, Eps):
 2      from math import fabs
 3      c = (a+b)/2.0
 4      Iters = 1
 5      while fabs(f(c)) >= Eps:
 6          if f(a)*f(c) < 0:
 7              b = c
 8          else:
 9              a = c
10          c = (a+b)/2
11          Iters += 1
12      return c, Iters
```

```
In [1]: f = lambda x: x**2 - 3
In [2]: a, b = 1., 2.
In [3]: Eps = 1e-8
In [4]: x, Iters = Bisection(f, a, b, Eps)
In [5]: print('Approximate root is:', x, '\nIterations:', Iters)
Approximate root is: 1.7320508062839508
Iterations: 25
```

## 4.1.2   The Newton-Raphson Method

If a function $f(x)$ is continuous in the neighbourhood of a point $x_0$, then it can be written in the form:

$$f(x) = f(x_0) + f'(x_0)(x - x_0) + \mathcal{O}((x - x_0)^2)$$

Now, if $x_1$ is a root for the function $f(x)$, then $f(x_1) = 0$. That is:

$$f(x_0) + f'(x_0)(x_1 - x_0) \approx f(x_1) = 0$$

From the above equation:

$$x_1 = x_0 - \frac{f(x_0)}{f'(x_0)}$$

provided that $f'(x_0) \neq 0$.

The Newton-Raphson method is an iterative method for finding an approximation to the closest root of the function $f(x)$ to an initial guess $x_0$. Starting from $x_0$, it generates a sequence of numbers $x_0, x_1, x_2, \ldots$, where,

$$x_{n+1} = x_n - \frac{f(x_n)}{f'(x_n)}$$

This sequence of numbers converges to the closest root of $f$ to $x_0$.

To find a root for the function $f(x) = x^2 - 3$, using the Newton-Raphson method, it is noticed that $f'(x) = 2x$, and therefore, the iterative method is of the form:

$$x^{(n+1)} = x^{(n)} - \frac{f(x^{(n)})}{f'(x^{(n)})} = x^{(n)} - \frac{(x^{(n)^2} - 3)}{2x^{(n)}}$$

The MATLAB function `NewtonRaphson.m`, implements the Newton-Raphson method:

```
1  function [x, Iter] = NewtonRaphson(f, fp, x0, Epsilon)
2      Iter = 0 ;
3      x1 = x0 - f(x0)/fp(x0) ;
4      while abs(f(x1)) ≥ Epsilon
5          x0 = x1 ;
6          x1 = x0 - f(x0)/fp(x0) ;
7          Iter = Iter + 1 ;
8      end
9      x = x1 ;
```

Calling the `NewtonRaphson` function from the command prompt:

```
>> format long
>> f = @(x) x^2 - 3 ;
>> fp = @(x) 2*x ;
>> Epsilon = 1e-8 ;
>> x0 = 1 ;
>>[x, Iterations] = NewtonRaphson(f, fp, x0, Epsilon)
x =
1.732050810014728
Iterations =
3
>> [x, Iterations] = NewtonRaphson(f, fp, -x0, Epsilon)
x =
-1.732050810014728
Iterations =
3
```

The code of the Python function `NewtonRaphson` is:

```
1  def NewtonRaphson(f, fp, x0, Eps):
2      from math import fabs
3      x = x0 - f(x0)/fp(x0)
4      Iters = 1
5      while fabs(f(x)) ≥ Eps:
6          x0 = x
7          x = x0 - f(x0)/fp(x0)
8          Iters += 1
9      return x, Iters
```

Running the above code with $x_0 = 1.0$ one time and $x_0 = -1.0$ another time:

```
In [6]: x, Iters = NewtonRaphson(f, fp, x0, Eps)
In [7]: print('Approximate root is:', x, '\nIterations:', Iters)
Approximate root is: 1.7320508100147276
Iterations: 4
In [8]: x, Iters = NewtonRaphson(f, fp, -x0, Eps)
In [9]: print('Approximate root is:', x, '\nIterations:', Iters)
Approximate root is: -1.7320508100147276
Iterations: 4
```

## 4.1.3    The Secant Method

The secant method has a close form as the Newton-Raphson method, but it does not require the analytical form of the derivative of $f(x)$ at $x_n$ ($f'(x_n)$). It replaces $f'(x_n)$ by the finite difference formula:

$$f'(x_n) \approx \frac{f(x_n) - f(x_{n-1})}{x_n - x_{n-1}}.$$

Hence, the secant method is of the form:

$$x_{n+1} = x_n - \frac{x_n - x_{n-1}}{f(x_n) - f(x_{n-1})} f(x_n)$$

Starting from some interval $[a,b]$ that contains a root for $f(x)$, the secant method iteration approaches the zero of $f(x)$ in $[a,b]$.

The MATLAB function Secant implements the secant method. It receives a function f, the limits of the interval $[a,b]$ that contains the root of $f$ and a tolerance $\varepsilon > 0$. It applies the secant method to return an approximate solution $x$ and the number of iterations *Iterations*

```
1  function [x, Iterations] = Secant(f, a, b, Eps)
2      x = b - ((b-a)*f(b))/(f(b)-f(a)) ;
3      Iterations = 1 ;
4      while fabs(f(x)) ≥ Eps
5          a = b ;
6          b = x ;
7          x = b - ((b-a)*f(b))/(f(b)-f(a)) ;
8          Iterations = Iterations + 1 ;
9      end
```

Calling the MATLAB function to find the approximate root of $x^2 - 3 = 0$ as follows.

```
>> f = @(x) x^2-3 ;
>> a = 1; b = 2 ;
>> Eps = 1e-8 ;
>> [x, Iterations] = Secant(f, a, b, Eps)
```

```
x =
1.732050807565499
Iterations =
5
```

The Python code of the function **Secant** is as follows.

```
1  def Secant(f, a, b, Eps):
2      from math import fabs
3      x = b - ((b-a)*f(b))/(f(b)-f(a))
4      Iterations = 1
5      while fabs(f(x)) >= Eps:
6          a, b = b, x
7          x = b - ((b-a)*f(b))/(f(b)-f(a))
8          Iterations += 1
9      return x, Iterations
```

To find the root of the equation $x^2 - 3 = 0$, the following Python instructions are used:

```
In [10]: f = lambda x: x**2-3
In [11]: a, b = 1., 2.
In [12]: Eps = 1e-8
In [13]: x, Iterations = Secant(f, a, b, Eps)
In [14]: print('Approximate root is:', x, '\nIterations:', Iterations)
Approximate root is: 1.732050807565499
Iterations: 5
```

## 4.1.4    The Iterative Method Towards a Fixed Point

A point $x^*$ is a fixed point for a function $g(x)$, if

$$g(x^*) = x^*$$

Suppose that the function $f(x)$ is differentiable and can be written as

$$f(x) = g(x) - x$$

If $x_1$ is a root of $f(x)$, then,

$$f(x_1) = g(x_1) - x_1 = 0 \Rightarrow x_1 = g(x_1)$$

That means $x_1$ is a fixed point for the function $g(x)$. The idea behind the iterative method towards a fixed point is to write the function $f(x)$ in the shape $g(x) - x$, and starting from some initial guess $x_0$, the method generates a sequence of numbers $x_0, x_1, x_2, \ldots$ that converges to the fixed point of the function $g(x)$, using the iterative rule:

$$x_{n+1} = g(x_n)$$

To show how the the iterative method works, it will be applied to find the roots of the function

$$f(x) = x^2 - 3$$

first the function $x^2 - 3$ is written in the form $g(x) - x$. One possible choice is to write:

$$f(x) = x^2 - 3 = x^2 + 2x + 1 - 2x - 4 = (x+1)^2 - 2x - 4$$

If $x^*$ is a root for $f(x)$, then:

$$f(x^*) = (x^* + 1)^2 - 2x^* - 4 = 0$$

from which

$$x^* = \frac{(x^* + 1)^2 - 4}{2}$$

Starting from an initial point $x_0$, the iterative method to find a root for $f(x)$ is:

$$x^{(n+1)} = \frac{(x^{(n)} + 1)^2 - 4}{2}$$

The following MATLAB code, computes a root of the function $f(x) = x^2 - 3$ using the iterative method:

```
1  function [x, Iterations] = IterativeMethod(g, x0, Epsilon)
2      Iterations = 0 ;
3      x = g(x0) ;
4      while abs(x-x0) ≥ Epsilon
5          x0 = x ;
6          x = g(x0) ;
7          Iterations = Iterations + 1 ;
8      end
```

From the command prompt, the following instructions are typed:

```
>> g = @(x)((x+1)^2-4)/2 ;
>> x0 = 1 ;
>> Epsilon = 1e-8 ;
>> [x, Iterations] = IterativeMethod(g, x0, Epsilon)
x =
-1.732050811416889
Iterations =
58
```

In Python, the code of the function IterativeMethod is:

```
1  def IterativeMethod(g, x0, Eps):
2      from math import fabs
3      x = g(x0)
4      Iters = 1
```

```
5        while fabs(x - x0) >= Eps:
6            x0 = x
7            x = g(x0)
8            Iters += 1
9        return x, Iters
```

Running the function `IterativeMethod` with $g(x) = ((x-1)^2 - 4)/2$ one time and $g(x) = (4 - (x+1)^2)/2$ another time:

```
In [15]: x0, Eps = 1.0, 1e-8
In [16]: g = lambda x: ((x+1)**2-4)/2.0
In [17]: x, Iters = IterativeMethod(g, x0, Eps)
In [18]: print('Approximate root is:', x, '\nIterations:', Iters)
Approximate root is: -1.732050811416889
Iterations: 59
In [19]: g = lambda x: (4-(x-1)**2)/2.0
In [20]: x, Iters = IterativeMethod(g, x0, Eps)
In [21]: print('Approximate root is:', x, '\nIterations:', Iters)
Approximate root is: 1.732050811416889
Iterations: 59
```

**Note:** It is worthy to notice that the selection of the function $g(x)$ is not unique. For example, for the function $f(x) = x^2 - 3$, the following iterative forms can be used:

1. $x^{(n+1)} = \frac{4 - (x^{(n)} - 1)^2}{2}$

2. $x^{(n+1)} = \frac{(3 - x^{(n)})(1 + x^{(n)})}{2}$

3. $x^{(n+1)} = \frac{(x^{(n)} + 3)(x^{(n)} - 1)}{2}$

The iteration $x_{n+1} = \frac{3}{x_n}$ cannot work.

## 4.1.5   Using the MATLAB and Python `solve` Function

The MATLAB solve function can be used to find all the roots of a nonlinear equation $f(x) = 0$. The MATLAB function solve belongs to the symbolic toolbox. It finds the roots of $f(x) = 0$ analytically.

To solve the nonlinear equations:

1. $x^2 - 3 = 0$

2. $e^{-x} = \sin x$

3. $x^3 + \cos x = \ln x$

we can use the MATLAB commands:

```
>> x = solve('x^2 - 3')
x =
3^(1/2)
```

```
-3^(1/2)
>> x = solve('exp(-x)-sin(x)')
x =
0.5885327439818610774324520457029
>> x = solve('x^3-cos(x)+log(x)')
x =
0.89953056480788905732035721409122
```

In Python, we use the `sympy.solve` function to solve the nonlinear equations:

```
In [22]: from sympy import *
In [23]: x = Symbol('x')
In [24]: solve(x**2-3, x)
[-sqrt(3), sqrt(3)]
```

Note: The Python symbolic library does not look as mature as the MAT-LAB symbolic toolbox. The second and third problems cannot be solved with Python, but MATLAB solves them. Python raises the exception:

```
In [25]: solve(exp(-x)-sin(x), x)
Traceback (most recent call last):
File "<ipython-input-43-7d4dd4404520>", line 1, in <module>
solve(exp(-x)-sin(x), x)
raise NotImplementedError('\n'.join([msg, not_impl_msg % f]))
NotImplementedError: multiple generators [exp(x), sin(x)]
No algorithms are implemented to solve equation -sin(x) + exp(-x)
In [26]: solve(x**3-cos(x)+log(x))
Traceback (most recent call last):
...
NotImplementedError: multiple generators [x, cos(x), log(x)]
No algorithms are implemented to solve equation x**3 + log(x) - cos(x)
```

## 4.2  Solving a System of Nonlinear Equations

In this section, a system of linear equations of the form is considered:

$$
\begin{aligned}
f_1(x_1, x_2, \ldots, x_n) &= 0 \\
f_2(x_1, x_2, \ldots, x_n) &= 0 \\
&\vdots \\
f_n(x_1, x_2, \ldots, x_n) &= 0
\end{aligned}
$$

where, $f_1, f_2, \ldots, f_n$ are nonlinear functions in the variables $x_1, x_2, \ldots, x_n$.

We can write the above system in the vector form, by writing:

$$\boldsymbol{x} = \begin{pmatrix} x_1 \\ x_2 \\ \vdots \\ x_n \end{pmatrix} \text{ and } \boldsymbol{f}(\boldsymbol{x}) = \begin{pmatrix} f_1(\boldsymbol{x}) \\ f_2(\boldsymbol{x}) \\ \vdots \\ f_n(\boldsymbol{x}) \end{pmatrix}$$

If $\boldsymbol{x}^*$ is any point in $\mathbb{R}^n$, the approximation by the first two terms of the Taylor expansion of $\boldsymbol{f}(\boldsymbol{x})$ around $\boldsymbol{x}^*$ is given by:

$$\boldsymbol{f}(\boldsymbol{x}) \approx \boldsymbol{f}(\boldsymbol{x}^*) + J(\boldsymbol{x}^*)(\boldsymbol{x} - \boldsymbol{x}^*)$$

where $J(\boldsymbol{x}^*)$ is the Jacobian matrix and the $ij$ component of the $J$ is defined by:

$$(J(\boldsymbol{x}^*))_{ij} = \left. \frac{\partial f_i(\boldsymbol{x})}{\partial x_j} \right|_{(\boldsymbol{x}=\boldsymbol{x}^*)}$$

Now, if we set $\boldsymbol{f}(\boldsymbol{x}) = 0$, we obtain the equation:

$$\boldsymbol{f}(\boldsymbol{x}^*) + J(\boldsymbol{x}^*)(\boldsymbol{x} - \boldsymbol{x}^*) = 0$$

from which we find that

$$\boldsymbol{x} = \boldsymbol{x}^* - J^{-1}(\boldsymbol{x}^*)\boldsymbol{f}(\boldsymbol{x}^*)$$

Starting from an initial guess $\boldsymbol{x}^{(0)}$ for the solution of $\boldsymbol{f}(\boldsymbol{x}) = 0$, the iteration:

$$\boldsymbol{x}^{(n+1)} = \boldsymbol{x}^{(n)} - J^{-1}(\boldsymbol{x}^{(n)})\boldsymbol{f}(\boldsymbol{x}^{(n)})$$

converges to the closest solution of $\boldsymbol{f}(\boldsymbol{x}) = 0$ to the initial point $\boldsymbol{x}^{(0)}$.

In MATLAB, the function 'jacobian' can be used to find the Jacobian matrix of the nonlinear system of equations $\boldsymbol{f}(\boldsymbol{x})$.

**Example 4.1** Use MATLAB to find a solution to the nonlinear system of equations:

$$\begin{aligned} x^2 + y^2 &= 30 \\ -x^2 + y^2 &= 24 \end{aligned}$$

We write:

$$\boldsymbol{f}(x, y) = [x^2 + y^2 - 30, -x^2 + y^2 - 24]^T$$

The Jacobian matrix for the nonlinear system is given by:

$$J(x, y) = \begin{pmatrix} 2x & 2y \\ -2x & 2y \end{pmatrix}$$

Starting from an initial guess $z^0 = (x^0, y^0)^T$, the iterative method is:

$$z^{(n+1)} = z^{(n)} - J^{-1}(z^{(n)})f(z^{(n)})$$
$$= \begin{pmatrix} x^{(n)} \\ y^{(n)} \end{pmatrix} - \begin{pmatrix} 2x^{(n)} & 2y^{(n)} \\ -2x^{(n)} & 2y^{(n)} \end{pmatrix}^{-1} \begin{pmatrix} x^{(n)2} + y^{(n)2} - 30 \\ -x^{(n)2} + y^{(n)2} - 24 \end{pmatrix}$$

The following MATLAB code implements the solution of the above given problem, starting from the initial guess $[1,1]^T$.

```
1        % SolveWithJacobi.m
2   function [z, Iterations] = Newton_sys(f, J, z0, Eps)
3        iJ = @(z) inv(J(z)) ;
4        z = z0 - iJ(z0)*f(z0) ;
5        Iters = 0 ;
6        while norm(z-z0, inf) ≥ Epsilon
7            z0 = z ;
8            z = z0 - iJ(z0)*f(z0) ;
9            Iters = Iters + 1 ;
10       end
```

The execution of the above code, will give the results:

```
>> f = @(z)[z(1)^2 + z(2)^2 - 30; z(1)^2 + z(2)^2 - 24] ;
>> J = @(z) [2*z(1), 2*z(2); -2*z(1), 2*z(2)] ;
>> z0 = [1; 1] ;
>> Eps = 1e-8 ;
>> [z, Iterations] = Newton_sys(f, J, z0, Eps) ;
>> fprintf('Iterations = %i\n', Iters) ;
fprintf('The solution of the system is given by:\n\n\t\t\t\t
x1 = %18.15f\n\t\t\t\t x2 = %18.15f\n\n', z(1), z(2)) ;
Iterations = 6
The solution of the system is given by:
x1 =   1.732050807568877
x2 =   5.196152422706632
```

The Python code for the function `SolveWithJacobi` is:

```
1   from numpy import array, inf
2   from numpy.linalg import norm, inv
3   def SolveWithJacobi(f, J, z0, Eps):
4        iJ = lambda z: inv(J(z))
5        z = z0 - iJ(z0)@f(z0)
6        Iterations = 1
7        while norm(z-z0, inf) ≥ Eps:
8            z0 = z
9            z = z0 - iJ(z0)@f(z0)
10           Iterations += 1
11       return z, Iterations
```

```
In [27]: f = lambda z: array([z[0]**2+z[1]**2-30,
-z[0]**2+z[1]**2-24])
In [28]: J = lambda z: array([[2*z[0], 2*z[1]], [-2*z[0],
2*z[1]]])
In [29]: Eps = 1e-8
In [30]: z0 = array([1., 1.])
In [31]: z, Iterations = SolveWithJacobi(f, J, z0, Eps)
In [32]: print('Approximate solution: (Iterations = ',
Iterations,') \nx =', z[0], '\ny =', z[1])
Out[33]: Approximate solution: (Iterations =  7 )
x = 5.196152422706632
y = 1.7320508075688774
```

It is also possible to use the MATLAB function 'solve' to find all the solutions of $f(x) = 0$, by using the MATLAB command:

```
[x1, x1, ..., xn] = solve([f1, f2, ..., fn], [x1, x2, ..., xn]);
```

For example:

```
>> [x, y] = solve('x^2+y^2-30', '-x^2+y^2-24', [x y])
x =
3^(1/2)
3^(1/2)
-3^(1/2)
-3^(1/2)
y =
3*3^(1/2)
-3*3^(1/2)
3*3^(1/2)
-3*3^(1/2)
```

Using the **sympy** library, the problem can be solved through the following code:

```
In [34]: from sympy import *
In [35]: x, y = symbols('x, y')
In [36]: solve([x**2+y**2-30, -x**2+y**2-24], [x, y])
Out[37]:
[(-sqrt(3), -3*sqrt(3)),
(-sqrt(3), 3*sqrt(3)),
(sqrt(3), -3*sqrt(3)),
(sqrt(3), 3*sqrt(3))]
```

**Example 4.2** We use Newton's method to solve the nonlinear system [47]:

$$3x_1 - \cos x_2 x_3 - 0.5 = 0$$
$$x_1^2 - 81(x_2 + 0.1)^2 + \sin x_3 + 1.06 = 0$$
$$e^{-x_1 x_2} + 20x_3 + \frac{10\pi - 3}{3} = 0$$

We write:

$$f = [3x_1 - \cos x_2 x_3 - 0.5, x_1^2 - 81(x_2+0.1)^2 + \sin x_3 + 1.06, e^{-x_1 x_2} + 20x_3 + \frac{10\pi - 3}{3}]^T$$

The Jacobian matrix associated to the given system is:

$$J(x_1, x_2, x_3) = \begin{pmatrix} 3 & x_3 \sin x_2 x_3 & x_2 \sin x_2 x_3 \\ 2x_1 & -162x_2 - 81/5 & \cos x_3 \\ -x_2 e^{-x_1 x_2} & -x_1 e^{-x_1 x_2} & 20 \end{pmatrix}$$

If Newton's iteration starts from an initial condition $z^{(0)} = (x_1^{(n)}, x_2^{(n)}, x_3^{(n)})^T$, then Newton's iteration is given by:

$$z^{(n+1)} = z^{(n)} - J^{-1}(z^{(n)}) f(z^{(n)})$$

The MATLAB code to compute the solution using the above iteration is given by:

```
>> clear ; clc ;
>> f = @(z)[3*z(1)-cos(z(2)*z(3)-1/2 ;
z(1)^2-81*(z(2)+0.1)^2+sin(z(3))+1.06 ;
exp(-z(1)*z(2))+20*z(3)+(10*pi-3)/3] ;
>> J = @(z) [3, z(3)*sin(z(2)*z(3)), z(2)*sin(z(2)*z(3));
2*z(1), - 162*z(2) - 81/5, cos(z(3));
-z(2)*exp(-z(1)*z(2)), -z(1)*exp(-z(1)*z(2)), 20] ;
>> z0 = [0.1; 0.1; -0.1] ;
>> Eps = 1e-8 ;
>> z, Iterations = Newton_sys(f, J, z0, Eps)
>> fprintf('Iterations = %i\n', Iterations) ;
Iterations = 4
>> fprintf('The solution of the system is given by:\n\n\t\t\t\t x1 =
%18.15f\n\t\t\t\t x2 = %18.15f\n\t\t\t\t x3 = %18.15f\n', z1(1), z1(2),
z1(3)) ;
The solution of the system is given by:
x1 =   0.500000000000000
x2 = -0.000000000000000
x3 = -0.523598775598299
```

Using Python, this example can be solved by calling the function Newton_sys as follows:

```
In [38]: f = lambda z: array([3*z[0]-cos(z[1]*z[2])-1/2,
z[0]**2-81*(z[1]+0.1)**2+sin(z[2])+1.06,
exp(-z[0]*z[1])+20*z[2]+(10*pi-3)/3])
In [39]: J = lambda z: array([[3, z[2]*sin(z[1]*z[2]), z[1]*sin(z[1]
*z[2])], [2*z[0], - 162*z[1] - 81/5, cos(z[2])],
[-z[1]*exp(-z[0]*z[1]), -z[0]*exp(-z[0]*z[1]), 20]])
In [40]: z0 = array([1., 1., 1.])
In [41]: z, Iterations = Newton_sys(f, J, z0, Eps)
```

```
In [42]: print('Approximate solution: (Iterations = ', Iterations,')
\nx =', z[0], '\ny =', z[1], '\nz =', z[2])
Approximate solution: (Iterations =  8 )
x = 0.49999999999999994
y = -1.6530395442910908e-17
z = -0.5235987755982988
```

# Part II

# Data Interpolation and Solutions of Differential Equations

# 5

## Data Interpolation

## Abstract

Data interpolation means to use a given set of $n+1$ data points to approximate a function $f(x)$ by a polynomial $P_n(x) = a_n x^n + a_{n-1} x^{n-1} + \ldots + a_1 x + a_0$ (of degree not exceeding $n$), such that $P_n(x_i) = f(x_i)$, $i = 0, \ldots, n$, where $a_0, \ldots, a_n$ are constants. The data points are given by the table:

| $x$ | $x_0$ | $x_1$ | $\ldots$ | $x_n$ |
|-----|-------|-------|----------|-------|
| $f(x)$ | $f(x_0)$ | $f(x_1)$ | $\ldots$ | $f(x_n)$ |

where $x_i \neq x_j$ for $i \neq j$ and $x_0 < x_1 < \cdots < x_n$.

This chapter discusses some of the interpolation methods and their implementation in MATLAB® and Python. It is divided into four sections. Section 1 discusses Lagrange interpolation and its implementation in MATLAB and Python. Section 2 discusses Newton's interpolation and the divided difference technique for finding the coefficients of Newton's interpolation. One-dimensional interpolations with MATLAB and Python are discussed in Sections 3 and 4.

## 5.1 Lagrange Interpolation

This section is divided into three sections. In the first section, construction and implementation of the Lagrange interpolating polynomial from a given data will be discussed. The proof of uniqueness of the Lagrange interpolating polynomial will be discussed in Section 2. In the last section, a formula of the interpolation error, using Lagrange interpolation will be presented.

### 5.1.1 Construction of Lagrange Interpolating Polynomial

In Lagrange interpolation, $n+1$ polynomials $L_0(x), L_1(x), \ldots, L_n(x)$ are constructed. Each polynomial $L_j(x), (j = 0, \ldots, n)$ is of degree $n$, such that

$$L_i(x_j) = \begin{cases} 1, & i = j \\ 0, & i \neq j \end{cases}$$

Then, the interpolating polynomial $P_n(x)$ is defined by the formula:

$$P_n(x) = L_0(x)y_0 + L_1(x)y_1 + \ldots + L_n(x)y_n = \sum_{i=0}^{n} L_i(x)y_i \qquad (5.1)$$

The polynomial $P_n(x)$ is of degree $n$, because each $L_i(x)$ is an $n^{th}$-degree polynomial, for all $i = 0, \ldots, n$. Moreover, $P_n(x)$ satisfies:

$$P_n(x_j) = \sum_{i=0}^{n} L_i(x_j)y_i = L_j(x_j)y_j = 1 \cdot y_j = y_j, \; j = 0, \ldots, n$$

This proofs that $P_n(x)$ interpolates the given data.

To construct the $n+1$ polynomials $L_i(x)$, $i = 0, \ldots, n$, it is important to notice that the roots of

$$\psi(x) = (x - x_0)(x - x_1) \ldots (x - x_n)$$

are $x_0, x_1, \ldots, x_n$. If $\tilde{L}(x)$ is defined by

$$\tilde{L}(x) = \frac{\psi(x)}{x - x_i} = (x - x_0)(x - x_1) \ldots (x - x_{i-1})(x - x_{i+1}) \ldots (x - x_n),$$

then $x_i$ is not a root of $\tilde{L}(x)$.

Now, $L_i(x)$ can be obtained from $\tilde{L}(x)$ using the formula:

$$L_i(x) = \frac{\tilde{L}_i(x)}{\tilde{L}_i(x_i)} = \frac{(x - x_0)(x - x_1) \ldots (x - x_{i-1})(x - x_{i+1}) \ldots (x - x_n)}{(x_i - x_0)(x_i - x_1) \ldots (x_i - x_{i-1})(x_i - x_{i+1}) \ldots (x_i - x_n)}$$

$$(5.2)$$

The polynomials $L_i(x)$ can also be written as:

$$L_i(x) = \prod_{j=0}^{n} \frac{x - x_j}{x_i - x_j},$$

and the Lagrange polynomial $P_n(x)$ can be written as:

$$P_n(x) = \sum_{i=0}^{n} \prod_{j=0}^{n} \frac{x - x_j}{x_i - x_j} y_i$$

The following MATLAB function `LagrangeInterp.m` constructs the Lagrange interpolation polynomial $P_n(x)$ using the data points given be two vectors $\boldsymbol{x}$ and $\boldsymbol{y}$, each of length $n$. It evaluates $P_n(x)$ at $x = t$ and returns the interpolation result.

```
1  function p = LagrangeInterp(t, x, y)
2  n = length(x) ;
3  p = 0 ;
```

```
 4  for i = 1 : n
 5  s = 1 ;
 6  for j = 1 : n
 7  if j ≠ i
 8  s = s*(t-x(j))/(x(i)-x(j)) ;
 9  else
10  continue
11  end
12  end
13  p = p + s*y(i) ;
14  end
```

Executing the above MATLAB code, we obtain:

```
>> x = 0:pi/11:pi ;
>> y = sin(x) ;
>> p = LagrangeInterp(pi/6, x, y)
p =
0.499999999914085
```

A Python script `LagInterp.py` implements the Lagrange interpolating polynomials with the following code:

```
 1  # LagInterp.py
 2  import numpy as np
 3  def LagrangeInterp(t, x, y):
 4  n = len(x)
 5  p = 0.0
 6  for i in range(n):
 7  s = 1
 8  for j in range(n):
 9  if j != i:
10  s *= (t-x[j])/(x[i]-x[j])
11  else:
12  continue
13  p += s*y[i]
14  return p
15
16  x = np.linspace(0.0, np.pi, 11)
17  y = np.sin(x)
18  p = LagrangeInterp(np.pi/6, x, y)
19  print(p)
```

Executing the code, shows the following results:

```
runfile('D:/PyFiles/LagInterp.py', wdir='D:/PyFiles')
0.4999999997868132
```

## 5.1.2  Uniqueness of Lagrange Interplation Polynomial

In this section *the fundamental theorem of algebra* will be used to prove that the Lagrange interpolating polynomial is unique. It states that *any polynomial of degree n with complex coefficients has at most n zeros.* So, if a polynomial $p(x)$ of degree $n$ has more that $n$ roots, it must be the zero polynomial $p(x) = 0, \forall x \in \mathbb{C}$.

To prove the uniqueness of Lagrange interpolating polynomial, assume that $p_n(x)$ and $q_n(x)$ are two polynomials of degree $n$, that interpolate the given data $(x_i, y_i), i = 0, \ldots, n$. Then, $p_n(x_i) = q_n(x_i) = y_i$ for all $i = 0, \ldots, n$. Let $r(x) = p_n(x) - q_n(x)$ for all $x \in [x_0, x_n]$. Since both $p_n(x)$ and $q_n(x)$ are both of degrees at most $n$, then so is $r(x)$. Now, $r(x_i) = p_n(x_i) - q_n(x_i) = 0$ for $i = 0, \ldots, n$. That is $r(x)$ has $n+1$ roots in $[x_0, x_n]$, while it is of degree $n$. From the fundamental theorem of algebra, this cannot happen unless $r(x)$ is the zero polynomial in $[x_0, x_n]$, that is $r(x) = 0, \ \forall x \in [x_0, x_n]$. This proves that $p_n(x) = q_n(x)$, for $x \in [x_0, x_n]$.

### 5.1.3   Lagrange Interpolation Error

Given a function $f : [a, b] \to \mathbb{R}$ and $n+1$ points $x_0, \ldots, x_n$ in $[a, b]$. Let $p_n(x)$ be a polynomial of degree $n$ that interpolates the data $(x_0, f(x_0)), \ldots, (x_n, f(x_n))$. that is $p_n(x_j) = f(x_j)$ for all $j = 0, \ldots, n$.

From the Lagrange interpolation, $p_n(x)$ is given by:

$$p_n(x) = \sum_{i=0}^{n} L_i(x) f(x_i)$$

Define the interpolation error by:

$$E_n(x) = f(x) - p_n(x), \ x \in [a, b] \tag{5.3}$$

Assuming that $f$ is differentiable continuously, a formula for the function $E_n(x)$ will be derived.

The following theorem gives an estimate for Lagrange interpolation.

**Theorem 5.1** *Let $x_0, x_1, \ldots, x_n$ be $n+1$ distinct values and $t$ any real value. Let $a = \min\{x_0, \ldots, x_n\}$ and $b = \max\{x_0, \ldots, x_n\}$. Let $f$ be a real valued function on $[a, b]$ that is differentiable $n+1$ times. Let $p_n(x)$ be the Lagrange polynomial that interpolates the data $(x_0, f(x_0)), \ldots, (x_n, f(x_n))$. Then, there exists $\xi \in [a, b]$ such that*

$$E_n(x) = f(x) - p_n(x) = \frac{\psi(x)}{(n+1)!} f^{(n+1)}(\xi),$$

**Proof:**
It is clear that the result is true for $x = x_i, \forall i = 0, \ldots, n$. Assume that the result is true when $x \neq x_i$. Fix a value $t \in [a, b]$ and define:

$$G(x) = \frac{\psi(x)}{\psi(t)} E_n(t), \ x \in [a, b],$$

where $\psi(x) = \prod_{j=0}^{n} (x - x_j)$.

Since the first $n+1$ derivatives exist for both $\psi(x)$ and $E_n(x)$, then so is $G(x)$. Therefore, $G(x)$ has $n+2$ distinct roots in $[a,b]$. From the intermediate value theorem, $G'(x)$ has $n+1$ distinct roots in $(a,b)$ and $G''(x)$ has $n$ distinct roots in $(a,b)$. By the mathematical induction, $G^{(k)}$ has $n-k+2$ distinct roots in $(a,b)$. Hence, $G^{(n+1)}$ has at least one root in $(a,b)$. Let $\xi$ be a root for $G^{(n+1)}$ in $(a,b)$, that is $G^{(n+1)}(\xi) = 0$.

From the definitions of $E_n(x)$ and $\psi(x)$, $E_n^{(n+1)}(x) = f^{(n+1)}(x)$ and $\psi^{(n+1)} = (n+1)!$. Then,

$$G^{(n+1)}(x) = f^{(n+1)}(x) - \frac{(n+1)!}{\psi(t)} E_n(t)$$

At $x = \xi$,

$$G^{(n+1)}(\xi) = f^{(n+1)}(\xi) - \frac{(n+1)!}{\psi(t)} E_n(t) = 0,$$

from which,

$$E_n(t) = \frac{\psi(t)}{(n+1)} f^{(n+1)}(\xi)$$

## 5.2 Newton's Interpolation

Given a set of data points $(x_j, f(x_j))$, $j = 0, \ldots, n$, the problem is to find a polynomial $P_n(x)$ of degree not exceeding $n$ such that $P_n(x_j) = f(x_j)$ for all $j = 0, \ldots, n$.

The Lagrange interpolating polynomial defined by

$$P_n(x) = \sum_{j=0}^{n} L_j(x) f(x_j),$$

where

$$L_i(x) = \prod_{\substack{j=1 \\ j \neq i}}^{n} \frac{x - x_j}{x_i - x_j},$$

suffer a problem that it is not possible to obtain $L_{i+1}(x)$ from $L_i(x)$ (that is there is no iterative method to compute the polynomials $L_i(x), i = 0, \ldots, n$). This makes the complexity of the algorithm high. Hence, Newton's interpolating polynomials can be seen as an alternative to the Lagrange polynomials.

### 5.2.1 Description of the Method

In Newton's interpolation method, a polynomial $P_k(x)$ of degree $k$ where $1 \leq k \leq n$ is computed from $p_{k-1}$ using an iterative technique. It starts with a constant polynomial $P_0(x)$, where

$$P_0(x) = c_0 = f(x_0).$$

Then, $P_k$ is obtained from the iteration:

$$P_k(x) = P_{k-1}(x) + c_k \prod_{j=0}^{k-1}(x - x_j), \qquad (5.4)$$

where $c_k$ is a constant that is determined by substituting the data point $(x_k, f(x_k))$ in the equation of $P_k(x)$, such that it satisfies the relation $P_k(x_j) = f(x_j)$, $j = 0, \ldots, n$.

From the iterative formula of Newton's interpolation polynomial, the Newton's polynomial that interpolates the data $(x_0, f(x_0)), \ldots, (x_k, f(x_k))$ descibed by equation (5.4) is of the form:

$$P_k(x) = c_0 + c_1(x - x_0) + c_2(x - x_0)(x - x_1) + \cdots + c_k(x - x_0)\ldots(x - x_{k-1}), \qquad (5.5)$$

with $P_0(x) = c_0 = f(x_0)$.

For a given data points $(x_j, f(x_j))$, the MATLAB function ComputeNewton-Coefs.m returns the coeficients of Newton's polynomial.

```
1   function c = ComputeNewtonCoefs(x, f)
2       n = length(x) ;
3       c = zeros(size(x)) ;
4       c(1) = f(1) ;
5       for j = 2 : n
6           z = x(j) - x(1:j-1) ;
7           pm1 = c(1) ;
8           for k = 1 : j
9               pm1 = pm1+sum(c(k)*prod(z(1:k-1))) ;
10          end
11          c(j) = (f(j)-pm1)/prod(z(1:j-1)) ;
12      end
13  end
```

Another MATLAB function NewtonInterp.m calls the function Compute-Newtoncoefs to construct the corresponding Newton's polynomial. It receives a set of data points $(x_j, f(x_j)), j = 0, \ldots, n$ and a value $\xi$ to return the corresponding value $yi$ obtained from the Newton's interpolating polynomial. The code of the function NewtonInterp.m is:

```
1   function yy = NewtonInterp(x, y, xi)
2       c = ComputeNewtonCoefs(x, y) ;
3       n = length(x) ;
4       z = zeros(1, n) ;
5       w = zeros(1, n) ;
6       z(1) = 1 ;
7       z(2:end) = xi - x(1:end-1) ;
```

```
 8        for k = 1 : n
 9            w(k) = prod(z(1:k)) ;
10        end
11        yy = sum(c.*w) ;
12    end
```

To test whether the functions give true results, they have been tested to interpolate data sampled from the function

$$f(x) = \frac{\cos(4x)}{1+x}, 0 \le x \le 2.$$

A total of 10 equidistant data points $(x_j, f(x_j)), j = 0, \ldots, 9$ are used to generate a Newton's polynomial and then it is used to approximate the values of the function at 101 equidistant data points. The MATLABs script `NewtonInterpCos.m` is used to do the task and plot the figure. Its code:

```
 1    clear ; clc ;
 2    x = linspace(0, 2, 10) ;
 3    y = cos(4*x)./(1+x) ;
 4    xx = linspace(0, 2, 101) ;
 5    yy = zeros(size(xx)) ;
 6    for j = 1 : length(yy)
 7        yy(j) = NewtonInterp(x, y, xx(j)) ;
 8    end
 9    plot(x, y, 'bo', xx, yy, '-.m', 'LineWidth', 3) ;
10    xlabel('x') ;
11    ylabel('y') ;
12    legend('Data points (x, y)', 'Newton Polynomial') ;
13    axis([-0.1, 2.1, -0.8, 1.1]) ;
14    set(gca, 'fontweight', 'bold') ;
15    grid on ;
16    set(gca, 'XTick', 0:0.2:2) ;
17    set(gca, 'YTick', -0.8:0.2:1) ;
```

The result of executing the code is shown in Figure 5.1.

The corresponding Python functions `ComputeNewtonCoefs.py` and `Newton-Interp.py` are implemented within a Python script `NewtonInterpSin.py` that interpolates 10 data points sampled from the function

$$f(x) = \frac{\sin(4x)}{1+x}, 0 \le x \le 2$$

to approximate the values of the function at 101 equidistant data points. The full code of `NewtonInterpSin.py` is:

```
 1    from numpy import zeros, prod, sin, linspace, arange
 2    import matplotlib.pyplot as plt
 3    def ComputeNewtonCoefs(x, f):
 4        n = len(x)
 5        c = zeros(len(x), 'float')
```

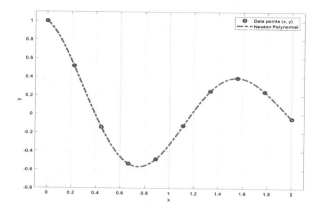

FIGURE 5.1: Approximation of the function $\cos(4x)/(1+x)$ at 101 points, using a Newton's interpolating polynomial constructed by 10 data points.

```
6      z = zeros(len(x), 'float')
7      c[0] = f[0]
8      for j in range(1, n):
9          z[0] = 1.0 ;
10         for k in range(j):
11             z[k+1] = x[j]-x[k]
12         pm1 = 0.0
13         w = zeros(j, 'float')
14         for k in range(j):
15             w[k] = prod(z[:k+1])
16             pm1 +=  (c[k]*w[k])
17         c[j] = (f[j]-pm1)/prod(z[:j+1])
18     return c
19
20 def NewtonInterp(x, y, xi):
21     c = ComputeNewtonCoefs(x, y)
22     n = len(x)
23     z = zeros(n, 'float') ;
24     w = zeros(n, 'float') ;
25     z[0] = 1 ;
26     for k in range(n-1):
27         z[k+1] = xi - x[k]
28     for k in range(n):
29         w[k] = prod(z[:k+1])
30     yy = sum(c*w)
31     return yy
32
33 x = linspace(0, 2, 10)
34 y = sin(4*x)/(1+x)
35 xx = linspace(0, 2, 101) ;
36 yy = zeros(len(xx), 'float') ;
37 for j in range(len(yy)):
38     yy[j] = NewtonInterp(x, y, xx[j]) ;
39
```

```
40  plt.plot(x, y, 'bo', label='Data points (x, y)', lw = 4)
41  plt.plot(xx, yy, '-.m', label='Newton Polynomial', lw= 3)
42  plt.xlabel('x', fontweight='bold')
43  plt.ylabel('y', fontweight='bold')
44  plt.legend()
45  plt.axis([-0.1, 2.1, -0.6, 1.0])
46  plt.grid(True, ls=':')
47  plt.xticks(arange(0, 2.2, 0.2), fontweight='bold')
48  plt.yticks(arange(-0.6, 1.0, 0.2), fontweight='bold') ;
```

By executing the code, the data points and Newton's interpolating polynomials are shown in Figure 5.2.

## 5.2.2   Newton's Divided Differences

Given a data table:

| $x$ | $x_0$ | $x_1$ | ... | $x_n$ |
|-----|-------|-------|-----|-------|
| $f(x)$ | $f(x_0)$ | $f(x_1)$ | ... | $f(x_n)$ |

and the problem is to find a Newton's polynomial

$$P_n(x) = c_0 + \sum_{k=1}^{n} c_k \prod_{j=0}^{k-1} (x - x_j),$$

that interpolates the given tabular data $(x_j, f(x_j))$, $j = 0, \ldots, n$.

The basis functions for the Newton's interpolating polynomials are the functions:

$$N_0(x) = 1, N_1(x) = x - x_0, N_2(x) = (x - x_0)(x - x_1), \ldots, N_n(x)$$
$$= (x - x_0) \ldots (x - x_{n-1}).$$

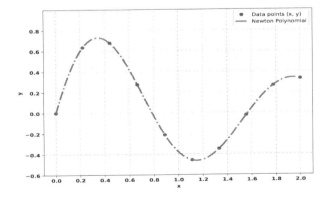

FIGURE 5.2: Approximation of the function $\sin(4x)/(1+x)$ at 101 points, using a Newton's interpolating polynomial constructed by 10 data points.

Therefore, computing a Newton's polynomial is restricted to computing the polynomial coefficients $c_0, \ldots, c_n$.

The method of divided differences is a short way for computing the coefficients of Newton's interpolating polynomials $c_j$, $j = 0, \ldots, n$. In a divided difference method, a divided differences table is constructed.

Before discussing the methods of constructing Newton's divided differences table, the following notations will be used:

$$
\begin{aligned}
f[x_k] &= f(x_k) = y_k, \ k = 0, \ldots, n \\
f[x_k, x_{k+1}] &= \frac{f[x_{k+1}] - f[x_k]}{x_{k+1} - x_k}, k = 0, \ldots, n-1 \\
f[x_k, x_{k+1}, x_{k+2}] &= \frac{f[x_{k+1}, x_{k+2}] - f[x_k, x_{k+1}]}{x_{k+2} - x_k}, k = 0, \ldots, n-2
\end{aligned}
$$

and generally,

$$
f[x_k, x_{k+1}, \ldots, x_{k+j+1}] = \frac{f[x_{k+1}, \ldots, x_{k+j+1}] - f[x_k, \ldots, x_{k+j}]}{x_{k+j+1} - x_k},
$$

$$
k = 0, \ldots, n - (j+1)
$$

Then, the Newton's divided difference table is defined by:

| $x_k$ | $f[x_k]$ | $f[x_{k-1}, x_k]$ | $\cdots$ | $f[x_0, \ldots, x_k]$ |
|---|---|---|---|---|
| $x_0$ | $f[x_0]$ | | | |
| $x_1$ | $f[x_1]$ | $f[x_0, x_1]$ | | |
| $\vdots$ | $\vdots$ | $\vdots$ | $\ddots$ | $\vdots$ |
| $x_n$ | $f[x_n]$ | $f[x_{n-1}, x_n]$ | $\cdots$ | $f[x_0, \ldots, x_n]$ |

The coefficients of Newton's interpolating polynomial are given by the following forms:

$$
\begin{aligned}
c_0 &= f(x_0) = f[x_0] \\
c_1 &= \frac{f[x_1] - f[x_0]}{x_1 - x_0} = f[x_0, x_1] \\
c_2 &= \frac{f[x_1, x_2] - f[x_0, x_1]}{x_2 - x_0} = f[x_0, x_1, x_2]
\end{aligned}
$$

and generally,

$$
c_k = \frac{f[x_1, \ldots, x_k] - f[x_0, \ldots, x_{k-1}]}{x_k - x_0} = f[x_0, \ldots, x_k], k = 1, \ldots, n.
$$

Hence, the coefficients of Newton's interpolating polynomials are the diagonal elements on the table of Newton's divided differences.

A MATLAB function `NewtonCoefs.m` receives data points $(x_j, f(x_j))$ and uses them to compute the Newton's divided difference table and hence the coefficients of Newton's interpolating polynomial. It returns both the divided difference table and coefficients. Its code is:

```
1  function [ndd, Coef] = NewtonCoefs(x, f)
2      % The function NewtonCoefs receives data points (x_j, ...
           f(x_j) and uses them
3      % to compute the Newton's divided difference table and ...
           hence the
4      % coefficients of Newton's interpolating polynomial. It ...
           returns both the
5      % divided difference table and coefficients.
6      Coef = zeros(1, length(x)) ;
7      ndd = zeros(length(x), length(x)+1) ;
8      ndd(:, 1) = x' ;
9      ndd(:, 2) = f' ;
10     for j = 3 : length(x)+1
11         for i = j-1 : length(x)
12             k = j - 2 ;
13             ndd(i, j) = (ndd(i, j-1)-ndd(i-1, ...
                   j-1))/(x(i)-x(i-k)) ;
14         end
15     end
16     for j = 1 : length(x)
17         Coef(j) = ndd(j, j+1) ;
18     end
19 end
```

To test the performance of the function `NewtonCoefs.m`, five data points $(x_j, f(x_j)), j = 0, \ldots, 4$ are considered, where $f(x) = 4e^{-x^2} \sin(2\pi x)$, $x \in [0, 1]$. The following MATLAB instructions are used to do the test:

```
>> x = linspace(0, 1, 5) ;
>> y = 4*exp(-x.^2).*sin(2*pi*x) ;
>> [ndd, Coefs] = NewtonCoefs(x, y)
ndd =
```

| | | | | | |
|---|---|---|---|---|---|
| 0 | 0 | 0 | 0 | 0 | 0 |
| 0.2500 | 3.7577 | 15.0306 | 0 | 0 | 0 |
| 0.5000 | 0.0000 | -15.0306 | -60.1224 | 0 | 0 |
| 0.7500 | -2.2791 | -9.1165 | 11.8282 | 95.9341 | 0 |
| 1.0000 | -0.0000 | 9.1165 | 36.4661 | 32.8506 | -63.0836 |

```
Coef =
0   15.0306   -60.1224   95.9341   -63.0836
```

The Newton's interpolating polynomial obtained by using these coefficients is shown in Figure 5.3.

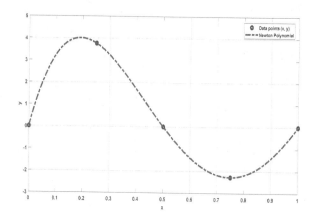

FIGURE 5.3: Approximation of the function $4e^{-x^2}\sin(2\pi x)$ at 101 points, using a Newton's interpolating polynomial constructed by 5 data points.

## 5.3 MATLAB's Interpolation Tools

MATLAB provides many functions for data interpolation. In this section, the MATLAB functions interp1, spline and pchip will be discussed.

### 5.3.1 Interpolation with the `interp1` Function

The MATLAB function `interp1` receives mainly two vectors $x$ and $y$ and a third argument which could be a scaler value or a vector. In this case, it interpolates the data $(x_j, y_j)$ with piecewise linear polynomials. To see this, 8 data points will be sampled from the function $y = e^{-x}\sin(x)$ in the interval $[0,3]$, then the constructed interpolating polynomial will be evaluated at 100 data points in $[0,3]$. This is can be done through the following MATLAB commands:

To construct the sample data points:

```
>> x = linspace(0, 3, 8) ;
>> y = exp(-x).*sin(x) ;
```

To define the points $s$ at which the interpolating polynomial will be evaluated:

```
>> s = linspace(0, 3) ;
```

The function `interp1` is used as follows:

```
>> P = interp1(x, y, s) ;
```

The function `interp1` joins any two points $(x_j, y_j)$ and $(x_{j+1}, y_{j+1})$ by straight lines. Hence, the interpolating polynomial is a piecewise-linear

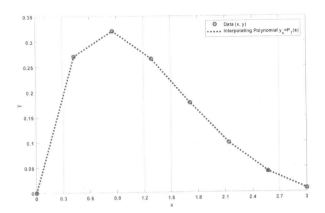

FIGURE 5.4: The original data $(x, y)$ and the data obtained by interpolation $(s, P(s))$.

polynomial consisting of the union of the straight lines joining between the points $(x_j, y_j)$, $j = 0, \ldots, 7$. At any given point $s_k, k = 0, \ldots, 99$ the polynomial is evaluated by determining $i$ such that $x_i \leq s \leq x_{i+1}$ and using the equation of the line joining between $(x_i, y_i)$ and $(x_{i+1}, y_{i+1})$. It finally returns the result in $P$.

In MATLAB the data and the interpolating polynomial are plotted by using:

```
>> plot(x, y, 'ro', s, P, ':b', 'LineWidth', 3) ;
>> xlabel('x') ;
>> ylabel('y') ;
>> legend('Data', '(s, P(s))')
```

The data points and interpolating polynomial are plotted in Figure 5.4.

## 5.3.2  Interpolation with the Spline Function

Interpolating with piecewise linear polynomials result in functions that are not differentiable at the interpolation points as seen in Figure 5.4. The MATLAB function spline computes a piecewise cubic polynomial such that the interpolating cubic polynomial and its first and second derivatives are smooth at the interpolation points $(x_j, y_j)$.

If the command interp1 is replaced by the command spline in the previous example, and replot the result:

```
>> P = spline(x, y, xx) ;
>> plot(x, y, 'ro', xx, P, ':b', 'LineWidth', 3) ;
>> xlabel('x') ;
>> ylabel('y') ;
>> legend('The data (x, y)', 'The spline P(x)')
```

The Figure 5.5 will be seen:

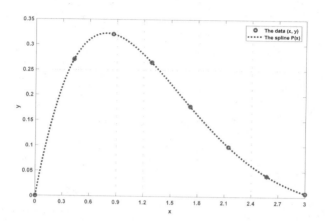

FIGURE 5.5: The original data $(x, y)$ and the data obtained by interpolation $(xx, P(xx))$.

By typing:

```
>> Q = spline(x, y)
```

The result is:

```
Q =
form: 'pp'
breaks: [0 0.4286 0.8571 1.2857 1.7143 2.1429 2.5714 3]
coefs: [7x4 double]
pieces: 7
order: 4
dim: 1
```

The structure $Q$ contains information about the interpolating polynomial. For 8 data points, 7 pieces of cubic splines are required to join between each two data points $(x_i, y_i)$ and $(x_{i+1}, y_{i+1}); i = 0, \dots, 7$. The field "breaks" contains the $x - coordinates$ of the data points. The field "coefs" contain the coeficients of each cubic spline piece between the two data points $(x_i, y_i)$ and $(x_{i+1}, y_{i+1}); i = 0, \dots, 7$. The error in approximating the original function by the interpolating spline polynomial is $h^5$, where $h = \max x_{i+1} - x_i; i = 1, \dots, 7$. Therefore, the order of convergence is 4.

Then, $Q$ can be evaluated at the points of the vector $xx$ by using the ppval function. This is done as follows:

```
>> S = ppval(Q, xx) ;
```

Now, $S$ and $P$ are identical. This can be seen through using the command:

```
>> Error = norm(P-S, inf)
Error =
0
```

### 5.3.3   Interpolation with the Function pchip

The function `pchip` stands for piecewise cubic Hermite interpolating polynomials. It employs sets of the cubic Hermite polynomials to approximate the function between any pair of data points.

The function `pchip` works in a similar way to the function `spline`. If the command `spline` is replaced by the command `pchip` in the previous example, by typing:

```
>> P = pchip(x, y, xx) ;
>> plot(x, y, 'ro', xx, P, ':b', 'LineWidth', 2) ;
>> xlabel('x') ;
>> ylabel('y') ;
>> legend('The data (x, y)', 'The pchip P(x)')
```

then Figure 5.6 will be obtained.

Again, if $Q$ is defined by:

```
>> Q = pchip(x, y)
```

the following result would be obtained:

```
Q =
form: 'pp'
```

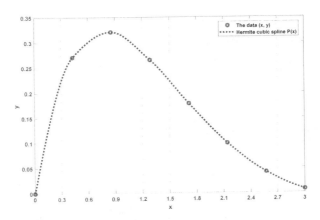

FIGURE 5.6: The original data $(x, y)$ and the data obtained by interpolation $(xx, P(xx))$ using the pchip function.

```
breaks: [0 0.4286 0.8571 1.2857 1.7143 2.1429 2.5714 3]
coefs: [7x4 double]
pieces: 7
order: 4
dim: 1
```

It is similar to what was obtained by using the spline command. The difference is in the values of the coefficients, and therefore, in the equations defining the 7 pieces of the interpolating polynomial.

The function ppval can be used to evaluate the interpolating polynomial $Q$ at the components of $xx$.

```
>> S = ppval(Q, xx) ;
>> Error = norm(P - S, inf)
Error =
0
```

### 5.3.4    Calling the Functions spline and pchip from interp1

The MATLAB function interp1 can perform the interpolation by using the cubic splines or the piecewise cubic Hermite polynomials instead of using a piecewise linear interpolation.

The MATLAB command:

```
>> P = interp1(x, y, xx, 'spline') ;
```

will interpolate the data $(x, y)$ by piecewise cubic splines similar to the command spline does, and the MATLAB command:

```
>> P = interp1(x, y, xx, 'pchip') ;
```

will interpolate the data $(x, y)$ by piecewise cubic Hermite polynomials as same as the command pchip does.

If the MATLAB command:

```
>> P = interp1(x, y, xx, 'cubic') ;
```

is executed, then it can do the same job as the function pchip.

## 5.4    Data Interpolation in Python

The Python's scipy.interpolate library contains various kinds of functions for data interpolation [18]. Some of these functions are CubicHermiteSpline, CubicSpline, LinearNDInterpolator, NearestNDInterpolator, interp1d, lagrange, pade, pchip, pchip_interpolate. There are many other functions for multiple dimension interpolation. In this section some of the Python's interpolation functions will be discussed.

## 5.4.1   The Function `interp1d`

The Python's function `interp1d` has a similar syntax to the MATLAB function `interp1`. If no kind of interpolation is received, it interpolates the given data points with linear piecewise polynomials. If it receives the kind (quadratic, cubic, etc.), then it interpolates the given data with that kind of piecewise continuous polynomial.

For example, ten equidistant sample points from the function

$$f(x) = \frac{\cos(4x)}{1+x}$$

will be used to interpolate the function with linear, quadratic and cubic piecewise polynomials, based on the `interp1d` function. The Python script `interpwp1d.py` implements the interpolation with `interp1d`:

```python
from numpy import cos, linspace, arange
from scipy.interpolate import interp1d
import matplotlib.pyplot as plt
x = linspace(0, 3, 10)
y = cos(4*x)/(1+x)
xx = linspace(0, 3, 101) ;
Lin1d = interp1d(x, y)
Quad1d = interp1d(x, y, kind = 'quadratic')
Cubic1d = interp1d(x, y, kind = 'cubic')
yy1 = Lin1d(xx)
yy2 = Quad1d(xx)
yy3 = Cubic1d(xx)

plt.subplot(211)
plt.plot(x, y, 'bo', label='Data points (x, y)', lw = 4)
plt.plot(xx, yy1, ls='-.', color='orangered', label='Linear', ...
    lw= 2)
plt.plot(xx, yy2, ls=':', color='purple', label='Quadratic', ...
    lw= 2)
plt.xlabel('x', fontweight='bold')
plt.ylabel('y', fontweight='bold')
plt.legend()
plt.grid(True, ls=':')
plt.xticks(arange(0, 3.3, 0.3), fontweight='bold')
plt.yticks(arange(-0.6, 1.2, 0.2), fontweight='bold')

plt.subplot(212)
plt.plot(x, y, 'bo', label='Data points (x, y)', lw = 4)
plt.plot(xx, yy3, ls='--', color='crimson', label='Cubic', lw= 2)
plt.xlabel('x', fontweight='bold')
plt.ylabel('y', fontweight='bold')
plt.legend()
plt.grid(True, ls=':')
plt.xticks(arange(0, 3.3, 0.3), fontweight='bold')
plt.yticks(arange(-0.6, 1.2, 0.2), fontweight='bold')
```

Executing this code shows in the graphs in Figure 5.7.

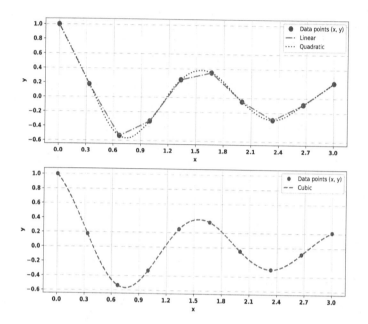

FIGURE 5.7: Interpolation of 10 equidistant sample points taken from $f(x) = cos(4x)/(1+x), 0 \le x \le 3$, with linear, quadratic and cubic polynomials.

## 5.4.2   The Functions `pchip_interpolate` and `CubicSpline` [18]

The two functions `pchip_interpolate` and `CubicSpline` have the same functions as the MATLAB's functions `pchip` and `spline`. They receive three arguments $(x, y, ps)$, the first two arguments represent the interpolation data points, and the last argument is the set of points at which the function is to be approximated.

The Python code `interppchipspline.py` interpolates the function $f(x) = sin(4x)/(1+x)$ at 10 data points in $[0.0, 3.0]$ and uses the interpolation polynomial to approximate the function at 101 data points.

```
1  from numpy import sin, linspace, arange
2  from scipy.interpolate import pchip_interpolate, CubicSpline
3  import matplotlib.pyplot as plt
4  x = linspace(0, 3, 10)
5  y = sin(4*x)/(1+x)
6  xx = linspace(0, 3, 101) ;
7
8  yy1 = pchip_interpolate(x, y, xx)
9  yy2 = CubicSpline(x, y)(xx)
```

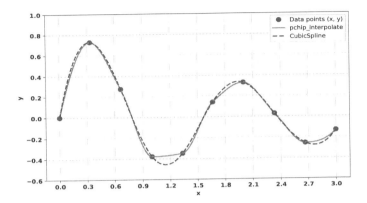

FIGURE 5.8: Interpolation of 10 equidistant sample points taken from $f(x) = sin(4x)/(1+x), 0 \le x \le 3$, with the functions `pchip_interpolate` and `CubicSpline`.

```
10
11  plt.plot(x, y, 'bo', label='Data points (x, y)', lw = 4)
12  plt.plot(xx, yy1, ls='-', color='orangered', ...
        label='pchip_interpolate', lw= 2)
13  plt.plot(xx, yy2, ls='--', color='purple', ...
        label='CubicSpline', lw= 2)
14  plt.xlabel('x', fontweight='bold')
15  plt.ylabel('y', fontweight='bold')
16  plt.legend()
17  plt.grid(True, ls=':')
18  plt.xticks(arange(0, 3.3, 0.3), fontweight='bold')
19  plt.yticks(arange(-0.6, 1.2, 0.2), fontweight='bold')
```

Executing the code shows in Figure 5.8.

### 5.4.3  The Function `lagrange`

The Python interpolation function `lagrange` interpolates a given data points, based on Lagrange interpolation. It receives two vectors $(x, y)$ representing the coordinates of the interpolation points to create a polynomial of degree $n$, where $n + 1$ is the number of interpolation points.

The Python script `LagrangeInterp.py` uses Lagrange interpolation to construct a ninth degree polynomial that interpolates data sampled from the function $f(x) = (cos(5x) + sin(5x))/(2(1 + x^2))$ based on 10 data points in $[0, 3]$.

```
1  from numpy import sin, cos, linspace, arange
2  from scipy.interpolate import lagrange
3  import matplotlib.pyplot as plt
4  x = linspace(0, 3, 10)
```

```
 5   y = (sin(5*x)+cos(5*x))/(2*(1+x**2))
 6   xx = linspace(0, 3, 101) ;
 7   LG = lagrange(x, y)
 8   print(LG)
 9   yy1 = LG(xx)
10
11   plt.plot(x, y, 'bo', label='Data points (x, y)', markersize = 8)
12   plt.plot(xx, yy1, ls='-', color='purple', label='Lagrange', ...
         lw= 2)
13
14   plt.xlabel('x', fontweight='bold')
15   plt.ylabel('y', fontweight='bold')
16   plt.legend()
17   plt.grid(True, ls=':')
18   plt.xticks(arange(0, 3.3, 0.3), fontweight='bold')
19   plt.yticks(arange(-0.6, 1.2, 0.2), fontweight='bold')
```

By running the code, the following output is obtained:

```
runfile('D:/PyFiles/LagrangeInterp.py', wdir='D:/PyFiles')
            9          8        7         6         5        4          3
-0.4591 x + 6.308 x - 35.79 x + 107.5 x - 180.9 x + 164 x - 66.05 x
         2
+ 2.858 x + 1.864 x + 0.5
```

Figure 5.9 shows the graph of the function $f(x)$ approximated at 101 points using Lagrange interpolation.

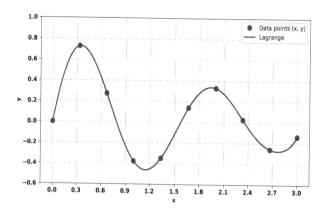

FIGURE 5.9: Lagrange interpolation the function $f(x) = (cos(5x) + sin(5x))/(2(1+x^2))$ based on 10 data points in $[0,3]$.

# 6

---

## Numerical Differentiation and Integration

---

### Abstract

This chapter discusses the numerical methods for approximating derivative and integrations of functions. The chapter is divided into two sections: the first section discusses the numerical differentiation of functions based on finite difference formulas and the second discusses the numerical integration based on Newton-Cotes and Gauss methods. Such numerical differentiation or integration algorithms are implemented using both MATLAB® and Python.

## 6.1   Numerical Differentiation

### 6.1.1   Approximating Derivatives with Finite Differences

Let $x_0 \in (a,b) \subset \mathbb{R}$ and $f \in \mathcal{C}^{n+1}[a,b]$. Then, for $h > 0$ and $k \leq n$ the Taylor expansion of $f$ around $x_0$ is given by:

$$f(x_0 + h) = f(x_0) + hf'(x_0) + \frac{h^2}{2!}f''(x_0) + \cdots + \frac{h^k}{k!}f^{(k)}(x_0)$$

$$+ \frac{h^{k+1}}{(k+1)!}f^{(k+1)}(\xi), \xi \in [x_0, x_0 + h] \tag{6.1}$$

and

$$f(x_0 + h) = f(x_0) - hf'(x_0) + \frac{h^2}{2!}f''(x_0) - \cdots + \frac{(-h)^k}{k!}f^{(k)}(x_0)$$

$$+ \frac{(-h)^{k+1}}{(k+1)!}f^{(k+1)}(\eta), \eta \in [x_0 - h, x_0] \tag{6.2}$$

By setting $k = 1$ in Equation (6.1) and solving for $f'(x_0)$ the first derivative of $f(x)$ at $x = x_0$ is given by:

$$f'(x_0) = \frac{f(x_0 + h) - f(x_0)}{h} + \frac{h}{2} f''(\xi), \xi \in [x_0, x_0 + h] \qquad (6.3)$$

Hence, by taking as small possible value of $h \geq \varepsilon > 0$, $f'(x)$ can be approximated by:

$$f'(x_0) \approx \frac{f(x_0 + h) - f(x_0)}{h} \qquad (6.4)$$

where $\varepsilon$ denotes the machine precision. If $h$ is taken to be less than $\varepsilon$, round-off error can affect the accuracy of approximate derivative at $x_0$.

The approximate formula of the derivative $f'(x)'$ in Equation (6.4) is called the `forward difference formula`.

By setting $k = 1$ in Equation (6.2) and solving for $f'(x_0)$ the first derivative of $f(x)$ at $x = x_0$ is given by:

$$f'(x_0) = \frac{f(x_0 + h) - f(x_0)}{h} - \frac{h}{2} f''(\eta), \eta \in [x_0 - h, x_0] \qquad (6.5)$$

from which $f'(x)$ can be approximated by:

$$f'(x_0) \approx \frac{f(x_0 + h) - f(x_0)}{h} \qquad (6.6)$$

where $\varepsilon$ denotes the machine precision. If $h$ is taken to be less than $\varepsilon$, round-off error can affect the accuracy of approximate derivative at $x_0$.

The approximate formula of the derivative $f'(x)'$ in Equation (6.4) is called the `forward difference formula`. Also, the approximate formula of $f'(x)$ in Equation (6.6) is called the `backward difference formula` [31].

The last terms of equations (6.3)

$$R_2^f(x) = \frac{h}{(2)!} f''(x), \ x \in [x_0, x_0 + h]$$

and (6.6)

$$R_2^b(t) = -\frac{h}{2} f''(t), \ t \in [x_0 - h, x_0]$$

give the `remainders` at points $x \in [x_0, x_0 + h]$ and $x_0 - h, x_0$, respectively. Ignoring the remainder while approximating some function is known as the `truncation error`. If $f''(x)$ is bounded by a constant $M_1 \in \mathbb{R}^+$ in $[t_0, t_0 + h]$ and by $M_2 \in \mathbb{R}^+$ in $[x_0 - h, x_0]$, then:

$$|R_2^f(x)| \ \leq \ M_1 \cdot h, for \ x \in [x_0, x_0 + h] \qquad (6.7)$$
$$|R_2^b(t)| \ \leq \ M_2 \cdot h, for \ t \in [x_0 - h, x_0]. \qquad (6.8)$$

This indicates that the truncation error in both the forward and backward difference formulas are $\mathcal{O}(h)$.

**Example 6.1** In this example, Equation (6.4) will be used to find approximation of derivative of $f(x) = e^{-\sin(x^3)/4}$ for $h = 10^{-1}, \ldots, 10^{-15}$. The exact derivative of $f(x)$ is

$$f'(x) = -\frac{3x^2 \cos(x^3)}{4} e^{-\sin(x^3)/4}$$

The error for each value of $h$ will also be shown.

The Python code is:

```
1  #fpapproxim.py
2  import numpy as np
3  f = lambda x: np.exp(-np.sin(x**3)/4)
4  fp = lambda x: -3*x**2*np.cos(x**3)/4*f(x)
5  h = 0.1
6  fpapprox = []
7  Eps = np.spacing(1.0)
8  while h >= Eps:
9      fp1 = (f(1.+h)-f(1.))/h
10     fpapprox.append([h, fp1, np.abs(fp(1.)-fp1)])
11     h /= 10
12 print('-----------------------------------------------------------')
13 print('   h', '\t\t\t', 'Approx Der', '\t\t   ', 'Approx Error')
14 print('-----------------------------------------------------------')
15 for x in fpapprox:
16     print('{0:1.3e}'.format(x[0]), '\t', ...
          '{0:1.15e}'.format(x[1]), '\t\t', ...
          '{0:1.15e}'.format(x[2]))
```

Executing this code will give the results:

```
runfile('D:/PyFiles/fpapproxim.py', wdir='D:/PyFiles')
```

| h | Approx Der | Approx Error |
|---|---|---|
| 1.000e-01 | -2.589437504200143e-01 | 6.940588094341621e-02 |
| 1.000e-02 | -3.231226286809608e-01 | 5.227002682469728e-03 |
| 1.000e-03 | -3.278426595997308e-01 | 5.069717636996818e-04 |
| 1.000e-04 | -3.282990900810301e-01 | 5.054128240039590e-05 |
| 1.000e-05 | -3.283445787927164e-01 | 5.052570714092486e-06 |
| 1.000e-06 | -3.283491261107940e-01 | 5.052526365068033e-07 |
| 1.000e-07 | -3.283495819683679e-01 | 4.939506259571402e-08 |
| 1.000e-08 | -3.283496252670658e-01 | 6.096364635332918e-09 |
| 1.000e-09 | -3.283496807782170e-01 | 4.941478654041376e-08 |
| 1.000e-10 | -3.283495697559146e-01 | 6.160751592210190e-08 |
| 1.000e-11 | -3.283484595328899e-01 | 1.171830540547258e-06 |
| 1.000e-12 | -3.284039706841212e-01 | 5.433932069076608e-05 |
| 1.000e-13 | -3.286260152890463e-01 | 2.763839256157974e-04 |
| 1.000e-14 | -3.330669073875469e-01 | 4.717276024116424e-03 |
| 1.000e-15 | -4.440892098500626e-01 | 1.157395784866321e-01 |

The MATLAB code is:

```
1  %fpapproxim.m
2  f = @(x) exp(-sin(x^3)/4) ;
3  fp = @(x) -3*x^2*cos(x^3)/4*f(x) ;
4  fpapp = zeros(15, 3) ;
5  h = 1e-1 ;
6  j = 1 ;
7  while h >= eps
8      fp1 = (f(1.0+h)-f(1.0))/h ;
9      fpapp(j,:) = [h, fp1, abs(fp(1)-fp1)] ;
10      h = h / 10 ;
11      j = j + 1 ;
12  end
13  fprintf('-------------------------------------------------------\n') ;
14  fprintf('     h \t\t\t\t Approx Der \t\t\t   Approx Error\n') ;
15  fprintf('-------------------------------------------------------\n') ;
16  for j = 1 : 15
17      fprintf('%5.3e\t\t%16.15e\t\t%16.15e\n', fpapp(j,1), ...
            fpapp(j,2), fpapp(j,3)) ;
18  end
```

Executing the MATLAB code gives similar results as Python code.

Figure 6.1 shows the relationship between the value of $h$ and the corresponding error. It shows that taking smaller value of $h$ improves the derivative approximation up to some limit, after that the approximate derivative gets worse.

The implementation of a Python/MATLAB code to approximate $f'(1)$ using formula (6.6) is by replacing the line:

```
fp1 = (f(1.0+h)-f(1.0))/h
```

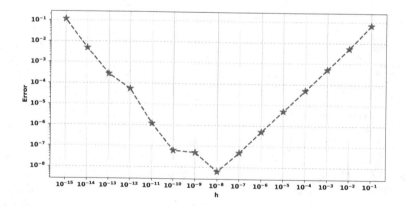

FIGURE 6.1: Approximation errors of $f'(1)$ versus different values of $h$ in the loglog scale, using the forward difference formula 6.4.

by the line:

```
fp1 = (f(1.0)-f(1.0-h))/h
```

in the codes of Example (6.1). Similar results can be obtained by running a code `fpbdfapprox.py` that is based on Equation (6.6):

```
runfile('D:/PyFiles/fpbdfapprox.py', wdir='D:/PyFiles')
```

| h | Approx Der | Approx Error |
|---|---|---|
| 1.000e-01 | -3.631033359769975e-01 | 3.475370461356703e-02 |
| 1.000e-02 | -3.332305430411631e-01 | 4.880911677732580e-03 |
| 1.000e-03 | -3.288531422600549e-01 | 5.035108966244262e-04 |
| 1.000e-04 | -3.284001380376989e-01 | 5.050667426836908e-05 |
| 1.000e-05 | -3.283546835874951e-01 | 5.052224064605593e-06 |
| 1.000e-06 | -3.283501365247687e-01 | 5.051613382045517e-07 |
| 1.000e-07 | -3.283496807782171e-01 | 4.941478659592491e-08 |
| 1.000e-08 | -3.283496363692961e-01 | 5.005865610918647e-09 |
| 1.000e-09 | -3.283495697559146e-01 | 6.160751592210190e-08 |
| 1.000e-10 | -3.283495697559146e-01 | 6.160751592210190e-08 |
| 1.000e-11 | -3.283484595328899e-01 | 1.171830540547258e-06 |
| 1.000e-12 | -3.282929483816587e-01 | 5.668298177174957e-05 |
| 1.000e-13 | -3.275157922644211e-01 | 8.338390990093592e-04 |
| 1.000e-14 | -3.219646771412953e-01 | 6.384954222135142e-03 |
| 1.000e-15 | -2.220446049250313e-01 | 1.063050264383992e-01 |

By setting $k = 2$ in equations (6.1) and (6.5), subtracting Equation (6.5) from (6.1) and solving for $f'(x)$ gives:

$$f'(x_0) = \frac{f(x_0+h) - f(x_0-h)}{2h} + \frac{h^2}{3} f'''(\tau), \; \tau \in [x_0 - h, x_0 + h] \qquad (6.9)$$

from which a third formula of the derivative is obtained

$$f'(x_0) = \frac{f(x_0+h) - f(x_0-h)}{2h} \qquad (6.10)$$

The formula in Equation (6.10) is called the **central difference formula**. If $|f'''(x)| \leq M \in \mathbb{R}^+$, then the approximation error of the central difference formula is bounded by $M/3 \cdot h^2$ which means that the central difference formula is $\mathcal{O}(h^2)$ method.

The Python/MATLAB code for implementing the central difference formula, is obtained by replacing the line:

```
fp1 = (f(1.0+h)-f(1.0))/h
```

in Example 6.1, by the line:

```
fp1 = (f(1.0+h)-f(1.0-h))/(2*h)
```

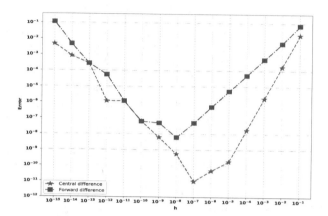

FIGURE 6.2: Comparing the forward differencing errors to the central difference errors in the loglog scale.

In Figure 6.2 the approximation errors obtained by the forward difference formula are compared to those obtained by the central difference formula in the loglog scale, for different values of $h$.

From Figure 6.2 it is clear that the central difference formula has better approximation of the first derivative than the forward difference formula, since it produces smaller errors.

In Figure 6.3 the graph of exact derivative of $f(x) = e^{-\sin(x^3)}$ is plotted in the interval $[0, \pi/2]$ against the graph of approximate derivative obtained by a forward difference formula.

If the interval $[0.0, \pi/2]$ is partitioned into $N$ sub-intervals $[x_0, x_1], \ldots,$ $[x_{N-1}, x_N]$, then the Python function diff in the numpy library can be used to compute a vector $df = [f(x_1) - f(x_0), f(x_2) - f(x_1), \ldots, f(x_{N+1}) - f(x_N)]$. If $dx = \pi/(2N)$, then $df/dx$ returns approximate values of the derivatives at the points $x_0, \ldots, x_{N-1}$. The following Python commands use the numpy function diff to graph $f'(x)$:

```
1  import numpy as np
2  import matplotlib
3  import matplotlib.pyplot as plt
4  matplotlib.rc('text', usetex=True)
5  matplotlib.rcParams['text.latex.preamble'] ...
       =[r"\usepackage{amsmath}"]
6  a, b = 0.0, 2.0
7  N = 1000
8  x = np.linspace(a, b, N+1)
9  f = lambda x: np.exp(-np.sin(x**3)/4)
10 df = np.diff(f(x))
```

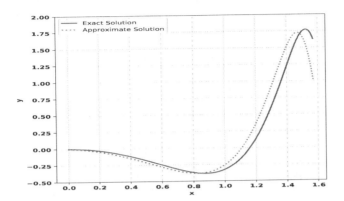

FIGURE 6.3: Graph of exact derivative of $f(x) = e^{-\sin(x^3)}$ is plotted in the interval $[0, \pi/2]$ against the graph of approximate derivative obtained by a forward difference formula.

```
11  dx = (b-a)/N
12  dfdx = df/dx
13  plt.plot(x[:N], dfdx, '-m', ...
        lw=2,label=r'$\mathbf{-\frac{3x^2\cos(x^3)}{4} ...
        e^{-\frac{\sin(x^3)}{4}}}$')
14  plt.xlabel('x', fontweight='bold')
15  plt.ylabel('y', fontweight='bold')
16  plt.grid(True, ls=':')
17  plt.legend(loc='upper left')
```

Figure 6.4 shows the graph of $f'(x), x \in [0, \frac{\pi}{2}]$ using the function `diff`.

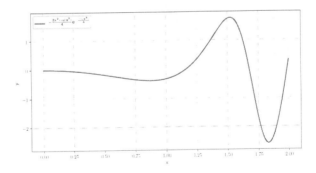

FIGURE 6.4: Graph of $f'(x), x \in [0, \frac{\pi}{2}]$ using the function `diff`.

To find an approximation of $f''(x)$ at $x = x_0$, Equation (6.1) is added to (6.5) and solving the resulting quantity for $f''(x)$, giving:

$$f''(x) = \frac{f(x_0 - h) - 2f(x_0) + f(x_0 + h)}{h^2} - \frac{h^2}{12}f^{(4)}(\xi), \ \xi \in [x_0 - h, x_0 + h]$$

from which,

$$f''(x) \approx \frac{f(x_0 - h) - 2f(x_0) + f(x_0 + h)}{h^2} \tag{6.11}$$

If $|f^{(4)}(x)| \leq M \in \mathbb{R}^+$, then $\frac{h^2}{12}|f^{(4)}(x)| \leq \frac{M}{12}h^2$, hence the truncation error in Equation (6.11) is of $\mathcal{O}(h^2)$.

The second derivative of $f(x) = e^{-x^2}$ is $f''(x) = 2(1 - 2x^2)e^{-x^2}$. The following MATLAB commands approximate the second derivative of $f(x), x \in [-3, 3]$ and graph it.

```
>> x = linspace(-3, 3, 601) ;
>> f = @(x) -exp(-x.^2) ;
>> y2 = diff(f(x), 2) ;
>> plot(x(2:end-1), y2, '-b', 'LineWidth', 2)
>> xlabel('x') ; ylabel('y') ;
>> legend('d^2f(x)/dx^2 = 2(1-2x^2)e^{-x^2}') ;
>> grid on
```

In Figure 6.5 the graph of $f''(x) = 2(1 - 2x^2)e^{-x^2}$ is plotted in the interval $[-3, 3]$.

It is very important to notice that if $h^2 < \varepsilon$ the resulting approximation errors get too large. For example, if we let $h$ take the values $10^{-j}$,

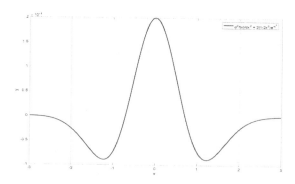

FIGURE 6.5: Graph of $f''(x) = 2(1 - 2x^2)e^{-x^2}$ in the period $[-3, 3]$.

$j = 1, \ldots, 10$, the corresponding approximate values and errors of $f''(x)$ will be as follows:

| h | Approx Der | Approx Error |
|---|---|---|
| 1.000e-01 | -7.296463309943767e-01 | 6.112551348507966e-03 |
| 1.000e-02 | -7.356975709826852e-01 | 6.131136019948968e-05 |
| 1.000e-03 | -7.357582692546494e-01 | 6.130882352906042e-07 |
| 1.000e-04 | -7.357588760470435e-01 | 6.295841181724882e-09 |
| 1.000e-05 | -7.357581210953866e-01 | 7.612474980378536e-07 |
| 1.000e-06 | -7.358003095703222e-01 | 4.142722743749605e-05 |
| 1.000e-07 | -7.271960811294772e-01 | 8.562801213407467e-03 |
| 1.000e-08 | -1.665334536937734e+00 | 9.295756545948495e-01 |
| 1.000e-09 | 1.110223024625156e+02 | 1.117580613448585e+02 |
| 1.000e-10 | 5.551115123125780e+03 | 5.551850882008122e+03 |
| 1.000e-11 | 5.551115123125780e+05 | 5.551122480714604e+05 |
| 1.000e-12 | 1.110223024625156e+08 | 1.110223031982745e+08 |

The table above shows that as the step size $h$ gets less than $10^{-8}$, the error becomes larger and larger. This is because of the approximation formula requires division by $h^2 \leq 10^{-16} < \varepsilon$.

## 6.2   Numerical Integration

If $f$ is a function that is defined over a real interval $[a, b]$, then the area under the curve of $f$ from $x = a$ to $x = b$ is given by the definite integral:

$$I = \int_a^b f(x) \, dx$$

If $F(x)$ is the anti-derivative of $f(x)$ in $[a, b]$, then from the fundamental theorem of calculus

$$I = \int_a^b f(x) \, dx = F(b) - F(a)$$

However, there are many functions defined over finite intervals whose anti-derivatives are unknown, although they do exist. Such examples include:

$$I_1 = \int_0^\pi \frac{\sin(x)}{x} \, dx$$

and

$$I_2 = \int_0^3 e^{-x^2} \, dx$$

The numerical integration methods work to find definite integral of a function $f$ in an interval $[a,b]$, through first partitioning the interval $[a,b]$ by points $x_0, x_1, \ldots, x_N$, such that $a \le x_0 < x_1 < \ldots < x_N \le b$, then approximate $I$ by a sum of the form:

$$I = \int_a^b f(x)\, dx \approx \sum_{j=0}^N w_j f(x_j) \tag{6.12}$$

where the different numerical integration methods differ from each other in the way by which the points $x_j$ are selected and coefficients $w_j$ are calculated.

Equation (6.12) is called `numerical quadrature`. Hence, the term numerical quadrature points to a form, in which an integration formula is approximated by a finite sum.

In this chapter, we discuss two classes of numerical integration methods, namely: the Newton-Cotes and Gauss quadratures.

## 6.2.1   Newton-Cotes Methods

In Newton-Cotes methods, the points $x_0, \ldots, x_N$ are chosen such that $x_0 = a < x_1 < \ldots < x_n = b$, with $h = (b-a)/n$. The idea behind Newton-Cotes methods to approximate $I = \int_a^b f(x)dx$ is first to approximate $f(x)$ by a Lagrange polynomial $P_n(x)$ of degree $n$. Then, to approximate $I$ by the integral of $P_n(x)$ [43, 47]. That is:

$$\int_a^b f(x)dx \approx \int_a^b P_n(x)dx = \int_a^b \sum_{j=0}^n L_j(x)f(x_j)dx = \sum_{j=0}^n \left(\int_a^b L_j(x)dx\right) f(x_j) \tag{6.13}$$

Hence, the quadrature weights $w_j$ in Newton-Cotes methods are given by:

$$w_j = \int_a^b L_j(x)dx$$

The numerical integration formulas for $n=1$ and $n=2$ will be derived and discussed.

(I) $n=1$: when $n=1$, $h=(b-a)/1 = b-a$ and the given data points are $(a, f(a))$ and $(b, f(b))$. The Lagrange interpolating polynomial is

$$P_1(x) = \frac{x-b}{a-b}f(a) + \frac{x-a}{b-a}f(b) = \frac{1}{h}\left(-(x-b)f(a) + (x-a)f(b)\right),$$

hence,

$$\int_a^b f(x)dx \approx \frac{1}{h}\int_a^b \left[-(x-b)f(a) + (x-a)f(b)\right] dx = \frac{h}{2}(f(b) + f(a)) \tag{6.14}$$

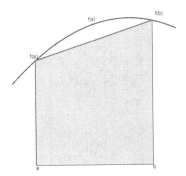

FIGURE 6.6: Approximating the area under curve by the trapezium area.

The formula in equation (6.14) gives the trapezoidal rule, where the area under the $f(x)$ curve from $x = a$ to $x = b$ is approximated by area of a trapezium whose bases are $f(a)$ and $f(b)$ and height is $b - a$.

In Figure 6.6 the area under the curve of $f(x)$ is approximated by the area of the trapezium (shaded area).

To find an estimate of the integral error, it is convenient to start from the formula:

$$f(x) = P_1(x) + \frac{(x-a)(x-b)}{2} f''(\xi), \xi \in [a, b]$$

from which the approximation error is given by:

$$\left| \int_a^b (f(x)dx - P_1(x)) \, dx \right| = \left| \int_a^b \frac{(x-a)(x-b)}{2} f''(\xi) dx \right|$$

$$\leq \frac{(b-a)^3}{12} |f''(\xi)| = \frac{h^3}{12} |f''(\xi)|$$

If $f''(x)$ is bounded by $M \in \mathbb{R}^+$ in $[a, b]$, then

$$\left| \int_a^b (f(x)dx - P_1(x)) \, dx \right| \leq \int_a^b |f(x) - P_1(x)| \, dx \leq \frac{M}{12} h^3 \qquad (6.15)$$

The error formula 6.15 shows that large value of $h = b - a$ causes large error.

**Example 6.2** In this example, the trapezoidal rule will be used to approximate the integral

$$I = \int_0^1 \frac{1}{1+x^2} dx$$

Here, $a = 0.0$, $b = 1.0$ and $f(x) = 1/(1+x^2)$. The following MATLAB commands can be used to evaluate the integral:

```
>> a = 0.0; b = 1.0 ; f = @(x) 1/(1+x^2) ;
>> I = (b-a)/2*(f(a)+f(b)) ;
>> disp(I)
   0.7500
```

In Python:

```
In [1]: a, b, f = 0.0, 1.0, lambda x: 1/(1+x**2)
In [2]: I = (b-a)/2*(f(b)+f(a))
In [3]: print(I)
0.75
```

The exact value of the integral in Example 6.2 is $\frac{\pi}{4} \approx 0.7853981633974483$. The approximation error is $\pi/4 - 0.75 = 0.03539816339744828$.

(II) $n = 2$: when $n = 2$, $h = (b-a)/2$ the given data points are $(a, f(a))$, $((a+b)/2, f((a+b)/2))$ and $(b, f(b))$. The Lagrange interpolating polynomial is:

$$P_2(x) = \frac{(x-c)(x-b)}{(a-c)(a-b)} f(a) + \frac{(x-a)(x-b)}{(c-a)(c-b)} f(c) + \frac{(x-a)(x-c)}{(b-a)(b-c)} f(b),$$

where $c = (a+b)/2$.

Let $x = a + th$, $0 \le t \le 2$ be a parametric representation of $x$ in terms of $t$. Then, $P_2(x)$ can be rewritten as:

$$P_2(x) = \frac{(t-1)(t-2)}{2} f(a) + t(t-2)f(c) + \frac{t(t-1)}{2} f(b)$$

Then,

$$\int_a^b f(x)dx \approx h \int_0^2 \left[ \frac{(t-1)(t-2)}{2} f(a) + t(t-2)f(c) + \frac{t(t-1)}{2} f(b) \right] dt$$

$$= h \left( \frac{1}{3} f(a) + \frac{4}{3} f\left( \frac{a+b}{2} \right) + \frac{1}{3} f(b) \right)$$

$$= \frac{b-a}{6} \left( f(a) + 4f\left( \frac{a+b}{2} \right) + f(b) \right) \tag{6.16}$$

The numerical quadrature (6.16) is called Simpson's rule.

The leading error term in approximating $\int_a^b f(x)dx$ by $\int_a^b P_2(x)dx$ in Equation (6.16) is:

$$\left| \int_a^b (f(x) - P_2(x)dx \right| \leq \int_a^b |f(x) - P_2(x)|\,dx$$

$$= \int_0^2 \left| \frac{t(t-1)(t-2)}{6} f'''(\xi) \right| = 0$$

where $a \leq \xi \leq b$. Therefore, it is necessary to move to the next leading term of the error. Hence,

$$\left| \int_a^b (f(x)dx - \int_a^b P_2(x)dx \right| \leq \int_0^2 \frac{t(t-1)(t-2)(t-4)}{120} |f^{(4)}(\eta)|$$

$$= \frac{h^5}{90} |f^{(4)}(\eta)|, \ a \leq \eta \leq b \qquad (6.17)$$

Applying the Simpson's rule to the problem of Example 6.2, with $a = 0.0, b = 1.0$ and $h = (b-a)/2 = 0.5$ Using MATLAB:

```
>> a = 0.0 ; b = 1.0; f = @(x) 1/(1+x^2) ;
>> h = (b-a)/2 ;
>> c = (a+b)/2 ;
>> I_s = h/3*(f(a)+4*f(c)+f(b))
I_s =
0.7833
>> format short e
>> disp(abs(pi/4-I_s))
2.0648e-03
```

In Python:

```
In [4]: import numpy as np
In [5]: a, b, f = 0.0, 1.0, lambda x: 1/(1+x**2)
In [6]: c, h = (a+b)/2, (b-a)/2
In [7]: I_s = h/3.0*(f(a)+4*f(c)+f(b))
In [8]: print(I_s)
0.7833333333333333
In [9]: print('{0:1.4e}'.format(np.abs(np.pi/4-I_s)))
2.0648e-03
```

The errors in approximating $\int_a^b f(x)dx$ by either the trapezoidal rule (6.14) or Simpson's rule (6.16) become large as $h = (b-a)/n$ gets large. To obtain better approximations of $\int_a^b f(x)dx$ are composite versions of Newton-Cotes method, which partition the interval $[a,b]$ into $N$-subintervals ($N \in \mathbb{Z}^+$) by points $x_0 = a < x_1 < \ldots < x_N = b$ where $x_{j+1} - x_j = h = (b-a)/N$. Then, in each sub-interval $[x_j, x_{j+1}]$ the integral $\int_a^b f(x)dx$ is approximated by a

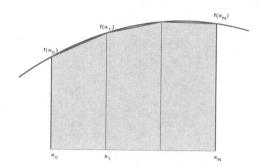

FIGURE 6.7: The composite trapezoidal rule.

quadrature. Two composite rules derived from the Newton-Cotes methods will be discussed.

(i) **The composite trapezoidal rule:** the derivation of the formula starts from the formula:

$$\int_a^b f(x)dx = \int_{x_0}^{x_N} f(x)dx = \int_{x_0}^{x_1} f(x)dx + \int_{x_1}^{x_2} f(x)dx + \ldots + \int_{x_{N-1}}^{x_N} f(x)dx,$$

then, in each sub-interval $[x_j, x_{j+1}]$ the integral $\int_{x_j}^{x_{j+1}} f(x)dx$ is approximated by $h/2(f(x_j) + f(x_{j+1}))$. This gives the formula:

$$
\begin{aligned}
\int_a^b f(x)dx &\approx \frac{h}{2}(f(x_0) + f(x_1)) + \frac{h}{2}(f(x_1) + f(x_2)) + \ldots \\
&\quad + \frac{h}{2}(f(x_{N-1}) + f(x_N)) \\
&= \frac{h}{2}(f(x_0) + 2f(x_1) + \ldots + 2f(x_{N-1}) + f(x_N)) \quad (6.18)
\end{aligned}
$$

Figure 6.7 shows the areas under trapeziums, which are in total very close to the area under the function's curve.

The following MATLAB code uses the composite trapezoidal rule with $N = 10 \times 2^k$, $k = 0, \ldots, 14$ to approximate $\int_0^1 1/(1+x^2)dx$:

```
1  %comptrapz.m
2  clear ; clc ;
3  format short e
4  a = 0.0; b = 1.0 ; f = @(x) 1./(1+x.^2) ;
5  N = 10 ; Approx = [] ;
6  fprintf('------------------------------------------------------\n')
7  fprintf('    N\t\t Approx Int.\t Error\n') ;
8  fprintf('------------------------------------------------------\n')
9  while N <= 200000
```

```
10      h = (b-a)/N ;
11      x = linspace(a, b, N+1) ;
12      I = h/2*(f(x(1))+f(x(N+1))) ;
13      I = I + h*sum(f(x(2:N))) ;
14      Err = abs(pi/4-I) ;
15      Approx = [Approx; [N I Err]] ;
16      fprintf('%6i\t%1.12f\t%1.12e\n', N, I, Err) ;
17      N = 2*N ;
18  end
19  fprintf('-------------------------------------------------------\n')
```

Running the above code, gives:

```
------------------------------------------------------------
   N          Approx Int.              Error
------------------------------------------------------------
    10      0.784981497227     4.166661706586083e-04
    20      0.785293996739     1.041666589161050e-04
    40      0.785372121731     2.604166654529561e-05
    80      0.785391652981     6.510416664773366e-06
   160      0.785396535793     1.627604166443142e-06
   320      0.785397756496     4.069010419716079e-07
   640      0.785398061672     1.017252599933016e-07
  1280      0.785398137966     2.543131505383656e-08
  2560      0.785398157040     6.357829041014895e-09
  5120      0.785398161808     1.589458453743475e-09
 10240      0.785398163000     3.973615880781267e-10
 20480      0.785398163298     9.934308931036640e-11
 40960      0.785398163373     2.483746541770415e-11
 81920      0.785398163391     6.210698622055588e-12
163840      0.785398163396     1.555644502104769e-12
------------------------------------------------------------
```

In Python, the code is:

```python
1  # compsimp.py
2  import numpy as np
3  a, b, f = 0.0, 1.0, lambda x: 1/(1+x**2)
4  ApproxTable = []
5  N = 10
6  print('-------------------------------------------------------')
7  print('    N\t Approx Int.\t Error')
8  print('-------------------------------------------------------')
9  while N <= 200000:
10     x, h = np.linspace(a, b, N+1), (b-a)/N
11     I = h/2*(f(x[0])+f(x[-1]))
12     I = I + h*sum(f(x[1:N]))
13     Err = np.abs(np.pi/4-I)
14     print('{0:6.0f}'.format(N), '\t', ...
           '{0:1.12f}'.format(I), '\t', '{0:1.12e}'.format(Err))
```

```
15      N = 2*N
16   print('-------------------------------------------------------------------------')
```

Executing the code gives the results:

```
runfile('D:/PyFiles/compsimp.py', wdir='D:/PyFiles')
-----------------------------------------------------------
     N     Approx Int.                        Error
-----------------------------------------------------------
    10    0.784981497227              4.166661706586e-04
    20    0.785293996739              1.041666589161e-04
    40    0.785372121731              2.604166654530e-05
    80    0.785391652981              6.510416664773e-06
   160    0.785396535793              1.627604166443e-06
   320    0.785397756496              4.069010419716e-07
   640    0.785398061672              1.017252601043e-07
  1280    0.785398137966              2.543131505384e-08
  2560    0.785398157040              6.357829041015e-09
  5120    0.785398161808              1.589458453743e-09
 10240    0.785398163000              3.973615880781e-10
 20480    0.785398163298              9.934308931037e-11
 40960    0.785398163373              2.483757644001e-11
 81920    0.785398163391              6.210698622056e-12
163840    0.785398163396              1.558975171179e-12
-----------------------------------------------------------
```

(ii) **The composite Simpson's rule**: Let $N \in 2\mathbb{Z}^+$ (even positive integer) with $N = 2k$, $h = (b-a)/N$ and $x_j = a + jh$. In an interval $[x_{2j}, x_{2(j+1)}]$, $j = 0,\ldots,k$ the composite Simpson's rule approximates $\int_{x_{2j}}^{x_{2(j+1)}} f(x)dx$ by $(h/3)(f(x_{2j}) + 4f(x_{2j+1}) + f(x_{2(j+1)}))$. Hence,

$$\int_a^b f(x)dx = \int_{x_0}^{x_2} f(x)dx + \ldots + \int_{x_{2k-2}}^{x_{2k}} f(x)dx$$

$$\approx \frac{h}{3}(f(x_0) + 4f(x_1) + f(x_2)) + \ldots + \frac{h}{3}(f(x_{2k-2})$$
$$+ 4f(x_{2k-1}) + f(x_{2k}))$$

$$= \frac{h}{3}(f(x_0) + f(x_N)) + \frac{4h}{3}\sum_{j=1}^{k} f(x_{2j-1}) + \frac{2h}{3}\sum_{j=1}^{k-1} f(x_{2j}) \quad (6.19)$$

The following MATLAB code uses the composite Simpson's rule to compute $\int_0^1 1/(1+x^2)dx$ with $N = 10 \times 2^k, k = 0,\ldots,9$

```
1   %compsimp.m
2   clear ; clc ;
3   format short e
4   a = 0.0; b = 1.0 ; f = @(x) 1./(1+x.^2) ; IExact = pi/4 ;
5   N = 10 ; Approx = [] ;
6   fprintf('----------------------------------------------------\n')
7   fprintf('   N\t\t Approx Int.\t\t Error\n') ;
8   fprintf('----------------------------------------------------\n')
9   while N ≤ 10000
10      [I, Err] = CompSimpson(f, a, b, N, IExact) ;
11      Approx = [Approx; [N I Err]] ;
12      fprintf('%6i\t\t%1.12f\t\t%1.15e\n', N, I, Err) ;
13      N = 2*N ;
14  end
15  fprintf('----------------------------------------------------\n')
16
17  function [I, Error] = CompSimpson(f, a, b, N, IExact)
18      h = (b-a)/N ;
19      x = linspace(a, b, N+1) ;
20      I = h/3*(f(x(1))+f(x(N+1))) ;
21      for j = 2 : N
22          if mod(j, 2) == 0
23              I = I + 4*h/3*f(x(j)) ;
24          else
25              I = I + 2*h/3*f(x(j)) ;
26          end
27      end
28      Error = abs(IExact-I) ;
29  end
```

Executing the above code gives:

| N | Approx Int. | Error |
|---|---|---|
| 10 | 0.785398153485 | 9.912644483023314e-09 |
| 20 | 0.785398163242 | 1.550021222485043e-10 |
| 40 | 0.785398163395 | 2.422284595127167e-12 |
| 80 | 0.785398163397 | 3.785860513971784e-14 |
| 160 | 0.785398163397 | 3.330669073875470e-16 |
| 320 | 0.785398163397 | 1.110223024625157e-16 |
| 640 | 0.785398163397 | 3.330669073875470e-16 |
| 1280 | 0.785398163397 | 2.220446049250313e-16 |
| 2560 | 0.785398163397 | 1.110223024625157e-16 |
| 5120 | 0.785398163397 | 1.887379141862766e-15 |

In Python, the code is:

```
1   #CompositeSimpson.py
2   import numpy as np
3   def CompSimpson(f, a, b, N, IExact):
```

```
4        x, h = np.linspace(a, b, N+1),  (b-a)/N
5        I = h/3.0*(f(x[0])+f(x[-1]))
6        for j in range(1, N):
7            if j%2 == 1:
8                I += 4.*h/3.*f(x[j])
9            else:
10               I += 2.*h/3.*f(x[j])
11       Error = np.abs(IExact-I)
12       return I, Error
13
14   a, b, f = 0.0, 1.0, lambda x: 1/(1+x**2)
15   ApproxTable = []
16   IExact = np.pi/4
17   N = 10
18   print('-----------------------------------------------------------')
19   print('    N\t Approx Int.\t Error')
20   print('-----------------------------------------------------------')
21   while N <= 10000:
22       I, Err = CompSimpson(f, a, b, N, IExact)
23       print('{0:6.0f}'.format(N), '\t', ...
              '{0:1.12f}'.format(I), '\t', '{0:1.12e}'.format(Err))
24       N = 2*N
25   print('-----------------------------------------------------------')
```

Executing this code gives:

```
runfile('D:/PyFiles/CompositeSimpson.py', wdir='D:/PyFiles')
----------------------------------------------------------
   N   Approx Int.                      Error
----------------------------------------------------------
   10  0.785398153485    9.912644594046e-09
   20  0.785398163242    1.550021222485e-10
   40  0.785398163395    2.422284595127e-12
   80  0.785398163397    3.796962744218e-14
  160  0.785398163397    3.330669073875e-16
  320  0.785398163397    1.110223024625e-16
  640  0.785398163397    3.330669073875e-16
 1280  0.785398163397    2.220446049250e-16
 2560  0.785398163397    1.110223024625e-16
 5120  0.785398163397    1.887379141863e-15
----------------------------------------------------------
```

Both MATLAB and Python contain functions to compute a definite integral $\int_a^b f(x)dx$ based on trapezoidal and Simpson's rules. The MATLAB function **trapz** receives two vectors $x$ (of discrete values in $[a,b]$) and $f(x)$ the corresponding values of $f$ to $x$. It applies the trapezoidal rule to return the integral value.

```
>> x = linspace(0, 1, 101) ;
>> f = exp(-x.^2) ;
>> format long g
```

```
>> I = trapz(x, f)
I =
0.74681800146797
```

Python also contains a `trapz` as a part of the `scipy.integrate` library. It differs from the MATLAB's trapz function by receiving the vector of $f$ before vector $x$:

```
In [10]: from scipy.integrate import trapz
In [11]: import numpy as np
In [12]: x = np.linspace(0, np.pi, 500)
In [13]: g = np.zeros_like(x)
In [14]: g[0], g[1:] = 1.0, np.sin(x[1:])/x[1:]
In [15]: I = trapz(g, x)
In [16]: print(I)
1.8519360005832526
```

Also, the MATLAB's function `quad` applies the Simpson's rule to evaluate $\int_a^b f(x)dx$. It receives a function handle to $f$, $a$ and $b$ and then it returns the value of the integral:

```
>> x = linspace(0, pi, 501) ;
>> h = @(x) sin(x)./x ;
>> I = quad(h, 0, pi)
I =
1.8519e+00
```

The Python's `quad` function located in the `scipy.integrate` library receives same arguments as MATLAB's quad function, but it returns two values: the integration value and the approximation error.

```
In [17]: from scipy.integrate import quad
In [18]: h = lambda x: np.exp(-x**2)
In [19]: I = quad(h, 0.0, 1.0)
In [20]: print('Integral = ',I[0], ',  Error = ', I[1])
Integral =  0.7468241328124271 ,  Error =  8.291413475940725e-15
```

### 6.2.2   The Gauss Integration Method

On approximating the integral

$$I = \int_a^b f(x)dx,$$

a Lagrange polynomial of degree $n$ that interpolates the integrand function $f(x)$ at equally spaced points to derive a Newton-Cotes quadrature. Such a Newton-Cotes quadrature is expected to be exact for a polynomial of degree not more than $n-1$.

Unlike Newton-Cotes methods, a Gauss integration method does not pre-determine the positions of the interpolation points $x_j$, $j = 1, \ldots, n$ in $[a,b]$, nor the corresponding weights $w_j$ with the purpose of finding a quadrature that is exact for polynomials of degrees up to $2n - 1$ [47]. In other words, the derivation of a Gauss quadrature is based on finding the set of points locations (Gauss points) and corresponding weights such that the quadrature is exact for polynomials of degrees $2n - 1$ or less.

All the Gauss points lie in the interval $[-1, 1]$, therefore the first step to start from is to use a variable transform

$$x = \frac{b - a}{2}s + \frac{b + a}{2}, \quad -1 \leq s \leq 1$$

to convert the integral problem from

$$I = \int_a^b f(x)\,dx$$

into the form

$$I = \frac{b - a}{2} \int_{-1}^{1} f\left(\frac{b - a}{2}s + \frac{b + a}{2}\right) ds.$$

Through this variable transform, the point $x = a$ is mapped into $s = -1$ and $x = b$ to $s = 1$. Values $a < x < b$ are mapped to $-1 < s < 1$.

The next step is to choose a positive integer $n$ and assuming that $f(x)$ is a polynomial of degree $2n - 1$. That is:

$$f(x) = c_0 + c_1 x + \cdots + c_{2n-1}x^{2n-1}, \tag{6.20}$$

where $c_1, \ldots, c_{2n-1}$.

The purpose is to find optimal sets of points $a \leq x_1 < \ldots < x_n \leq b$ (through finding $-1 \leq s_1 < s_2 < \ldots < s_{n-1} < s_n \leq 1$) and weights $w_1, w_2, \ldots, w_n$ such that

$$\int_a^b f(x)\,dx = \frac{b - a}{2} \int_{-1}^{1} f\left(\frac{b - a}{2}s + \frac{b + a}{2}\right) ds \approx \frac{b - a}{2} \sum_{j=1}^{n} w_j f\left(\frac{b - a}{2}s_j + \frac{b + a}{2}\right)$$

$$\tag{6.21}$$

is exact for $f(x) = c_0 + c_1 x + \cdots + c_{2n-1}x^{2n-1}$.

Integrating Equation (6.20) from $a$ to $b$ gives:

$$\int_a^b f(x)\,dx = \int_a^b \left(c_0 + c_1 x + \cdots + c_{2n-1}x^{2n-1}\right) dx$$

$$= c_0(b - a) + c_1 \frac{b^2 - a^2}{2} + \cdots + c_{2n-1}\frac{b^{2n} - a^{2n}}{2n} \tag{6.22}$$

From Equation (6.20):

$$f(x_j) = c_0 + c_1 x_j + \cdots + c_{2n-1}x_j^{2n-1}, \quad j = 1, \ldots, n$$

hence,

$$
\begin{aligned}
w_1 f(x_1) + \cdots + w_n f(x_n) &= w_1 \left( c_0 + \cdots + c_{2n-1} x_1^{2n-1} \right) \\
&\quad + w_2 \left( c_0 + \cdots + c_{2n-1} x_2^{2n-1} \right) \\
&\quad + \cdots + w_n \left( c_0 + \cdots + c_{2n-1} x_n^{2n-1} \right) \\
&= c_0 \left( w_1 + \ldots + w_n \right) + c_1 \left( w_1 x_1 + \ldots + w_n x_n \right) \\
&\quad + \cdots + c_{2n-1} \left( w_1 x_1^{n-1} + \cdots + w_n x_n^{n-1} \right) \quad (6.23)
\end{aligned}
$$

Comparing the terms of Equation (6.22) and (6.23) gives $2n$ nonlinear equations in the $2n$ unknowns $x_j$ and $w_j$, $j = 1, \ldots, n$.

$$
\begin{aligned}
w_1 + w_2 + \cdots + w_n &= b - a \\
w_1 x_1 + w_2 x_2 + \cdots + w_n x_n &= \frac{b^2 - a^2}{2} \\
w_1 x_1^2 + w_2 x_2^2 + \cdots + w_n x_n^2 &= \frac{b^3 - a^3}{3} \qquad (6.24) \\
\vdots \quad &= \quad \vdots \\
w_1 x_1^{n-1} + w_2 x_2^{n-1} + \cdots + w_n x_n^{n-1} &= \frac{b^n - a^n}{n}
\end{aligned}
$$

The points $x_1, \ldots, x_n$ and weights $w_1, \ldots, w_n$ are found by solving the system of nonlinear equations (6.24).

Following is the derivation of the Gauss quadratures for $n = 1$ and $n = 2$:

1. **One point Gauss quadrature:** In the case that $n = 1$, there is one point $x_1$ and one weight $w_1$ to be found. Since $2n - 1 = 1$, it is assumed that $f(x) = c_0 + c_1 x$. The equations in $x_1$ and $w_1$ are:

$$
\begin{aligned}
w_1 &= b - a \\
w_1 x_1 &= \frac{b^2 - a^2}{2}
\end{aligned}
$$

Solving this system of equations gives:

$$
w_1 = b - a \text{ and } x_1 = \frac{b + a}{2}.
$$

The Gauss quadrature is:

$$
\int_a^b f(x) dx = (b - a) f\left( \frac{b + a}{2} \right) \qquad (6.25)
$$

2. **Two points Gauss quadrature:** For $n = 2$, there are two points $x_1, x_2$ and weights $w_1, w_2$ to be found. Since $2n - 1 = 3$, the function $f(x)$ is

assumed to be cubic (*i.e.* $f(x) = c_0 + c_1 x + c_2 x^2 + c_3 x^3$). The system of equations in $x_1, x_2, w_1$ and $w_2$ is:

$$w_1 + w_2 = b - a$$

$$w_1 x_1 + w_2 x_2 = \frac{b^2 - a^2}{2}$$

$$w_1 x_1^2 + w_2 x_2^2 = \frac{b^3 - a^3}{3}$$

$$w_1 x_1^3 + w_2 x_2^3 = \frac{b^4 - a^4}{4}$$

The following Python code can by used to find the solution of the nonlinear system:

```
In [21]: import sympy as smp
In [22]: from sympy.solvers import solve
In [23]: w1, w2, x1, x2, a, b = smp.symbols('w1, s2, x1, x2,
         a, b', cls=smp.Symbol)
In [24]: Syst = [w1+w2-(b-a), w1*x1+w2*x2-(b**2-a**2)/2, \
   ...: w1*x1**2+w2*x2**2-(b**3-a**3)/3,
         w1*x1**3+w2*x2**3-(b**4-a**4)/4]
In [25]: vrs = [w1, w2, x1, x2]
In [26]: Sol = solve(Syst, vrs)
In [27]: print(Sol[0])
(-(a - b)/2, -(a - b)/2, -sqrt(3)*a/6 + a/2
+ sqrt(3)*b/6 + b/2, a/2 + b/2 + sqrt(3)*(a - b)/6)
In [28]: print(Sol[1])
(-(a - b)/2, -(a - b)/2, sqrt(3)*a/6 + a/2
- sqrt(3)*b/6 + b/2, a/2 + b/2 - sqrt(3)*(a - b)/6)
```

The Python symbolic library gives two solutions of the nonlinear system, in which the values of $x_1$ and $x_2$ exchange. If we put a condition $x_1 < x_2$ the solution of the nonlinear system is:

$$w_1 = w_2 = \frac{b-a}{2}, x_1 = \frac{b+a}{2} - \frac{b-a}{6}\sqrt{3} \text{ and } x_2 = \frac{b+a}{2} + \frac{b-a}{6}\sqrt{3}$$

The two-points Gauss quadrature is:

$$\int_a^b f(x)dx = \frac{b-a}{2}\left(f\left(\frac{b+a}{2} - \frac{b-a}{6}\sqrt{3}\right) + f\left(\frac{b+a}{2} + \frac{b-a}{6}\sqrt{3}\right)\right)$$

$$(6.26)$$

**Example 6.3** In this example, the one- and two-points Gauss quadratures will be used to evaluate the integral:

$$\int_0^2 \frac{xdx}{1+2x^2}$$

The interval $[0, 2]$ will be divided into $N$ sub-intervals ($N$ is a positive integer) by points $0 = x_0 < x_1 < \ldots < x_N = 2$. Each sub-interval has a length $h = 2/N$. Then, in each subinterval $[x_j, x_{j+1}]$ the one- or two points Gauss quadrature will be used to approximate $\int_{x_j}^{x_{j+1}} f(x)dx$.

The 1-point Gauss quadrature approximates $\int_{x_j}^{x_{j+1}} f(x)dx$ as

$$\int_{x_j}^{x_{j+1}} f(x)dx \approx hf\left(\frac{x_j + x_{j+1}}{2}\right)$$

Then,

$$\int_a^b f(x)dx \approx \sum_{j=0}^{N-1} \int_{x_j}^{x_{j+1}} f(x)dx = h\sum_{j=0}^{N-1} f\left(\frac{x_j + x_{j+1}}{2}\right)$$

The 2-points Gauss quadrature approximates $\int_{x_j}^{x_{j+1}} f(x)dx$ as

$$\int_{x_j}^{x_{j+1}} f(x)dx \approx \frac{h}{2}\left(f\left(\frac{x_j + x_{j+1}}{2} - \frac{x_{j+1} - x_j}{6}\sqrt{3}\right)\right.$$
$$\left. + f\left(\frac{x_j + x_{j+1}}{2} + \frac{x_{j+1} - x_j}{6}\sqrt{3}\right)\right)$$

Then,

$$\int_a^b f(x)dx \approx \sum_{j=0}^{N-1} \int_{x_j}^{x_{j+1}} f(x)dx$$
$$= \frac{h}{2}\sum_{j=0}^{N-1}\left(f\left(\frac{x_j + x_{j+1}}{2} - \frac{x_{j+1} - x_j}{6}\sqrt{3}\right)\right.$$
$$\left. + f\left(\frac{x_j + x_{j+1}}{2} + \frac{x_{j+1} - x_j}{6}\sqrt{3}\right)\right)$$

The following MATLAB code shows the approximations and errors of the 1- and 2-points Gauss quadratures for different values of $N(= 10 \times 2^k, k = 0, \ldots, 10)$.

```
1  clear ; clc ;
2  f = @(x) x./(1+2*x.^2) ;
3  a = 0.0 ; b = 2.0 ;
4  N = 10 ;
5  IExact = log(9)/4 ;
6  GL1Approx = [] ; GL2pApprox = [] ;
7
8  while N < 20000
9      x = linspace(a, b, N+1) ;
10     h = (b-a)/N ;
11     I1 = h*sum(f((x(1:N)+x(2:N+1))/2)) ;
```

```
12      Err1 = abs(I1-IExact) ;
13      GL1Approx = [GL1Approx; [N I1 Err1]] ;
14      I2 = ...
            h/2*sum(f((x(1:N)+x(2:N+1))/2-(x(2:N+1)-x(1:N))*sqrt(3)/6)
15      +...f((x(1:N)+x(2:N+1))/2+(x(2:N+1)-x(1:N))*sqrt(3)/6)) ;
16      Err2 = abs(I2-IExact) ;
17      GL2pApprox = [GL2pApprox; [N I2 Err2]] ;
18      N = 2*N ;
19  end
20
21  fprintf('Integration with 1-point Gauss quadrature\n')
22  fprintf('-------------------------------------------------------\n')
23  fprintf('   N\t\t Approx Int.\t\t Error\n') ;
24  fprintf('-------------------------------------------------------\n')
25  [m, L] = size(GL1Approx) ;
26  for j = 1 : m
27      fprintf('%6i\t\t%1.12f\t\t%1.15e\n', GL1Approx(j,1), ...
28      GL1Approx(j,2), GL1Approx(j,3)) ;
29  end
30  fprintf('-------------------------------------------------------\n\n\n')
31
32  fprintf('Integration with 2-points Gauss quadrature\n')
33  fprintf('-------------------------------------------------------\n')
34  fprintf('   N\t\t Approx Int.\t\t Error\n') ;
35  fprintf('-------------------------------------------------------\n')
36  [m, L] = size(GL2pApprox) ;
37  for j = 1 : m
38      fprintf('%6i\t\t%1.12f\t\t%1.15e\n', GL2pApprox(j,1), ...
39      GL2pApprox(j,2), GL2pApprox(j,3)) ;
40  end
41  fprintf('-------------------------------------------------------\n\n\n')
```

By executing the code, the following results are obtained:

```
Integration with 1-point Gauss quadrature
```

| N | Approx Int. | Error |
|---|---|---|
| 10 | 0.551141201821 | 1.835057487226677e-03 |
| 20 | 0.549760289497 | 4.541451628828908e-04 |
| 40 | 0.549419404209 | 1.132598753686986e-04 |
| 80 | 0.549334442201 | 2.829786668079315e-05 |
| 160 | 0.549313217734 | 7.073400441592881e-06 |
| 320 | 0.549307912618 | 1.768283512948443e-06 |
| 640 | 0.549306586401 | 4.420667167881476e-07 |
| 1280 | 0.549306254850 | 1.105164195713826e-07 |
| 2560 | 0.549306171963 | 2.762908835052258e-08 |
| 5120 | 0.549306151241 | 6.907272531719855e-09 |
| 10240 | 0.549306146061 | 1.726818577019174e-09 |

Integration with 2-points Gauss quadrature

```
--------------------------------------------------------
    N          Approx Int.                 Error
--------------------------------------------------------
    10       0.549301461604        4.682729895844062e-06
    20       0.549305863669        2.806649841424758e-07
    40       0.549306126963        1.737151811287419e-08
    80       0.549306143251        1.083119705036495e-09
   160       0.549306144266        6.765477067460779e-11
   320       0.549306144330        4.227507233167671e-12
   640       0.549306144334        2.640110352558622e-13
  1280       0.549306144334        1.576516694967722e-14
  2560       0.549306144334        1.554312234475219e-15
  5120       0.549306144334        6.661338147750939e-16
 10240       0.549306144334        4.440892098500626e-16
--------------------------------------------------------
```

The Python code is:

```python
 1  import numpy as np
 2  f = lambda x: x/(1.0+2.0*x**2)
 3  a, b = 0.0, 2.0
 4  N = 10
 5  IExact = np.log(9)/4
 6  GL1Approx, GL2Approx = [], []
 7  while N < 20000:
 8      I1, I2 = 0.0, 0.0
 9      x = np.linspace(a, b, N+1)
10      h = (b-a)/N
11      for j in range(N):
12          I1 += h*(f((x[j]+x[j+1])/2.0))
13          I2 += ...
               h/2*(f((x[j]+x[j+1])/2-(x[j+1]-x[j])*np.sqrt(3.)/6.)+\
14          f((x[j]+x[j+1])/2+(x[j+1]-x[j])*np.sqrt(3.)/6.))
15      Err1 = abs(I1-IExact) ;
16      GL1Approx.append([N, I1, Err1])
17      Err2 = abs(I2-IExact)
18      GL2Approx.append([N, I2, Err2])
19      N = 2*N
20
21  print('Integration with 1-point Gauss quadrature\n')
22  print('---------------------------------------------------\n')
23  print('   N\t\t Approx Int.\t\t Error\n') ;
24  print('---------------------------------------------------\n')
25  m = len(GL1Approx)
26  for j in range(m):
27      print('{0:6.0f}'.format(GL1Approx[j][0]), '\t', \
28          '{0:1.12f}'.format(GL1Approx[j][1]), '\t', \
29          '{0:1.12e}'.format(GL1Approx[j][2]))
30  print('---------------------------------------------------\n\n\n')
31
32  print('Integration with 2-points Gauss quadrature\n')
```

```
33   print('--------------------------------------------------------\n')
34   print('     N\t\t Approx Int.\t\t Error\n') ;
35   print('--------------------------------------------------------\n')
36   m = len(GL2Approx) ;
37   for j in range(m):
38       print('{0:6.0f}'.format(GL2Approx[j][0]), '\t', \
39            '{0:1.12f}'.format(GL2Approx[j][1]), '\t', \
40            '{0:1.12e}'.format(GL2Approx[j][2]))
41   print('-----------------------------------------------\n\n\n')
```

Since the number of subintervals is doubled in the above example, the rates of convergences for the 1- and 2-points Gauss quadratures can be computed using the rule:

$$R_N = \log_2\left(\frac{Error_N}{Error_{2N}}\right)$$

The following Python code can be used to see the rates of convergences of the two methods:

```
1    RConv1 = []
2    print('Rates of convergence for 1-point Gauss quadrature: \n')
3    print('----------------------\n')
4    print('  N\t Conv. Rate\n')
5    print('----------------------\n')
6    for k in range(m-1):
7        RConv1.append(np.log2(GL1Approx[k][2]/GL1Approx[k+1][2]))
8        print('{0:4.0f}'.format(GL1Approx[k][0]), '\t ...
              {0:2.3f}'.format(RConv1[-1]))
9    print('----------------------\n')
10
11   RConv2 = []
12   print('Rates of convergence for 2-points Gauss quadrature: \n')
13   print('----------------------\n')
14   print('  N\t Conv. Rate\n')
15   print('----------------------\n')
16   for k in range(m-1):
17       RConv2.append(np.log2(GL2Approx[k][2]/GL2Approx[k+1][2]))
18       print('{0:4.0f}'.format(GL2Approx[k][0]), '\t ...
              {0:2.3f}'.format(RConv2[-1]))
19   print('----------------------\n')
```

Executing the code gives:

```
Rates of convergence for 1-point Gauss quadrature:
----------------------
  N       Conv. Rate
----------------------
   10      2.015
   20      2.004
   40      2.001
   80      2.000
  160      2.000
```

| | |
|---|---|
| 320 | 2.000 |
| 640 | 2.000 |
| 1280 | 2.000 |
| 2560 | 2.000 |
| 5120 | 2.000 |

----------------------

Rates of convergence for 2-points Gauss quadrature:
----------------------

| N | Conv. Rate |
|---|---|
| 10 | 4.060 |
| 20 | 4.014 |
| 40 | 4.003 |
| 80 | 4.001 |
| 160 | 4.000 |
| 320 | 3.998 |
| 640 | 4.010 |
| 1280 | 4.402 |
| 2560 | 0.000 |
| 5120 | -0.652 |

----------------------

The above results show that the 1-point Gauss quadrature behaves as a second order method and the 2-point Gauss quadrature behaves as a fourth-order method.

Higher-order Gauss quadrature methods can be obtained for $n \geq 3$. In fact, in $n$-points Gauss quadrature method, the points $x_1, x_2, \ldots, x_n$ are the roots of the Legendre polynomial $P_n(x)$. Hence, this class of points is referred to as Legendre-Gauss points. The Python function legendre located in scipy.special library receives a parameter $n$ and returns the coefficients of the corresponding $n^{th}$-degree Legendre polynomial [5, 44].

```
In [29]: from scipy.special import legendre
In [30]: legendre(4)
Out [30]: poly1d([ 4.37500000e+00,   4.85722573e-16,
-3.75000000e+00,   2.42861287e-16,   3.75000000e-01])
In [31]: print(legendre(4))
       4          3         2
4.375 x + 4.857e-16 x - 3.75 x + 2.429e-16 x + 0.375
In [32]: np.roots(legendre(4))
Out[32]: array([ 0.86113631, -0.86113631,   0.33998104,
-0.33998104])
```

The last line np.roots(legendre(4)) gives the point locations of the 4-points Legendre-Gauss quadrature.

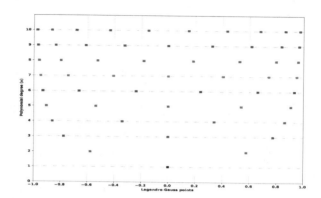

FIGURE 6.8: The positions of Legendre-Gauss points for polynomials of degrees $n = 1, \ldots, 10$.

Figure 6.8 shows the distribution of Legendre-Gauss points for polynomials of degrees 1 to 10.

Having the $n$ coordinates of the Legendre-Gauss points $x_1, \ldots, x_n$, the corresponding weights $w_1, \ldots, w_n$ can be found by solving the linear system:

$$
\begin{bmatrix}
1 & 1 & 1 & \cdots & 1 \\
x_1 & x_2 & x_3 & \cdots & x_n \\
x_1^2 & x_2^2 & x_3^2 & \cdots & x_n^2 \\
\vdots & \vdots & \vdots & \ddots & \vdots \\
x_1^{n-1} & x_2^{n-1} & x_3^{n-1} & \cdots & x_n^{n-1}
\end{bmatrix}
\begin{bmatrix}
w_1 \\
w_2 \\
w_3 \\
\vdots \\
w_n
\end{bmatrix}
=
\begin{bmatrix}
2 \\
0 \\
\frac{2}{3} \\
\vdots \\
\frac{1-(-1)^n}{n}
\end{bmatrix}
\tag{6.27}
$$

The matrix at the left-hand side of Equation (6.27) is the transpose matrix of the Vandermonde matrix of type $n \times n$, constructed from the vector of points $x_1, \ldots, x_n$.

The following Python code (`FindLGParams.py`) includes a function `LGpw` that receives a parameter `n` and returns the corresponding Legendre-Gauss points and weights.

```
1   #FindLGParams.py
2
3   from scipy.special import legendre
4   import numpy as np
5
6   def LGpw(n):
7       s = list(np.sort(np.roots(legendre(n))))
8       X = (np.fliplr(np.vander(s))).T
9       Y = np.array([(1-(-1)**j)/j for j in range(1, n+1)])
10      w = np.linalg.solve(X, Y)
11      return s, w
```

```
12
13   s, w = LGpw(3)
14   print('n = 3:\n')
15   print('Points are: ', s, '\n Weights are:', w)
16
17   print('n = 6:\n')
18   s, w = LGpw(6)
19   print('Points are: ', s, '\n Weights are:', w)
```

By executing the code the following results are obtained:

```
runfile('D:/PyFiles/FindLGParams.py', wdir='D:/PyFiles')
n = 3:
Points are:  [-0.7745966692414835, 0.0, 0.7745966692414834]
Weights are: [0.55555556 0.88888889 0.55555556]


n = 6:
Points are:  [-0.9324695142031514, -0.6612093864662644,
-0.23861918608319704, 0.23861918608319688, 0.6612093864662634,
0.9324695142031519]
Weights are: [0.17132449 0.36076157 0.46791393 0.46791393
0.36076157 0.17132449]
```

**Example 6.4** This example uses the function LGpw to find the points and weights of a 5-points Legendre-Gauss quadrature, then applies it to find the integration in Example 6.3:

$$\int_0^2 \frac{x}{1+2x^2}\,dx$$

The Python code LG5Approx.py implements the method:

```
1    from scipy.special import legendre
2    import numpy as np
3
4    def LGpw(n):
5        s = list(np.sort(np.roots(legendre(n))))
6        X = (np.fliplr(np.vander(s))).T
7        Y = np.array([(1-(-1)**j)/j for j in range(1, n+1)])
8        w = np.linalg.solve(X, Y)
9        return s, w
10
11   s, w = GLpw(5)
12
13   a, b = 0.0, 2.0
14   f = lambda x: x/(1.0+2.0*x**2)
15   N = 5
16   IExact = np.log(9.0)/4.0
17   GL5Approx = []
18   while N < 1000:
19       x, h = np.linspace(a, b, N+1), (b-a)/N
```

```
20        I = 0.0
21        for j in range(N):
22            xm = (x[j]+x[j+1])/2.0
23            I += h/2.0*sum([w[k]*f(xm+h/2*s[k]) for k in ...
                  range(len(s))])
24        Err = abs(I-IExact)
25        GL5Approx.append([N, I, Err])
26        N = 2*N
27
28    print('Integration with 5-point Gauss quadrature\n')
29    print('-----------------------------------------------------------\n')
30    print('   N\t\t Approx Int.\t\t Error\n') ;
31    print('-----------------------------------------------------------\n')
32    m = len(GL5Approx)
33    for j in range(m):
34        print('{0:6.0f}'.format(GL5Approx[j][0]), '\t', \
35            '{0:1.12f}'.format(GL5Approx[j][1]), '\t', \
36            '{0:1.12e}'.format(GL5Approx[j][2]))
37    print('-----------------------------------------------------------\n\n\n')
```

Executing the Python code gives the following results:

```
runfile('D:/PyFiles/LG5Approx.py', wdir='D:/PyFiles')
Integration with 5-point Gauss quadrature
```

| N | Approx Int. | Error |
|---|---|---|
| 5 | 0.549306144553503 | 3.995001813766e-10 |
| 10 | 0.549306144334415 | 6.564653283378e-13 |
| 20 | 0.549306144334055 | 6.063411283909e-16 |
| 40 | 0.549306144334055 | 2.021137094636e-16 |
| 80 | 0.549306144334055 | 2.021137094636e-16 |
| 160 | 0.549306144334055 | 4.042274189272e-16 |
| 320 | 0.549306144334055 | 4.042274189272e-16 |
| 640 | 0.549306144334054 | 1.414795966245e-15 |

The MATLAB code `LGApprox` is:

```
1  n = 5 ;
2  [s, w] = LGpw(n) ;
3  a = 0.0 ; b = 2.0 ; f = @(x) x./(1+2*x.^2) ;
4  N = 5 ;
5  IExact = log(9.0)/4 ;
6  LG5Approx = [] ;
7  while N < 1000
8      x = linspace(a, b, N+1) ;
9      h = (b-a)/N ;
10     I = 0.0 ;
11     for j = 1 : N
12         xm = (x(j)+x(j+1))/2 ;
```

```
13              I = I + h/2.0*sum(w.*f(xm+h/2*s)) ;
14      end
15      Err = abs(I - IExact) ;
16      LG5Approx = [LG5Approx; [N, I, Err]] ;
17      N = 2*N ;
18 end
19
20 fprintf('Integration with 5-point Gauss quadrature\n')
21 fprintf('------------------------------------------------------------\n')
22 fprintf('   N\t\t Approx Int.\t\t Error\n') ;
23 fprintf('------------------------------------------------------------\n')
24 m = length(LG5Approx) ;
25 for j = 1 : m
26      fprintf('%4i\t%1.15f\t%1.12e\n', LG5Approx(j, 1), ...
27      LG5Approx(j, 2), LG5Approx(j, 3)/IExact) ;
28 end
29 fprintf('------------------------------------------------------------\n')
30
31 function [s, w] = LGpw(n)
32      s = sort(roots(legendreV(n))) ;
33      X = fliplr(vander(s))' ;
34      Y = zeros(n, 1) ;
35      for j = 1 : n
36          Y(j) = (1-(-1)^j)/j ;
37      end
38      w = X\Y ;
39 end
40
41 function Legn = legendreV(n)
42
43      %legendreV.m
44      %
45      % This function receives a parameter n representing the ...
                polynomial degree
46      % and returns the cofficients of Legendre polynomial of ...
                degree n using the
47      % recursive relation
48      %            P_n(x) = ...
                ((2n-1)/n)xP_{n-1}(x)-((n-1)/n)P_{n-2}(x)
49      %
50      % Written by Eihab B.M. Bashier, ebashier@du.edu.om
51
52      if n == 0
53          Legn = [0, 1] ;
54      elseif n == 1
55          Legn = [1, 0] ;
56      else
57          L1 = conv([1, 0], legendreV(n-1)) ;
58          L2 = legendreV(n-2) ;
59          if length(L2) < length(L1)
60              L2 = [zeros(1, length(L1)-length(L2)) L2] ;
61          end
62          Legn = (2*n-1)/n*L1-(n-1)/n*L2 ;
63      end
64 end
```

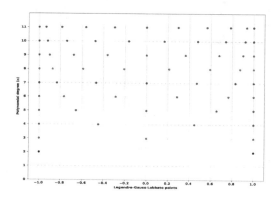

FIGURE 6.9: The distribution of Legendre-Gauss-Lobbatto points in the interval $[-1,1]$ for $n = 2, 3, \ldots, 11$.

The numerical quadratures based on Legendre-Gauss points are open. That is they are not including the boundary points $-1.0$ and $1.0$ (hence the integration boundaries $a$ and $b$). There is another class of Gauss quadratures that imposes the choice of the boundary points $-1.0$ and $1.0$. So, in addition to $x_1 = -1.0$ and $x_n = 1.0$, an $n$-points closed Gauss quadrature may choose the roots of $(1 - x^2)P'_{n-1}(x)$ where $P'_{n-1}(x)$ is the derivative of the Legendre polynomial of degree $n - 1$ with $n \geq 2$. This kind of points is referred as `Legendre-Gauss-Lobbatto (LGL)` points.

Figure 6.9 shows the Legendre-Gauss-Lobbatto points for different values of $n$.

The corresponding weights to the $n$-LGL points can be obtained from Equation (6.27). For $n = 2$ and $n = 3$, the resulting quadratures coincide with the trapezoidal and Simpson's rules. For $n \geq 4$ the resulting quadratures are different from those obtained by Newton-Cotes methods.

Because the boundary points are enforced to be included among the $n$ LGL-points, the accuracy of the method is reduced by 2. Hence, instead of being accurate up to degree $2n - 1$ they are accurate up to degree $2n - 3$.

The Python code `LGLParams` includes a function `LGLpw` that receives a positive integer $n \geq 2$ and returns the corresponding $n$-LGL points $-1 = x_1 < x_2 < \cdots < x_n = 1$ and weights $w_1, \ldots, w_n$. It then approximates the definite integral

$$I = \int_0^2 \frac{x}{1 + 2x^2},$$

using 4-LGL points.

```
1  import numpy as np
2  from scipy.special import legendre
3  def LGLpw(n):
4      s = list(np.sort(np.roots([-1, 0, ...
              1]*np.polyder(legendre(n-1)))))
5      X = (np.fliplr(np.vander(s))).T
6      Y = np.array([(1-(-1)**j)/j for j in range(1, n+1)])
7      w = np.linalg.solve(X, Y)
8      return s, w
9
10 n = 4
11 s, w = LGLpw(n)
12
13 a, b = 0.0, 2.0
14 f = lambda x: x/(1.0+2.0*x**2)
15 N = 5
16 IExact = np.log(9.0)/4.0
17 LGL5Approx = []
18 while N < 1000:
19     x, h = np.linspace(a, b, N+1), (b-a)/N
20     I = 0.0
21     for j in range(N):
22         xm = (x[j]+x[j+1])/2.0
23         I += h/2.0*sum([w[k]*f(xm+h/2*s[k]) for k in ...
                  range(len(s))])
24     Err = abs(I-IExact)
25     LGL5Approx.append([N, I, Err])
26     N = 2*N
27
28 print('Integration with 5-point Gauss quadrature\n')
29 print('----------------------------------------------------\n')
30 print('  N\t\t Approx Int.\t\t Error\n') ;
31 print('----------------------------------------------------\n')
32 m = len(LGL5Approx)
33 for j in range(m):
34     print('{0:6.0f}'.format(LGL5Approx[j][0]), '\t', \
35         '{0:1.15f}'.format(LGL5Approx[j][1]), '\t', \
36         '{0:1.12e}'.format(LGL5Approx[j][2]/IExact))
37 print('----------------------------------------------------\n\n\n')
38
39 RConv5 = []
40 print('Rates of convergence for '+str(n)+'-points ...
          Legendre-Gauss-Lobbato quadrature: \n')
41 print('------------------------\n')
42 print('  N\t Conv. Rate\n')
43 print('------------------------\n')
44 for k in range(m-1):
45     RConv5.append(np.log2(LGL5Approx[k][2]/LGL5Approx[k+1][2]))
46     print('{0:4.0f}'.format(LGL5Approx[k][0]), '\t ...
          {0:2.3f}'.format(RConv5[-1]))
47 print('------------------------\n')
```

Executing the code gives the following results:

```
runfile('D:/PyFiles/LGLParams.py', wdir='D:/PyFiles')
Integration with 5-point Gauss quadrature
```

| N | Approx Int. | Error |
|---|---|---|
| 5 | 0.549303859258719 | 4.159930413361e-06 |
| 10 | 0.549306121271021 | 4.198575552244e-08 |
| 20 | 0.549306144007535 | 5.944232913986e-10 |
| 40 | 0.549306144329063 | 9.088447173451e-12 |
| 80 | 0.549306144333977 | 1.410753692056e-13 |
| 160 | 0.549306144334054 | 2.021137094636e-15 |
| 320 | 0.549306144334055 | 4.042274189272e-16 |
| 640 | 0.549306144334054 | 1.212682256782e-15 |

Rates of convergence for 4-points Legendre-Gauss-Lobbato quadrature:

| N | Conv. Rate |
|---|---|
| 5 | 6.631 |
| 10 | 6.142 |
| 20 | 6.031 |
| 40 | 6.009 |
| 80 | 6.125 |
| 160 | 2.322 |
| 320 | -1.585 |

The table of rates of convergence, shows that the Gauss quadrature based on 4-LGL points behaves as order 6 integration method.

The MATLAB code is:

```
1   n = 4 ;
2   [s, w] = LGLpw(n) ;
3   a = 0.0 ; b = 2.0 ; f = @(x) x./(1+2*x.^2) ;
4   N = 5 ;
5   IExact = log(9.0)/4 ;
6   LGL4Approx = [] ;
7   while N < 1000
8       x = linspace(a, b, N+1) ;
9       h = (b-a)/N ;
10      I = 0.0 ;
11      for j = 1 : N
12          xm = (x(j)+x(j+1))/2 ;
13          I = I + h/2.0*sum(w.*f(xm+h/2*s)) ;
14      end
15      Err = abs(I - IExact) ;
16      LGL4Approx = [LGL4Approx; [N, I, Err]] ;
17      N = 2*N ;
18  end
```

```
19
20   fprintf('Integration with 4-point Gauss quadrature\n')
21   fprintf('---------------------------------------------------------------\n')
22   fprintf('   N\t\t Approx Int.\t\t Error\n') ;
23   fprintf('---------------------------------------------------------------\n')
24   m = length(LGL4Approx) ;
25   for j = 1 : m
26       fprintf('%4i\t%1.15f\t%1.12e\n', LGL4Approx(j, 1), ...
27           LGL4Approx(j, 2), LGL4Approx(j, 3)/IExact) ;
28   end
29   fprintf('---------------------------------------------------------------\n')
30
31   RConv4 = [] ;
32   fprintf('Rates of convergence for 4-points Gauss quadrature: \n')
33   fprintf('-------------------------\n')
34   fprintf('  N\t Conv. Rate\n')
35   fprintf('-------------------------\n')
36   for k = 1:m-1
37       RConv4 =[RConv4;(log2(LGL4Approx(k, 3)/LGL4Approx(k+1, ...
             3)))] ;
38       fprintf('%4i\t%1.12e\n', LGL4Approx(k, 1), RConv4(end)) ;
39   end
40   fprintf('-------------------------\n')
41
42   function [s, w] = LGLpw(n)
43       s = sort(roots(conv([-1, 0, 1],polyder(legendreV(n-1))))) ;
44       X = fliplr(vander(s))' ;
45       Y = zeros(n, 1) ;
46       for j = 1 : n
47           Y(j) = (1-(-1)^j)/j ;
48       end
49       w = X\Y ;
50   end
51
52   function Legn = legendreV(n)
53       if n == 0
54           Legn = [0, 1] ;
55       elseif n == 1
56           Legn = [1, 0] ;
57       else
58           L1 = conv([1, 0], legendreV(n-1)) ;
59           L2 = legendreV(n-2) ;
60           if length(L2) < length(L1)
61               L2 = [zeros(1, length(L1)-length(L2)) L2] ;
62           end
63           Legn = (2*n-1)/n*L1-(n-1)/n*L2 ;
64       end
65   end
```

A quadrature based on $n$ Legendre-Gauss points is accurate for polynomials of degree $2n-1$ or less. Imposing the inclusion of the two boundary points as Legendre-Gauss-Lobbato points reduces the accuracy by 2 to $2n-3$.

Another class of Gauss points is the Legendre-Gauss-Radau (LGR) points in which one boundary point is selected among the $n$ points. The LGR-points are the roots of $P_{n-1}(x) + P_n(x)$, where $P_n(x)$ is the Legendre polynomial

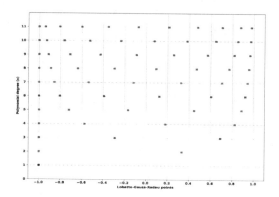

FIGURE 6.10: The distribution of Legendre-Gauss-Radau points in the interval $[-1,1]$ for $n = 1,3,\ldots,11$.

of degree $n$. The positions of the $n$ LGR-points are shown in Figure 6.10 for $n = 1,\ldots,11$.

A MATLAB function `PolyAdd` the receives vectors of coefficients of two polynomials $P_1$ of degree $n$ and $P_2(x)$ of degree $m$ and returns the vectors of coefficients of $P_1(x) + P_2(x)$ is described by the following code:

```
1  function P = PolyAdd(P1, P2)
2      n1 = length(P1) ; n2 = length(P2) ;
3      if n1 == n2
4          P = P1 + P2 ;
5      elseif n1 > n2
6          P = P1 + [zeros(1, n1-n2) P2] ;
7      else
8          P = [zeros(1, n2-n1) P1]+P2 ;
9      end
10 end
```

The MATLAB function `LGRpw` receives a positive integer $n$ and uses the function `PolyAdd` to return the LGR points and weights.

```
1  function [s, w] = LGRpw(n)
2      s = sort(roots(poly_add(legendreV(n-1), legendreV(n)))) ;
3      X = fliplr(vander(s))' ;
4      Y = zeros(n, 1) ;
5      for j = 1 : n
6          Y(j) = (1-(-1)^j)/j ;
7      end
8      w = X\Y ;
9  end
```

Running the MATLAB code:

```matlab
% LGRApp.m
clear ; clc ;
n = 3 ;
[s, w] = LGRpw(n) ;
a = 0.0 ; b = 2.0 ; f = @(x) x./(1+2*x.^2) ;
N = 5 ;
IExact = log(9.0)/4 ;
LGRApprox = [] ;
while N < 1000
    x = linspace(a, b, N+1) ;
    h = (b-a)/N ;
    I = 0.0 ;
    for j = 1 : N
        xm = (x(j)+x(j+1))/2 ;
        I = I + h/2.0*sum(w.*f(xm+h/2*s)) ;
    end
    Err = abs(I - IExact) ;
    if Err < eps
        break ;
    end
    LGRApprox = [LGRApprox; [N, I, Err]] ;
    N = 2*N ;
end

fprintf('%s%i%s\n', 'Integration with  Gauss quadrature based ...
    on', n, '-LGR points:')
fprintf('---------------------------------------------------------\n')
fprintf('   N\t\t Approx Int.\t\t Error\n') ;
fprintf('---------------------------------------------------------\n')
m = length(LGRApprox) ;
for j = 1 : m
    fprintf('%4i\t%1.15f\t%1.12e\n', LGRApprox(j, 1), ...
    LGRApprox(j, 2), LGRApprox(j, 3)/IExact) ;
end
fprintf('---------------------------------------------------------\n')

RConv4 = [] ;
fprintf('%s%i%s\n', 'Rates of convergence for  Gauss ...
    quadrature based on ', n, '-LGR points')
fprintf('------------------------\n')
fprintf('  N\t Conv. Rate\n')
fprintf('------------------------\n')
for k = 1:m-1
    RConv4 =[RConv4;(log2(LGRApprox(k, 3)/LGRApprox(k+1, 3)))] ;
        fprintf('%4i\t%3.3f\n', LGRApprox(k, 1), RConv4(end)) ;
end
fprintf('------------------------\n')
```

gives the results:

```
Integration with  Gauss quadrature based on 3-LGR points:
----------------------------------------------------------
    N           Approx Int.                  Error
----------------------------------------------------------
    5       0.549307676087172        2.788523546236e-06
   10       0.549306137628403        1.220749463804e-08
   20       0.549306144241099        1.692233412952e-10
   40       0.549306144332712        2.444363202253e-12
   80       0.549306144334036        3.375298948042e-14
  160       0.549306144334055        4.042274189272e-16
  320       0.549306144334055        4.042274189272e-16
  640       0.549306144334054        1.414795966245e-15
----------------------------------------------------------
```

```
Rates of convergence for  Gauss quadrature based on 3-LGR points
------------------------------
    N         Conv. Rate
------------------------------
    5           7.836
   10           6.173
   20           6.113
   40           6.178
   80           6.384
  160           0.000
  320          -1.807
------------------------------
```

The Python function LGRpw(n) receives an integer $n$ and returns the corresponding $n$-LGR points and weights:

```
1  from scipy.special import legendre
2  import numpy as np
3  def LGRpw(n):
4      s = list(np.sort(np.roots(legendre(n-1)+legendre(n))))
5      X = (np.fliplr(np.vander(s))).T
6      Y = np.array([(1-(-1)**j)/j for j in range(1, n+1)])
7      w = np.linalg.solve(X, Y)
8      return s, w
```

Applying to the integral

$$I = \int_0^2 \frac{x}{1+2x^2} dx,$$

with $n = 4$ gives the following results:

```
Integration with Gauss quadrature based on 4-LGR points
-------------------------------------------------------
  N          Approx Int.                 Error
-------------------------------------------------------
  5       0.549306013521088       2.381421874512e-07
 10       0.549306144353480       3.536282517630e-11
 20       0.549306144334126       1.291506603473e-13
 40       0.549306144334055       4.042274189272e-16
 80       0.549306144334055       2.021137094636e-16
-------------------------------------------------------
```

```
Rates of convergence for Gauss quadrature based on 4-LGR points:
----------------------
  N    Conv. Rate
----------------------
  5    12.717
 10     8.097
 20     8.320
 40     1.000
----------------------
```

# 7

## Solving Systems of Nonlinear Ordinary Differential Equations

### Abstract

Differential equations have wide applications in modelling real phenomena [61, 62]. Analytical solutions of differential equations can be found for few special and simple cases. Hence, numerical methods are more appropriate in finding solutions of such differential equations.

This chapter discusses some of the numerical methods for solving a system of initial value problems:

$$\frac{d\boldsymbol{y}}{dx} = \boldsymbol{f}(x, \boldsymbol{y}), \ \boldsymbol{y}(a) = \boldsymbol{y}_0, \tag{7.1}$$

where $\boldsymbol{y} : \mathbb{R} \to \mathbb{R}^n$, $\boldsymbol{f} : \mathbb{R} \times \mathbb{R}^\varkappa \to \mathbb{R}^n$ and $x \in [a, b]$. Here, we assume that the functions $f_j(x, \boldsymbol{y})$ are Lipschitz continuous in $[a, b]$ for all $j = 1, \ldots, n$ and at least one function $f_k(x, \boldsymbol{y})$ is nonlinear for some $k \in \{1, 2, \ldots, n\}$.

It is divided into five sections. The first, second and third sections discuss the general idea of Runge-Kutta methods, explicit and implicit Runge-Kutta methods. The fourth section discusses the MATLAB® built-in functions for solving systems of differential equations. The fifth section discusses the scipy functions and gekko Python methods for solving initial value problems.

## 7.1 Runge-Kutta Methods

An M-stage Runge-Kutta method for solving Equation (7.1), is any numerical method characterized by a triplet $A = (a)_{ij} \in R^{M \times M}, \boldsymbol{b} = (b)_i \in R^M$ and $\boldsymbol{c} = (c)_i \in R^M$, $i, j = 1, \ldots, M$ [13]. The triplet $(A, \boldsymbol{b}, \boldsymbol{c})$ satisfies the Butcher array

$$
\begin{array}{c|c}
\boldsymbol{c} & A \\
\hline
& \boldsymbol{b}^T
\end{array}
=
\begin{array}{c|ccc}
c_1 & a_{11} & \cdots & a_{1M} \\
\vdots & \vdots & \ddots & \vdots \\
c_M & a_{M1} & \cdots & a_{MM} \\
\hline
& b_1 & \cdots & b_M
\end{array}
$$

The parameters $c_j$, $j = 1, \ldots, M$ are such that $0 \leq c_1 \leq c_2 \leq \ldots \leq c_M \leq 1$. The parameters $a_{ij}$, $i, j = 1, \ldots, M$ satisfy

$$\sum_{j=1}^{M} a_{ij} = c_i,$$

and the parameters $b_i$, $i = 1, \ldots, M$ satisfy

$$\sum_{i=1}^{M} b_i = 1.$$

A M-stage Runge-Kutta method, for solving (7.1) is described below. Divide the time interval $[a, b]$ into $N$ sub-intervals $[x_i, x_{i+1}], i = 0, \ldots, N$, each is of length $h_i = x_{i+1} - x_i$. Then, $a = x_0 < x_1 < \cdots < x_N = b$.

Given $y(x_i)$, we compute $y(x_{i+1})$, by integrating equation (7.1) on $[x_i, x_{i+1}]$, as

$$y(x_{i+1}) = y(x_i) + \int_{x_i}^{x_{i+1}} f(x, y(x))dx, \qquad (7.2)$$

$i = 0, \ldots, N - 1$.

On each sub-interval $[x_i, x_{i+1}]$ let $x_i^{(j)} = x_i + c_j h_i$ for $j = 1, \ldots, M$. The slope is then evaluated as $f(x_i^{(j)}, y(x_i^{(j)}))$, and the weighted sum of the slope over the sub-interval $[x_i^{(1)}, x_i^{(j)}]$; $j = 1, \ldots, M$ is given by

$$f(x_i^{(j)}, y(x_i^{(j)})) \approx \sum_{l=1}^{M} a_{jl} f(x_i^{(l)}, y(x_i^{(l)}))$$

where $\sum_{l=1}^{M} a_{jl} = c_j$. Then

$$\begin{aligned} y(x_i^{(j)}) &= y(x_i^{(1)}) + \int_{x_i^{(1)}}^{x_i^{(j)}} f(x, y(x))dx \\ &\approx y(x_i^{(1)}) + h_i \sum_{l=1}^{M} a_{jl} f(x_i^{(l)}, y(x_i^{(l)})) \end{aligned} \qquad (7.3)$$

The weighted average of the slope over the interval $[x_i, x_{i+1}]$ is then a convex combination of the slopes $f(x_i^{(j)}, y(x_i^{(j)})); j = 1, \ldots, M$, given by

$$\sum_{l=1}^{M} b_j f(x_i^{(j)}, y(x_i^{(j)})), \quad \sum_{i=1}^{M} b_i = 1$$

Equation (7.2) finally becomes

$$y(x_i^{(M)}) \approx y(x_i^{(1)}) + h_i \sum_{j=1}^{M} b_j f(x_i^{(j)}, y(x_i^{(j)})) \qquad (7.4)$$

The formulas (7.3) and (7.4) define the M-stage Runge-Kutta method. If the coefficients $a_{jl} = 0$ for all $l > j$ the resulting Runge-Kutta method is `explicit`. If $a_{jl} = 0$, for all $l > j$, but $a_{jj} \neq 0$ for some $j$, the method is `semi-explicit`. Otherwise, it is `implicit` [14].

The quantity $\boldsymbol{\xi}_i = \boldsymbol{y}(x_i^{(1)}) + h_i \sum_{j=1}^{M} b_j \boldsymbol{f}(x_i^{(j)}, \boldsymbol{y}(t_i^{(j)})) - \boldsymbol{y}(x_i^{(M)})$ defines the *local truncation error* of the M-stage Runge-Kutta method. The M-stage Runge-Kutta method is said to be of `order` $P \in \mathbb{N}$, if its local truncation error is of $\mathcal{O}(h^{P+1})$, where $h = \max_{i=0,\dots,N-1} \{h_i\}$.

The order of the M-stage Runge-Kutta method is achieved by satisfying conditions on the coefficients $(a)_{ij}, b_i$ and $c_i$, $i,j \in \{1,\dots,M\}$. Butcher [14] stated the number of conditions on an implicit M-stage Runge-Kutta method, to be of order $P$ as in table (7.1).

TABLE 7.1: The number of restrictions on a $P$-th order implicit Runge-Kutta method

| Order $(P)$ | 1 | 2 | 3 | 4 | 5 | 6 | 7 | 8 |
|---|---|---|---|---|---|---|---|---|
| Number of Restrictions | 1 | 2 | 4 | 8 | 17 | 37 | 85 | 200 |

In Table (7.2) we state the conditions on the coefficients $(A, b, c)$ for the M-stage Runge-Kutta methods up to order four. Here a method can of order $P$ if it satisfies all the conditions from 1 to $P$.

TABLE 7.2: The relationships between the order of the M-stage Runge-Kutta method, and the conditions on the triplet $(A, b, c)$

| Order | Conditions | |
|---|---|---|
| 1 | $\sum_{i=1}^{M} b_i = 1$ | |
| 2 | $\sum_{i=1}^{M} \sum_{j=1}^{M} b_i a_{ij} = \frac{1}{2}$ | |
| 3 | $\sum_{i=1}^{M} b_i c_i^2 = \frac{1}{3}$ | $\sum_{i=1}^{M} \sum_{j=1}^{M} b_i a_{ij} c_j = \frac{1}{6}$ |
| 4 | $\sum_{i=1}^{M} b_i c_i^3 = \frac{1}{4}$ | $\sum_{i=1}^{M} \sum_{j=1}^{M} b_i a_{ij} c_j = \frac{1}{12}$ |
| | $\sum_{i=1}^{M} \sum_{j=1}^{M} b_i c_i a_{ij} c_j = \frac{1}{8}$ | $\sum_{i=1}^{M} \sum_{j=1}^{M} (\sum_{l=1}^{M} b_l a_{li}) a_{ij} c_j = \frac{1}{24}$ |

## 7.2 Explicit Runge-Kutta Methods

An M-stage Runge-Kutta method is explicit if its Butcher table satisfies $a_{ij} = 0$ for all $j \geq i$. That is

$$\frac{c \ \big|\ A}{\ \big|\ b^T} = \begin{array}{c|ccccc} 0 & 0 & 0 & \cdots & 0 & 0 \\ c_2 & a_{21} & 0 & \cdots & 0 & 0 \\ c_3 & a_{31} & a_{32} & \ddots & 0 & 0 \\ \vdots & \vdots & \vdots & \ddots & \ddots & \vdots \\ c_M & a_{M1} & a_{M2} & \cdots & a_{MM-1} & 0 \\ \hline & b_1 & b_2 & \cdots & b_{M-1} & b_M \end{array}$$

In an explicit Runge-Kutta method, the function slope at a point $x_i^{(\ell)}$ depends only on the slopes at previous points $x_i^{(1)}, \ldots, x_i^{(\ell-1)}$. The most important explicit Runge-Kutta methods are the `Euler's methods`, `Heun's method` and the `classical fourth-order Runge-Kutta method` [16].

### 7.2.1   Euler's Method

The simplest explicit Runge-Kutta method is the Euler method, whose Butcher tabular is:

$$\frac{c \ \big|\ A}{\ \big|\ b^T} = \frac{0 \ \big|\ 0}{\ \big|\ 1}$$

and is derived from:

$$\boldsymbol{y}(x_{i+1}) = \boldsymbol{y}(x_i) + h_i f(x_i, \boldsymbol{y}(x_i)) + \mathcal{O}(h^2)$$

as

$$\boldsymbol{y}^{i+1} = \boldsymbol{y}^i + h_i \boldsymbol{f}(x_i, \boldsymbol{y}(x_i)) \tag{7.5}$$

where $\boldsymbol{y}^i \approx \boldsymbol{y}(x_i)$ and $h = \max\{h_i, i = 1, \ldots, N\}$.

Because the local truncation error in Euler's method is $\mathcal{O}(h^2)$, the approximation error of the method is $\mathcal{O}(h)$, as the summation of the local truncation errors of the $N$ subintervals is $\mathcal{O}(N \cdot h^2) = \mathcal{O}(h)$.

**Example 7.1 (solving initial value problem with Euler's method)** In this example, the Euler's method will be used to solve the initial value problem

$$\frac{dy}{dt} = t - 2ty, \ y(0) = 1, \ \text{for } t \in [0, 2].$$

To apply Euler's methods, the interval $[0, 2]$ is divided into $N$ subintervals (each of length $h = 2/N$) by the points $t_0 = 0 < t_1 < \ldots < t_N = 2$, where $t_j = j \cdot h$ and $t_{j+1} - t_j = h = 2/N$. At a point $t_j$, the solution function $y(t_j)$ is approximated by $y_j$.

Starting from the initial point $(t_0, y_0) = (0, 1)$, Euler's methods use the difference formula:

$$y_{j+1} = y_j + hf(t_j, y_j) = y_j + h(t_j - 2t_j y_j), \ j = 0, 1, \ldots, N - 1$$

to approximate $y$ at $t_{j+1}$ giving $y_{j+1}$.

The exact solution of the initial value problem

$$\dot{y}(t) = t - 2ty, \ y(0) = 1$$

is

$$y(t) = \frac{1}{2}\left(1 + e^{-t^2}\right)$$

and the error in approximating $y(t_j)$ by $y_j$ is

$$Error = \max\{|y(t_j) - y_j|, j = 0, 1, \ldots, N\}.$$

The Python code `SolveIVPWithEuler` solves the given initial value problem using Euler's method:

```python
1   # SolveIVPWithEuler.py
2   from numpy import linspace, zeros, exp, inf
3   from numpy.linalg import norm
4   def EulerIVP(f, a, b, N, y0):
5       t = linspace(a, b, N+1)
6       yexact = 0.5*(1+exp(-t**2))
7       f = lambda t, y: t-2*t*y
8       h = 2.0/N
9       y = zeros((N+1,), 'float')
10      y[0] = y0
11      for j in range(N):
12          y[j+1] = y[j] + h*f(t[j], y[j])
13      return t, y, norm(y-yexact, inf)
14
15  Rows = 15
16  f = lambda t, y: 2*t*y
17  a, b = 0.0, 2.0
18  y0 = 1.0
19  EulerErrors = zeros((Rows, 2), 'float')
20  print('------------------------------------------')
21  print('     N\t\t Error')
22  print('------------------------------------------')
23  for j in range(Rows):
24      N = 2**(4+j)
25      [t, y, Error] = EulerIVP(f, a, b, N, y0)
26      EulerErrors[j, 0] =  N
27      EulerErrors[j, 1] = Error
28      print('{0:8.0f}'.format(N), '\t', '{0:1.12e}'.format(Error))
29  print('------------------------------------------')
30
31  RatesOfConvergence = zeros((Rows-1, 2), 'float')
32  print('Rates of convergence of Eulers method:')
33  print('------------------------------------------')
34  print('     N\t\t Conv. Rate')
35  print('------------------------------------------')
36  for j in range(Rows-1):
37      RatesOfConvergence[j, 0] = EulerErrors[j, 0]
38      RatesOfConvergence[j, 1] = log2(EulerErrors[j, ...
39          1]/EulerErrors[j+1, 1])
40      print('{0:6.0f}'.format(RatesOfConvergence[j, 0]), '\t\t',\
41      '{0:1.3f}'.format(RatesOfConvergence[j, 1]))
42  print('------------------------------------------')
```

By executing the code the following results are obtained.

```
runfile('D:/PyFiles/SolveIVPWithEuler.py', wdir='D:/PyFiles')
------------------------------------------
     N              Error
------------------------------------------
     16          2.212914947992e-02
     32          1.061531476368e-02
     64          5.201658072192e-03
    128          2.570740890148e-03
    256          1.278021559005e-03
    512          6.371744244492e-04
   1024          3.181315727203e-04
   2048          1.589510709855e-04
   4096          7.944690307737e-05
   8192          3.971629161525e-05
  16384          1.985635490032e-05
  32768          9.927729750725e-06
  65536          4.963752954890e-06
 131072          2.481848491165e-06
 262144          1.240917262835e-06
------------------------------------------


Rates of convergence of Eulers method:
--------------------------------------
     N              Conv. Rate
--------------------------------------
     16          1.060
     32          1.029
     64          1.017
    128          1.008
    256          1.004
    512          1.002
   1024          1.001
   2048          1.001
   4096          1.000
   8192          1.000
  16384          1.000
  32768          1.000
  65536          1.000
 131072          1.000
--------------------------------------
```

From the table of order of convergence, it can be seen that Euler method is a first-order method ($\mathcal{O}(h)$).

The MATLAB code is:

```
1   clear ; clc ;
2   f = @(t, y) t-2*t*y ;
3   a = 0.0 ; b = 2.0 ; y0 = 1 ;
4   yexact = @(t) 0.5*(1+exp(-t.^2)) ;
5   Rows = 15 ;
6   EulerErrors = zeros(Rows, 2) ;
7   fprintf('-------------------------------------------\n') ;
8   fprintf('       N\t\t\t Error\n') ;
9   fprintf('-------------------------------------------\n') ;
10  for j = 1 : Rows
11      N = 2^(3+j) ;
12      [t, y, Error] = EulerIVP(f, a, b, N, y0) ;
13      EulerErrors(j, 1) =  N  ;
14      EulerErrors(j, 2) = Error ;
15      fprintf('%8i\t%1.10e\n', N, Error) ;
16  end
17  fprintf('-------------------------------------------\n') ;
18
19  fprintf('-------------------------------------------\n\n') ;
20  RatesOfConvergence = zeros(Rows-1, 2) ;
21  fprintf('Rates of convergence of Eulers method:\n') ;
22  fprintf('-------------------------------------------\n') ;
23  fprintf('       N\t\t Conv. Rate\n') ;
24  fprintf('-------------------------------------------\n') ;
25  for j = 1 : Rows - 1
26      RatesOfConvergence(j, 1) = EulerErrors(j, 1) ;
27      RatesOfConvergence(j, 2) = log2(EulerErrors(j, ...
            2)/EulerErrors(j+1, 2)) ;
28      fprintf('%8i\t %1.12e\n', RatesOfConvergence(j, 1), ...
29      RatesOfConvergence(j, 2)) ;
30  end
31  fprintf('-------------------------------------------\n') ;
32
33  function [t, y, Error] = EulerIVP(f, a, b, N, y0)
34      t = linspace(a, b, N+1) ;
35      h = (b-a)/N ;
36      y = zeros(1, N+1) ;
37      yexact = 0.5*(1+exp(-t.^2)) ;
38      y(1) = y0 ;
39      for j = 1 : N
40          y(j+1) = y(j) + h*f(t(j), y(j)) ;
41      end
42      Error = norm(y-yexact, inf) ;
43  end
```

## 7.2.2  Heun's Method

Heun's method has the Butcher tabular:

$$\frac{c \mid A}{\quad b^T} = \frac{\begin{array}{c|cc} 0 & 0 & 0 \\ 1 & 1 & 0 \end{array}}{\quad\; \frac{1}{2} \quad \frac{1}{2}}$$

Hence, the Heun's method is of the form:

$$y^{i+1} = y^i + \frac{h}{2}\left(f\left(x_i, y^i\right) + f\left(x_{i+1}, y^i + h_i f(x_i, y^i)\right)\right)$$

The following MATLAB code solves Example 7.1 using Heun's method:

```matlab
1  clear ; clc ;
2  f = @(t, y) t-2*t*y ;
3  a = 0.0 ; b = 2.0 ; y0 = 1 ;
4  yexact = @(t) 0.5*(1+exp(-t.^2)) ;
5  Rows = 15 ;
6  HeunErrors = zeros(Rows, 2) ;
7  fprintf('------------------------------------------\n') ;
8  fprintf('        N\t\t\t Error\n') ;
9  fprintf('------------------------------------------\n') ;
10 for j = 1 : Rows
11     N = 2^(3+j) ;
12     [t, y, Error] = HeunIVP(f, a, b, N, y0) ;
13     HeunErrors(j, 1) = N  ;
14     HeunErrors(j, 2) = Error ;
15     fprintf('%8i\t%1.10e\n', N, Error) ;
16 end
17 fprintf('------------------------------------------\n\n') ;
18 RatesOfConvergence = zeros(Rows-1, 2) ;
19 fprintf('Rates of convergence of Heuns method:\n') ;
20 fprintf('------------------------------------------\n') ;
21 fprintf('        N\t\t Conv. Rate\n') ;
22 fprintf('------------------------------------------\n') ;
23 for j = 1 : Rows - 1
24     RatesOfConvergence(j, 1) = HeunErrors(j, 1) ;
25     RatesOfConvergence(j, 2) = log2(HeunErrors(j, ...
26         2)/HeunErrors(j+1, 2)) ;
26     fprintf('%8i\t %1.12e\n', RatesOfConvergence(j, 1), ...
27     RatesOfConvergence(j, 2)) ;
28 end
29 fprintf('------------------------------------------\n') ;
30
31 function [t, y, Error] = HeunIVP(f, a, b, N, y0)
32 t = linspace(a, b, N+1) ;
33 h = (b-a)/N ;
34 y = zeros(1, N+1) ;
35 yexact = 0.5*(1+exp(-t.^2)) ;
36 y(1) = y0 ;
37 for j = 1 : N
38     k1 = f(t(j), y(j)) ;
39     k2 = f(t(j+1), y(j)+h*k1) ;
40     y(j+1) = y(j) + h/2*(k1+k2) ;
41 end
42 Error = norm(y-yexact, inf) ;
43 end
```

By executing this code, the following results are obtained.

```
------------------------------------
    N                   Error
------------------------------------
    16          1.6647420306e-03
    32          3.8083683419e-04
    64          9.1490315453e-05
   128          2.2439310337e-05
   256          5.5575532305e-06
   512          1.3830109670e-06
  1024          3.4496196688e-07
  2048          8.6142039946e-08
  4096          2.1523228200e-08
  8192          5.3792741372e-09
 16384          1.3446276315e-09
 32768          3.3613245520e-10
 65536          8.4030005176e-11
131072          2.1010082563e-11
262144          5.2486903712e-12
------------------------------------
```

```
Rates of convergence of Heuns method:
------------------------------------
    N                Conv. Rate
------------------------------------
    16          2.128053707149e+00
    32          2.057482078881e+00
    64          2.027590701520e+00
   128          2.013506569375e+00
   256          2.006637264733e+00
   512          2.003303382587e+00
  1024          2.001647915261e+00
  2048          2.000823009405e+00
  4096          2.000411061126e+00
  8192          2.000204811383e+00
 16384          2.000104948066e+00
 32768          2.000053370390e+00
 65536          1.999822741673e+00
131072          2.001052433088e+00
------------------------------------
```

The table of rates of convergence shows that Heun's method is a second-order method.

The Python code is:

```
from numpy import linspace, zeros, exp, inf, log2
from numpy.linalg import norm
def HeunIVP(f, a, b, N, y0):
    t = linspace(a, b, N+1)
    yexact = 0.5*(1+exp(-t**2))
    h = 2.0/N
    y = zeros((N+1,), 'float')
    y[0] = y0
    for j in range(N):
        k1 = f(t[j], y[j])
        k2 = f(t[j+1], y[j]+h*k1)
        y[j+1] = y[j] + h/2.0*(k1 + k2)
    return t, y, norm(y-yexact, inf)

Rows = 15
f = lambda t, y: t-2*t*y
a, b = 0.0, 2.0
y0 = 1.0
HeunErrors = zeros((Rows, 2), 'float')
print('---------------------------------------------')
print('     N\t\t Error')
print('---------------------------------------------')
for j in range(Rows):
    N = 2**(4+j)
    [t, y, Error] = HeunIVP(f, a, b, N, y0)
    HeunErrors[j, 0] =  N
    HeunErrors[j, 1] = Error
    print('{0:8.0f}'.format(N), '\t', '{0:1.12e}'.format(Error))
print('---------------------------------------------\n')

RatesOfConvergence = zeros((Rows-1, 2), 'float')
print('Rates of convergence of Heuns method:')
print('---------------------------------------------')
print('     N\t\t Conv. Rate')
print('---------------------------------------------')
for j in range(Rows-1):
    RatesOfConvergence[j, 0] = HeunErrors[j, 0]
    RatesOfConvergence[j, 1] = log2(HeunErrors[j, ...
        1]/HeunErrors[j+1, 1])
    print('{0:6.0f}'.format(RatesOfConvergence[j, 0]), '\t\t',\
    '{0:1.3f}'.format(RatesOfConvergence[j, 1]))
print('---------------------------------------------')
```

## 7.2.3   The Fourth-Order Runge-Kutta Method

The classical fourth-order Runge-Kutta method has the Butcher tabular:

$$
\frac{c \mid A}{\quad \mid b^T} =
\begin{array}{c|cccc}
0 & 0 & 0 & 0 & 0 \\
\frac{1}{2} & \frac{1}{2} & 0 & 0 & 0 \\
\frac{1}{2} & 0 & \frac{1}{2} & 0 & 0 \\
1 & 0 & 0 & 1 & 0 \\
\hline
 & \frac{1}{6} & \frac{2}{6} & \frac{2}{6} & \frac{1}{6}
\end{array}
$$

**Example 7.2 (Lotka-Voltera model)** In this example the fourth-order Runge-Kutta method will be used to solve the following predator-prey model:

$$\dot{x}(t) = 0.7x(t) - 1.3x(t)y(t), x(0) = 0.9$$
$$\dot{y}(t) = x(t)y(t) - y(t), \ y(0) = 0.1$$

with $t \in [0, 50]$.

The MATLAB code for solving the problem of this example is as follows:

```
1  % SolvePredPrey.m
2  clear ; clc ;
3  f = @(t, z) [0.7*z(1)-1.3*z(1)*z(2);
4  -z(2)+z(1)*z(2)] ;
5  a = 0.0 ; b = 50 ; y0 = [0.9; 0.1] ;
6  N = 500 ;
7  x = linspace(a, b, N+1) ;
8  h = (b-a)/N ;
9  y = zeros(2, N+1) ;
10 y(:, 1) = y0 ;
11 for j = 1 : N
12     k1 = f(x(j), y(:, j)) ;
13     k2 = f(x(j)+h/2, y(:, j)+h/2*k1) ;
14     k3 = f(x(j)+h/2, y(:, j)+h/2*k2) ;
15     k4 = f(x(j)+h, y(:, j)+h*k3) ;
16     y(:, j+1) = y(:, j) + h/6.0*(k1+2*k2+2*k3+k4) ;
17 end
18 figure(1) ;
19 plot(x, y(1, :), '-b', x, y(2, :), '-.m', 'LineWidth', 2)
20 xlabel('Time (t)') ;
21 ylabel('Population') ;
22 legend('Prey', 'Predator') ;
23 set(gca, 'FontSize', 12)
24 set(gca, 'Fontweight', 'bold') ;
25 grid on ;
26 set(gca, 'XTick', linspace(a, b, 11)) ;
27 set(gca, 'YTick', 0:0.3:3) ;
28 set(gca, 'GridLineStyle', ':')
29 grid on ;
```

Executing the code `SolvePredPrey` shows in the graph in Figure 7.1. The Python code is:

```
1  from numpy import linspace, zeros, arange, array
2  import matplotlib.pyplot as plt
3  f = lambda t, z: array([0.7*z[0]-1.3*z[0]*z[1], z[0]*z[1] - z[1]])
4  a, b, N = 0.0, 50.0, 500
5  h = (b-a)/N
6  x = linspace(a, b, N+1)
7  y = zeros((1+N, 2), 'float')
8  y[0, 0], y[0, 1] = 0.9, 0.1
9  for j in range(N):
10     k1 = f(x[j], y[j, :])
11     k2 = f(x[j]+h/2, y[j, :]+h/2*k1)
```

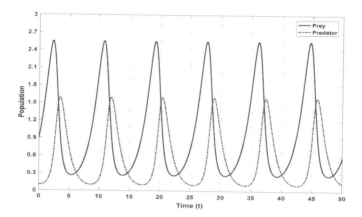

FIGURE 7.1: Dynamics of the predator and prey populations obtained by the fourth-order Runge-Kutta method.

```
12      k3 = f(x[j]+h/2, y[j, :]+h/2*k2)
13      k4 = f(x[j+1], y[j, :]+h*k3)
14      y[j+1, :] = y[j, :] + h/6.*(k1+2.*k2+2.*k3+k4)
15   plt.figure(1, figsize=(14, 14))
16   plt.plot(x, y[:, 0], color='b', ls='-', lw= 2, label='Prey')
17   plt.plot(x, y[:, 1], color='m',ls='-.', lw= 2, label='Predator')
18   plt.xticks(arange(a, b+0.5, 5), fontweight='bold')
19   plt.yticks(arange(0, 3.3, 0.3), fontweight='bold')
20   plt.xlabel('Time (t)', fontweight='bold')
21   plt.ylabel('Population', fontweight='bold')
22   plt.grid(True, ls=':')
23   plt.legend()
```

## 7.3   Implicit Runge-Kutta Methods

This section discusses some implicit and collocation Runge-Kutta methods for solving the initial value problems (7.1).

An M-stages Runge-Kutta method is implicit, if in its Butcher tablular $a_{ij} \neq 0$ for some $j > i$.

### 7.3.1   The Backward Euler Method

The simplest implicit Runge-Kutta method is the backward Euler method. The Butcher table for the backward Euler method is as follows:

$$\begin{array}{c|c} 1 & 1 \\ \hline & 1 \end{array}$$

which takes the form:

$$y^{n+1} = y^n + h\boldsymbol{f}(x_n, y^{n+1})$$

The backward Euler method is a first order method with a truncation error of $\mathcal{O}(h)$ where $h$ is the maximum step size between two mesh points $x_j$ and $x_{j+1}$.

It can be seen that the evaluation of the solution function $\boldsymbol{y}(x)$ at $x = x_{n+1}$ requires the evaluation of the slope function at the same point $(x = x_{n+1})$. To avoid this problem, Euler method can be used for giving an estimate of $\boldsymbol{y}^{n+1}$ for the right-hand side.

The following Python code solves example 7.1:

```python
from numpy import linspace, zeros, exp, inf, log2
from numpy.linalg import norm
def BackwardEulerIVP(f, a, b, N, y0):
    t = linspace(a, b, N+1)
    yexact = 0.5*(1+exp(-t**2))
    h = 2.0/N
    y = zeros((N+1,), 'float')
    y[0] = y0
    for j in range(N):
        y[j+1] = y[j] + h*f(t[j], y[j]+h*f(t[j], y[j]))
    return t, y, norm(y-yexact, inf)

Rows = 15
f = lambda t, y: t-2*t*y
a, b = 0.0, 2.0
y0 = 1.0
BackwardEulerErrors = zeros((Rows, 2), 'float')
print('------------------------------------------')
print('      N\t\t Error')
print('------------------------------------------')
for j in range(Rows):
    N = 2**(4+j)
    [t, y, Error] = BackwardEulerIVP(f, a, b, N, y0)
    BackwardEulerErrors[j, 0] =  N
    BackwardEulerErrors[j, 1] = Error
    print('{0:8.0f}'.format(N), '\t', '{0:1.12e}'.format(Error))
print('------------------------------------------\n')

RatesOfConvergence = zeros((Rows-1, 2), 'float')
print('Rates of convergence of BackwardEulers method:')
print('------------------------------------------')
print('      N\t\t Conv. Rate')
print('------------------------------------------')
for j in range(Rows-1):
    RatesOfConvergence[j, 0] = BackwardEulerErrors[j, 0]
    RatesOfConvergence[j, 1] = log2(BackwardEulerErrors[j, ...
        1]/BackwardEulerErrors[j+1, 1])
    print('{0:6.0f}'.format(RatesOfConvergence[j, 0]), '\t\t',\
        '{0:1.3f}'.format(RatesOfConvergence[j, 1]))
print('------------------------------------------')
import matplotlib.pyplot as plt
```

```
41  plt.figure(1, figsize=(10, 10))
42  plt.plot(t, y, color='purple',lw=2, label='$y(x) = ...
        0.5(1+e^{-x^2})$')
43  plt.xlabel('x', fontweight='bold')
44  plt.ylabel('y', fontweight='bold')
45  plt.grid(True, ls=':')
46  plt.xticks(arange(0.0, 2.2, 0.2), fontweight='bold')
47  plt.yticks(arange(0.5, 1.05, 0.05), fontweight='bold')
48  plt.legend()
```

By executing this code, the following results are obtained.

```
runfile('D:/PyFiles/SolveIVPWithBackwardEuler.py',
    wdir='D:/PyFiles')
```

```
-------------------------------------------
    N              Error
-------------------------------------------
    16          4.337814100757e-02
    32          2.037832192200e-02
    64          9.905832986266e-03
   128          4.886733400528e-03
   256          2.427510216728e-03
   512          1.209810748381e-03
  1024          6.039264582937e-04
  2048          3.017197337041e-04
  4096          1.507990196000e-04
  8192          7.538431169796e-05
 16384          3.768835804929e-05
 32768          1.884322987655e-05
 65536          9.421377654029e-06
131072          4.710629515681e-06
262144          2.355299925094e-06
-------------------------------------------
```

```
Rates of convergence of BackwardEulers method:
---------------------------------------
    N           Conv. Rate
---------------------------------------
    16          1.090
    32          1.041
    64          1.019
   128          1.009
   256          1.005
   512          1.002
  1024          1.001
  2048          1.001
  4096          1.000
  8192          1.000
```

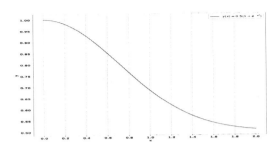

FIGURE 7.2: Solution of the initial value problem of Example 7.1, given by $y(x) = (1 + e^{-x^2})/2$.

| | |
|---|---|
| 16384 | 1.000 |
| 32768 | 1.000 |
| 65536 | 1.000 |
| 131072 | 1.000 |

------------------------------------

The graph of the problem solution is shown in Figure 7.2.
The MATLAB code is:

```
1   clear ; clc ;
2   f = @(t, y) t-2*t*y ;
3   a = 0.0 ; b = 2.0 ; y0 = 1 ;
4   yexact = @(t) 0.5*(1+exp(-t.^2)) ;
5   Rows = 15 ;
6   BackwardEulerErrors = zeros(Rows, 2) ;
7   fprintf('-------------------------------------------\n') ;
8   fprintf('       N\t\t\t Error\n') ;
9   fprintf('-------------------------------------------\n') ;
10  for j = 1 : Rows
11      N = 2^(3+j) ;
12      [t, y, Error] = BackwardEulerIVP(f, a, b, N, y0) ;
13      BackwardEulerErrors(j, 1) =  N  ;
14      BackwardEulerErrors(j, 2) = Error ;
15      fprintf('%8i\t%1.10e\n', N, Error) ;
16  end
17  fprintf('-------------------------------------------\n\n') ;
18
19  plot(t, y, 'm', 'LineWidth', 2) ;
20  xlabel('x') ; ylabel('y') ;
21  grid on ;
22  set(gca, 'XTick', linspace(a, b, 11)) ;
23  set(gca, 'YTick', 0.5:0.05:1.0) ;
24  set(gca, 'fontweight', 'bold') ;
25  legend('y(x) = (1+e^{-x^2})/2') ;
26
27  function [t, y, Error] = BackwardEulerIVP(f, a, b, N, y0)
```

```
28    t = linspace(a, b, N+1) ;
29    h = (b-a)/N ;
30    y = zeros(1, N+1) ;
31    yexact = 0.5*(1+exp(-t.^2)) ;
32    y(1) = y0 ;
33    for j = 1 : N
34        y(j+1) = y(j) + h*f(t(j), y(j)+h*f(t(j), y(j))) ;
35    end
36    Error = norm(y-yexact, inf) ;
37 end
```

## 7.3.2   Collocation Runge-Kutta Methods

Collocation methods are implicit Runge-Kutta methods, but not all implicit Runge-Kutta methods are collocation methods.

To advance the solution of the differential equation from a point $x_n$ to a point $x_{n+1}$, a collocation method constructs a polynomial $P_M(x)$ of degree $M$ that interpolates the solution of the ODE at collocation points $x_n + c_j h$, $j = 1,\ldots,M$, with $0 \le c_j \le 1$, $j = 1,\ldots,M$. That is

$$P'_M(x_n + c_j h) = f(x_n + c_j h, P_M(x_n + c_j h)), \; j = 1,\ldots,M$$

and satisfies,

$$P_M(x_i) = y^i, \; i = 0,1,\ldots,n.$$

Collocation Runge-Kutta methods are deeply related to the Gauss quadrature methods discussed in the past chapter. The collocation are obtained by transforming Gauss points (in $[-1,1]$) such as the LG, LGL, LGR, ...etc points to the interval $[0,1]$, using the transformation

$$x(s) = \frac{1}{2}(1+s).$$

### 7.3.2.1   Legendre-Gauss Methods

The first four Gauss-Legendre methods have the Butcher's tables:

(i) $M = 1$:

$$\begin{array}{c|c} \frac{1}{2} & \frac{1}{2} \\ \hline & 1 \end{array}$$

This method is called the `mid-point rule`.

(ii) $M = 2$:

$$\begin{array}{c|cc} \frac{1}{2}-\frac{\sqrt{3}}{6} & \frac{1}{4} & \frac{1}{4}-\frac{\sqrt{3}}{6} \\ \frac{1}{2}+\frac{\sqrt{3}}{6} & \frac{1}{4}+\frac{\sqrt{3}}{6} & \frac{1}{4} \\ \hline & \frac{1}{2} & \frac{1}{2} \end{array}$$

(iii) $M = 3$ :

$$
\begin{array}{c|ccc}
\frac{1}{2} - \frac{\sqrt{15}}{10} & \frac{5}{36} & \frac{2}{9} - \frac{\sqrt{15}}{10} & \frac{5}{36} - \frac{\sqrt{15}}{30} \\[2ex]
\frac{1}{2} & \frac{5}{36} + \frac{\sqrt{15}}{24} & \frac{2}{9} & \frac{5}{36} - \frac{\sqrt{15}}{24} \\[2ex]
\frac{1}{2} + \frac{\sqrt{15}}{10} & \frac{5}{36} + \frac{\sqrt{15}}{30} & \frac{2}{9} + \frac{\sqrt{15}}{15} & \frac{5}{36} \\[1ex]
\hline
 & \frac{15}{18} & \frac{4}{9} & \frac{5}{18}
\end{array}
$$

**Example 7.3** In this example the mid-point rule will be used to solve the logistic-growth model:

$$\dot{P}(t) = rP(t)(1 - P(t)), P(0) = P_0, t \in [0, T]$$

The exact solution of the model is

$$P(t) = \frac{P_0}{P_0 + (1 - P_0)e^{-rt}}$$

The Python code for solving the logistic growth model with $r = 0.2, P_0 = 0.1$ in $[0, 30]$ is:

```
1   from numpy import linspace, zeros, exp, inf, log2, arange
2   from numpy.linalg import norm
3   def GaussMidIVP(f, a, b, N, P0):
4       t = linspace(a, b, N+1)
5       Pexact = P0/(P0+(1-P0)*exp(-0.2*t))
6       h = (b-a)/N
7       P = zeros((N+1,), 'float')
8       P[0] = P0
9       for j in range(N):
10          Ph = P[j] + h/2*f(t[j], P[j])
11          P[j+1] = P[j] + h*f(t[j]+h/2, Ph)
12      return t, P, norm(P-Pexact, inf)
13
14  Rows = 12
15  r = 0.2
16  f = lambda t, P: r*P*(1-P)
17  a, b = 0.0, 30.0
18  P0 = 0.1
19  GaussMidErrors = zeros((Rows, 2), 'float')
20  print('------------------------------------------------')
21  print('     N\t\t Error')
22  print('------------------------------------------------')
23  N = 100
24  for j in range(Rows):
25      [t, P, Error] = GaussMidIVP(f, a, b, N, P0)
26      GaussMidErrors[j, 0] = N
27      GaussMidErrors[j, 1] = Error
28      print('{0:8.0f}'.format(N), '\t', '{0:1.12e}'.format(Error))
29      N *= 2
30  print('------------------------------------------------\n')
31
```

```
32  RatesOfConvergence = zeros((Rows-1, 2), 'float')
33  print('Rates of convergence of Gauss Midpoint method:')
34  print('--------------------------------------------')
35  print('    N\t\t Conv. Rate')
36  print('--------------------------------------------')
37  for j in range(Rows-1):
38      RatesOfConvergence[j, 0] = GaussMidErrors[j, 0]
39      RatesOfConvergence[j, 1] = log2(GaussMidErrors[j, ...
            1]/GaussMidErrors[j+1, 1])
40      print('{0:6.0f}'.format(RatesOfConvergence[j, 0]), '\t\t',\
41      '{0:1.3f}'.format(RatesOfConvergence[j, 1]))
42  print('--------------------------------------------')
43
44  import matplotlib.pyplot as plt
45  plt.figure(1, figsize=(10, 10))
46  plt.plot(t, P, color='purple',lw=2, label='$P(t) = ...
        P_0/(P_0+(1-P_0)e^{-rt})$')
47  plt.xlabel('Time (t)', fontweight='bold')
48  plt.ylabel('Population (P(t))', fontweight='bold')
49  plt.grid(True, ls=':')
50  plt.xticks(arange(a, b+(b-a)/10, (b-a)/10), fontweight='bold')
51  plt.yticks(arange(0.0, 1.05, 0.1), fontweight='bold')
52  plt.legend()
```

By executing the code, the errors and rates of convergence for different values of mesh points are shown below.

```
runfile('D:/PyFiles/SolveLogisticWithMidPoint.py',
    wdir='D:/PyFiles')
```

| N | Error |
|---|-------|
| 100 | 1.948817140496e-05 |
| 200 | 4.811971295204e-06 |
| 400 | 1.195559782174e-06 |
| 800 | 2.979672797387e-07 |
| 1600 | 7.437647475683e-08 |
| 3200 | 1.857971743124e-08 |
| 6400 | 4.643131656934e-09 |
| 12800 | 1.160556650781e-09 |
| 25600 | 2.901111573195e-10 |
| 51200 | 7.252876077501e-11 |
| 102400 | 1.812816563529e-11 |
| 204800 | 4.535261055594e-12 |
| 409600 | 1.134092819655e-12 |
| 819200 | 2.786659791809e-13 |
| 1638400 | 6.639133687258e-14 |

Rates of convergence of Gauss Midpoint method:

| N | Conv. Rate |
|---|---|
| 100 | 2.018 |
| 200 | 2.009 |
| 400 | 2.004 |
| 800 | 2.002 |
| 1600 | 2.001 |
| 3200 | 2.001 |
| 6400 | 2.000 |
| 12800 | 2.000 |
| 25600 | 2.000 |
| 51200 | 2.000 |
| 102400 | 1.999 |
| 204800 | 2.000 |
| 409600 | 2.025 |
| 819200 | 2.069 |

The MATLAB code is:

```matlab
clear ; clc ;
Rows = 15 ;
r = 0.25 ;
f = @(t, P) r*P*(1-P) ;
a = 0.0; b = 20.0 ;
P0 = 0.2 ;
GaussMidErrors = zeros(Rows, 2) ;
fprintf('-----------------------------------------\n') ;
fprintf('     N\t\t Error\n') ;
fprintf('-----------------------------------------\n') ;
N = 100 ;
for j = 1 : Rows
    [t, P, Error] = GaussMidIVP(f, a, b, N, P0) ;
    GaussMidErrors(j, 1) = N ;
    GaussMidErrors(j, 2) = Error ;
    fprintf('%8i\t%1.12e\n', N, Error) ;
    N = 2 * N ;
end
fprintf('-----------------------------------------\n')

figure(1)
plot(t, P, '-m', 'LineWidth', 2) ;
legend('P(t) = P_0/(P_0+(1-P_0)e^{-rt})') ;
xlabel('Time (t)') ;
ylabel('Population (P(t))') ;
grid on
set(gca, 'fontweight', 'bold') ;
set(gca, 'fontsize', 12) ;
set(gca, 'XTick', a:(b-a)/10:b) ;
```

```
30   set(gca, 'YTick', 0:0.1:1) ;
31
32   RatesOfConvergence = zeros(Rows-1, 2) ;
33   fprintf('Rates of convergence of Gauss Midpoint method:\n')
34   fprintf('----------------------------------------\n')
35   fprintf('     N\t\t Conv. Rate\n')
36   fprintf('----------------------------------------\n')
37   for j = 1 : Rows-1
38       RatesOfConvergence(j, 1) = GaussMidErrors(j, 1) ;
39       RatesOfConvergence(j, 2) = log2(GaussMidErrors(j, ...
             2)/GaussMidErrors(j+1, 2)) ;
40       fprintf('%6i\t\t%1.3f\n', RatesOfConvergence(j, 1), ...
             RatesOfConvergence(j, 2)) ;
41   end
42   fprintf('----------------------------------------\n') ;
43
44   function [t, P, Error] = GaussMidIVP(f, a, b, N, P0)
45       t = linspace(a, b, N+1) ;
46       h = (b-a)/N ;
47       P = zeros(1, N+1) ;
48       Pexact = P0./(P0+(1-P0)*exp(-0.25*t)) ;
49       P(1) = P0 ;
50       for j = 1 : N
51           Ph = P(j) + h/2*f(t(j), P(j)) ;
52           P(j+1) = P(j) + h*f(t(j)+h/2, Ph) ;
53       end
54       Error = norm(P-Pexact, inf) ;
55   end
```

The solution graph is explained in Figure 7.3.

### 7.3.2.2   Lobatto Methods

There are three kinds of Lobatto methods: Lobatto IIIA, Lobatto IIIB and
Lobatto IIIC methods. In this section only two Lobatto methods will be used
for solving the IVP (7.1), both are Lobatto IIIA methods.

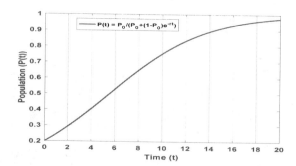

FIGURE 7.3: Solution of the logistic growth model using the Gauss mid-point
method.

The first method is the `implicit trapezoidal` method which has the Butcher table:

$$\begin{array}{c|cc} 0 & 0 & 0 \\ 1 & \frac{1}{2} & \frac{1}{2} \\ \hline & \frac{1}{2} & \frac{1}{2} \end{array}$$

and has the form:

$$y^{n+1} = y^n + \frac{1}{2}\left(f(t_n, y^n) + f(t_{n+1}, y^{n+1})\right)$$

The implicit trapezoidal method is of second order.

**Example 7.4** In this example, MATLAB and Python programmes based on the implicit trapezoidal rule will be used to solve a system of two ordinary differential equations, representing two competing species $x(t)$ and $y(t)$, $t \in [0,T]$, both are subject to logistic growth model, in the absence of the other population:

$$\begin{aligned} \frac{dx(t)}{dt} &= r_1 x(t)(1 - x(t)) - a x(t) y(t), \ x(0) = x_0 \\ \frac{dy(t)}{dt} &= r_2 y(t)(1 - y(t)) - b x(t) y(t), \ y(0) = y_0 \end{aligned}$$

The MATLAB code is

```
1   % ITSolCompSpec.m
2   clear ; clc ;
3   t0 = 0 ; T = 50.0 ; r1 = 0.3 ; r2 = 0.2 ; a = 1.0 ; b = 0.9 ;
4   f = @(t, z) [0.3*z(1)*(1-z(1))-z(1)*z(2); ...
         0.2*z(2)*(1-z(2))-0.9*z(1)*z(2)] ;
5   z0 = [0.5; 0.5] ; Dim = length(z0) ;
6   N = 100*ceil(T-t0) ;
7   t = linspace(t0, T, N+1) ;
8   h = (T-t0)/N ;
9   Epsilon = 1e-15 ;
10  n = length(t) ;
11  z = zeros(Dim, N+1) ;
12  YPrime = zeros(n, Dim) ;
13  z(:, 1) = z0 ;
14  i = 1 ;
15  while (i <= N)
16      k1 = f(t(i), z(:, i)) ;
17      k2 = feval(f, t(i+1),z(:, i)+h*k1) ;
18      YPrime(i, :) = k1' ;
19      yest = z(:, i) + h*k1 ;
20      z(:, i+1) = z(:, i) + h/2.0*(k1+k2) ;
21      while(norm(z(:, i+1)-yest)>=Epsilon)
22          yest = z(:, i+1) ;
23          k2 = feval(f, t(i+1),yest) ;
24          z(:, i+1) = z(:, i) + (h/2.0)*(k1+k2) ;
25      end
```

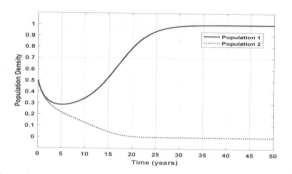

FIGURE 7.4: Solution of the competition model based on implicit trapezoidal rule solver.

```
26        i = i + 1 ;
27   end
28   x = z(1, :) ;
29   y = z(2, :) ;
30   plot(t, x, '-b', t, y, ':m', 'LineWidth', 2) ;
31   grid on
32   set(gca, 'fontweight', 'bold')
33   legend('Population 1', 'Population 2') ;
34   axis([t0, T, -0.1, 1.1])
35   xlabel('Time (years)') ; ylabel('Population Density') ;
36   set(gca, 'XTick', t0:(T-t0)/10:T) ;
37   set(gca, 'YTick', 0:0.1:1) ;
```

By executing the code, the solution of the competition model is shown in Figure 7.4.

The Python code is:

```
1   import numpy as np
2   t0 = 0 ; T = 50.0 ; r1 = 0.3 ; r2 = 0.2 ; a = 1.0 ; b = 0.9 ;
3   f = lambda t, z: np.array([0.3*z[0]*(1-z[0])-z[0]*z[1], ...
        0.2*z[1]*(1-z[1])-0.9*z[0]*z[1]])
4   z0 = [0.5, 0.5]
5   Dim = len(z0)
6   N = 100*int(T-t0)
7   t = np.linspace(t0, T, N+1)
8   h = (T-t0)/N
9   Epsilon = 1e-15
10  n = len(t)
11  z = np.zeros((Dim, N+1), 'float')
12  YPrime = np.zeros((n, Dim), 'float')
13  z[:, 0] = z0
14  i = 0
15  while i < N:
16      k1 = f(t[i], z[:, i])
17      k2 = f(t[i+1],z[:, i]+h*k1)
18      YPrime[i, :] = k1
```

```
19      yest = z[:, i] + h*k1
20      z[:, i+1] = z[:, i] + h/2.0*(k1+k2)
21      while np.linalg.norm(z[:, i+1]-yest)≥Epsilon:
22          yest = z[:, i+1] ;
23          k2 = f(t[i+1],yest) ;
24          z[:, i+1] = z[:, i] + (h/2.0)*(k1+k2)
25      i = i + 1
26
27  x, y = z[0, :], z[1, :]
28  y = z[1, :]
29  import matplotlib.pyplot as plt
30  plt.plot(t, x, '-b', lw=2, label='Population 1')
31  plt.plot(t, y, ':m', lw=2, label='Population 2')
32  plt.grid(True, ls=':')
33  #set(gca, 'fontweight', 'bold')
34  plt.legend()
35  plt.axis([t0, T, -0.1, 1.1])
36  plt.xlabel('Time (years)', fontweight='bold')
37  plt.ylabel('Population Density', fontweight='bold')
38  plt.xticks(np.arange(t0, T+(T-t0)/10, (T-t0)/10), ...
            fontweight='bold')
39  plt.yticks(np.arange(-0.1, 1.1, 0.1), fontweight='bold')
```

The second method is `Hermite-Simpson`'s method. The Butcher's table of the method is:

$$
\begin{array}{c|ccc}
0 & 0 & 0 & 0 \\[4pt]
\frac{1}{2} & \frac{5}{24} & \frac{1}{3} & -\frac{1}{24} \\[4pt]
1 & \frac{1}{6} & \frac{2}{3} & \frac{1}{6} \\[4pt]
\hline
 & \frac{1}{6} & \frac{2}{3} & \frac{1}{6}
\end{array}
$$

and has the form:

$$
y^{n+1} = y^n + \frac{1}{6}\left( f(t_n, y^n) + 4f(t_{n+1/2}, y^{n+1/2}) + f(t_{n+1}, y^{n+1}) \right)
$$

where $t_{n+1/2} = t_n + h/2$ and $y^{n+1/2} \approx y(t_n + h/2)$. The Hermite-Simpson method is a fourth order.

**Example 7.5** The Hermite-Simpson method will be used to solve an HIV model described by the following equations:

$$
\frac{dT(t)}{dt} = s - \mu_T T(t) + rT(t)(1 - (T(t) + I(t))/T_{max}) - k_1 T(t)V(t), \ T(0) = 1000
$$

$$
\frac{dI(t)}{dt} = k_2 T(t)V(t) - \mu_I I(t), \ I(0) = 0
$$

$$
\frac{dV(t)}{dt} = N_v \mu_b I(t) - k_1 T(t)V(t) - \mu_V V(t), V(0) = 10^{-3}
$$

where $s = 10, mu_T = 0.02, r = 0.03, T_{max} = 1500, k_1 = 2.4 \times 10^{-5}, k_2 = 2.0 \times 10^{-5}, \mu_I = 0.26, N_v = 850, \mu_b = 0.24, \mu_V = 2.4$ and $t \in [0, 200]$.

The MATLAB code is:

```
1   clear ;
2   t0 = 0 ; tf = 200 ;
3   s = 10 ;
4   mu_T = 0.02 ;
5   r = 0.03 ;
6   Tmax = 1500 ;
7   k1 = 2.4e-5 ;
8   k2 = 2.0e-5 ;
9   mu_I = 0.26 ;
10  Nv = 850 ;
11  NC = 200 ;
12  Q = 2 ;
13  mu_b = 0.24 ;
14  mu_V = 2.4 ;
15
16  f = @(t, x) [s-mu_T*x(1)+r*x(1)*(1-(x(1)+x(2))/Tmax)-k1*x(1)*x(3);
17  k2*x(1)*x(3)-mu_I*x(2) ;
18  Nv*mu_b*x(2)-k1*x(1)*x(3)-mu_V*x(3)] ;
19  x0 = [1e+3; 0; 1e-3] ;
20  N = 10*(tf-t0) ;
21  Epsilon = 1e-13 ;
22  x = zeros(1+N, 3) ;
23  t = linspace(t0, tf, 1+N) ;
24  h = (tf-t0)/N   ;
25  x(1,:) = x0 ;
26
27  for i = 1 : N
28      K1 = f(t(i), x(i, :)) ;
29      xest = x(i, :) + h*K1' ;
30      K3 = f(t(i+1), xest) ;
31      xmid = 0.5*(x(i, :)+xest)+h/8*(K1'-K3') ;
32      K2 = f(t(i)+h/2, xmid) ;
33      x(i+1,:) = x(i,:) + h/6*(K1'+4*K2'+K3') ;
34  end
35  T = x(:, 1) ; I = x(:, 2) ; V = x(:, 3) ;
36  figure(1) ;
37  subplot(3, 1, 1) ;
38  plot(t, T, '-b', 'LineWidth', 2) ;
39  xlabel('Time') ;ylabel('Susceptible CD4+ T-Cells') ;
40  grid on ;
41  set(gca, 'fontweight', 'bold') ;
42  set(gca, 'XTick', linspace(0, tf, 11)) ;
43  grid on ;
44  subplot(3, 1, 2) ;
45  plot(t, I, '--r', 'LineWidth', 2) ;
46  xlabel('Time') ;ylabel('Infected CD4+ T-Cells') ;
47  grid on ;
48  set(gca, 'fontweight', 'bold') ;
49  set(gca, 'XTick', linspace(0, tf, 11)) ;
50  grid on ;
51  subplot(3, 1, 3) ;
52  plot(t, V, '-.m', 'LineWidth', 2) ;
53  xlabel('Time') ;ylabel('Viral load') ;
54  grid on ;
```

```
55 set(gca, 'fontweight', 'bold') ;
56 set(gca, 'XTick', linspace(0, tf, 11)) ;
57 grid on ;
```

Executing the above code shows in the graph in Figure 7.5.
The Python code is:

```
1  import numpy as np
2  t0 = 0
3  tf = 200
4  s = 10
5  mu_T = 0.02
6  r = 0.03
7  Tmax = 1500
8  c1 = 2.4e-5
9  c2 = 2.0e-5
10 mu_I = 0.26
11 Nv = 850
12 mu_b = 0.24
13 mu_V = 2.4
14
15 f = lambda t, x: ...
     np.array([s-(mu_T+r*(1-(x[0]+x[1])/Tmax)-c1*x[2])*x[0],\
16       c2*x[0]*x[2]-mu_I*x[1],\
17       Nv*mu_b*x[1]-c1*x[0]*x[2]-mu_V*x[2]])
18 z0 = [1e+3, 0.0, 1e-3] ;
19 Dim = len(z0)
20 N = 100*int(tf-t0)
21 t = np.linspace(t0, tf, N+1)
22 h = (tf-t0)/N
23 Epsilon = 1e-15
24 n = len(t)
25 z = np.zeros((Dim, N+1), 'float')
26 zmid = np.zeros((Dim, N), 'float')
27 YPrime = np.zeros((n, Dim), 'float')
28 z[:, 0] = z0
29 i = 0
30 while i < N:
31     k1 = f(t[i], z[:, i])
32     zest = z[:, i]+h*k1
33     k3 = f(t[i+1], zest)
34     zmid[:, i] = 0.5*(z[:, i]+zest) + h/8*(k1-k3)
35     k2 = f(t[i]+h/2.0,zmid[:, i])
36     z[:, i+1] = z[:, i] + h/6.0*(k1+4*k2+k3)
37 i = i + 1
38
39 T, I, V = z[0, :], z[1, :], z[2, :]
40 import matplotlib.pyplot as plt
41 plt.figure(1)
42 plt.plot(t, T, color='orangered', lw=2)
43 plt.subplot(3, 1, 1)
44 plt.plot(t, T, color='orangered', lw=2)
45 plt.xlabel('Time')
46 plt.ylabel('Susceptible CD4+ T-Cells', fontweight='bold')
47 plt.grid(True, ls=':')
```

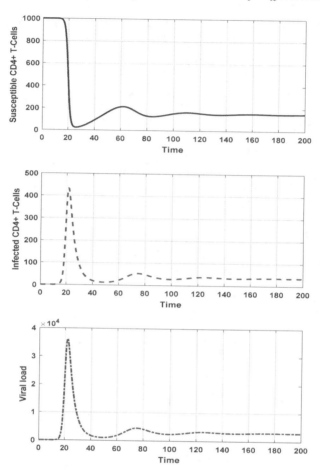

FIGURE 7.5: Solution of the HIV model obtained by the Hermite-Simpson's method.

```
48  plt.xticks(np.linspace(0, tf, 11), fontweight='bold')
49  plt.subplot(3, 1, 2) ;
50  plt.plot(t, I, color='purple', lw=2)
51  plt.xlabel('Time')
52  plt.ylabel('Infected CD4+ T-Cells', fontweight='bold')
53  plt.grid(True, ls=':')
54  plt.xticks(np.linspace(0, tf, 11))
55  plt.subplot(3, 1, 3)
56  plt.plot(t, V, color='darkred', lw=2)
57  plt.xlabel('Time', fontweight='bold')
58  plt.ylabel('Viral load', fontweight='bold')
59  plt.grid(True, ls=':')
60  plt.xticks(np.linspace(0, tf, 11), fontweight='bold')
```

## 7.4  MATLAB ODE Solvers

MATLAB provides many routines to solve systems of first order differential equations. In this lecture we will discuss how the MATLAB function ode45 can be used for solving a system of first order initial value problems (IVPs) of ordinary differential equations.

### 7.4.1  MATLAB ODE Solvers

There are two general classes of ODE solvers in MATLAB: solvers for non-stiff problems, and solvers for stiff problems.

(1) **Non-stiff problems solvers:**

MATLAB has three solvers for non-stiff problems: ode23, ode45 and ode113.

The ode23 is of low order that integrates a system of first-order differential equations using second and third order explicit Runge-Kutta methods.

The ode45 is of medium order that integrates a system of first-order odes using fourth and fifth order explicit Runge-Kutta methods. It is more accurate than ode23 in solving non-stiff problems.

The ode113 solver belongs to the Adams-Bashforth-Moulton predictor-corrector pairs of orders 1 to 13. Therefore, the ode113 is a low to medium ODE solver. In many problems it is more efficient than the ode45 solver.

(2) **Stiff problems solvers** MATLAB possesses five solvers for solving stiff problems. They are ode15s, ode23s, ode23t, ode23b and ode15i.

### 7.4.2  Solving a Single IVP

We consider an IVP of the form:

$$\frac{dy}{dt} = f(t, y(t)), \ t \in [t_0, t_1], \tag{7.6}$$

$$y(t_0) = y_0 \tag{7.7}$$

To solve the IVP (7.6)-(7.7), we have to write a function, which returns the right-hand side of Equation (7.6). This can be done as follows:

```
1  function z = fp(t, y)
2  z = f(t, y) ;
```

Now from the command window (or a MATLAB script), we can define the time period, the initial condition and call the ode45 function to solve the given IVP. We do this as follows:

```
1  tspan = [t_0, t_1] ;
2  IC = y0 ;
3  [t, y] = ode45(@fp, tspan, y0) ;
4  plot(t, y) ;
```

The symbol @ in front of the name of the function, informs MATLAB that what is follows is the name of the function which defines the slope.

**Example 7.6** Use MATLAB to solve the initial value problem:

$$\frac{dy}{dx} = 0.01y(t)\left(1 - y(t)\right),\ t \in [0, 700],\ y(0) = 0.2$$

To solve the above given IVP, we first define a function `LogisticGrowthSlope` as follows:

```
1       % LogisticGrowthSlope.m
2       function z = LogisticGrowthSlope(t, y)
3       z = 0.01*y*(1-y) ;
```

Then, a MATLAB script `SolveLogisticWithAlee.m` to solve the problem and show the results is as follows:

```
1  % SolveLogisticWithAlee.m
2  clear ; clc ;
3  tspan = [0, 700] ;
4  y0 = 0.3 ;
5  x0 = 0.2 ;
6  LogGrowth = @(t, y) 0.01*y*(1-y)*(4*y-1) ;
7  [t, y] = ode45(LogGrowth, tspan, y0) ;
8  [s, x] = ode45(LogGrowth, tspan, x0) ;
9  plot(t, y, '-b', s, x, '-.r', 'LineWidth', 2) ;
10 legend('y_0 = 0.3 > Threshold = 0.25', 'x_0 = 0.2 < Threshold ...
        = 0.25') ;
11 hold on ;
12 plot([0.0, 700], [0.25, 0.25], '--k') ;
13 xlabel('Time') ;
14 ylabel('Population') ;
15 grid on ;
16 axis([0, 700, 0, 1.1]) ;
17 set(gca, 'fontweight', 'bold') ;
18 set(gca, 'GridLineStyle', ':') ;
19 set(gca, 'YTick', 0:0.125:1) ;
```

executing the script `SolveLogisticWithAlee` will show the solution as in Figure 7.6:

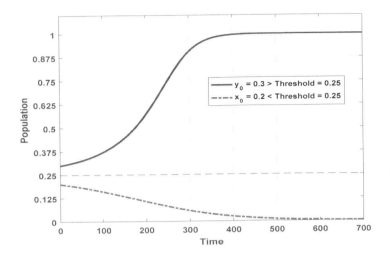

FIGURE 7.6: Solution of the logistic growth model with Alee effect using the ode45 MATLAB solver.

### 7.4.3 Solving a System of IVPs

In this section, we consider a system of initial value problems. The system is of the form:

$$
\begin{aligned}
\frac{dy_1}{dt} &= f_1(t, y_1(t), \ldots, y_n(t)), \ y_1(t_0) = y_1^0, \ t \in [t_0, t_1] \\
\frac{dy_2}{dt} &= f_2(t, y_1(t), \ldots, y_n(t)), \ y_2(t_0) = y_2^0, \ t \in [t_0, t_1] \\
\vdots &= \vdots \\
\frac{dy_n}{dt} &= f_n(t, y_1(t), \ldots, y_n(t)), \ y_n(t_0) = y_n^0, \ t \in [t_0, t_1]
\end{aligned}
\tag{7.8}
$$

The first step to solve the above system of IVPs is to write it in a vector form. To do that, we let:

$$
z = \begin{pmatrix} y_1(t) \\ y_2(t) \\ \vdots \\ y_n(t) \end{pmatrix}
$$

then,

$$
\frac{dz(t)}{dt} = \begin{pmatrix} f_1(t, z(t)) \\ f_2(t, z(t)) \\ \vdots \\ f_n(t, z(t)) \end{pmatrix} \text{ and } z(t_0) = \begin{pmatrix} y_1(t_0) \\ y_2(t_0) \\ \vdots \\ y_n(t_0) \end{pmatrix} = \begin{pmatrix} y_1^0 \\ y_2^0 \\ \vdots \\ y_n^0 \end{pmatrix}
$$

The second step is to write a function, which can evaluate the right-hand side of the system, where the output is an n-dimensional vector. This is done as follows:

```
1  function z = fp(t, z)
2  z = [ f_1(t, z) ;
3  f_2(t, z) ; ...
4
5  f_n(t, z) ;
6  ] ;
```

Finally, we write a script to define the initial condition, time space and calls the ode45 function to solve the system.

```
T  = [t_0, t_1] ;
z0 = [y_10; y_20; ...; y_n0] ;
[t, z] = ode45(@fp, T, z0) ;
```

The ode45 function discretizes the time space $[t_0, t_1]$ into $m$ (not equidistant) points and returns the result in $t$. Also, the matrix $z$ is of type $(m+1) \times (n+1)$. The first column of $z$ ($z(:,1)$) is the solution vector $y_1$, the second column ($z(:,2)$) is the solution vector $y_2$, ...etc.

**Example 7.7** Using MATLAB, solve:

$$\dot{S}(t) = -0.03SI + 0.02R, \ S(0) = 0.85, \ t \in [0, 300]$$
$$\dot{I}(t) = 0.03SI - 0.01I, \ I(0) = 0.15, \ t \in [0, 300]$$
$$\dot{R}(t) = 0.01I - 0.02R, \ R(0) = 0, \ t \in [0, 300]$$

We define a function SIRSSlope.m to evaluate the right-hand sides of the system equations.

```
1  % SIRSSlope
2  function z = SIRSSlope(t, x)
3      S = x(1) ; I = x(2) ; R = x(3) ;
4      z = [-0.03*S*I+0.02*R; 0.03*S*I-0.01*I; 0.01*I-0.02*R] ;
5  end
```

We write a MATLAB script 'SolveSIRS.m' to implement the solver:

```
1  % SolveSIRS.m
2  clear ; clc ;
3  tspan = [0, 300] ;
4  z0 = [0.85; 0.15; 0] ;
5  [t, z] = ode45(@SIRSSlope, tspan, z0) ;
6  S = z(:, 1) ; I = z(:, 2) ; R = z(:, 3) ;
7  plot(t, S, '--b', t, I, ':r', t, R, '-.m', 'LineWidth', 2) ;
8  grid on ;
9  xlabel('Time') ;
```

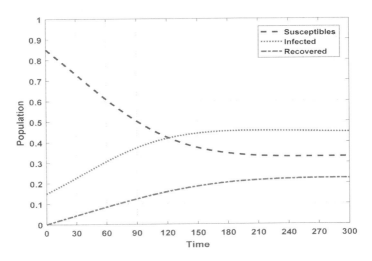

FIGURE 7.7: Solution of the SIRS model with the ode45 MATLAB solver.

```
10   ylabel('Population') ;
11   legend('Susceptibles', 'Infected', 'Recovered') ;
12   axis([0, 300, 0, 1]) ;
13   set(gca, 'fontweight', 'bold') ;
14   set(gca, 'XTick', linspace(0, 300, 11)) ;
15   set(gca, 'GridLineStyle', ':') ;
```

The output of executing the script SolveSIRS.m is Figure 7.7.

## 7.4.4   Solving Stiff Systems of IVPs

If the IVP system is stiff, the MATLAB function ode45 fails in finding a
solution for the problem. An example for such a system is the Vanderpool
oscillator, given by the second order differential equation:

$$y''(x) + \sigma(y^2(x) - 1)y'(x) + y(x) = 0, \ y(0) = 1, \ y'(0) = 2$$

If we let $y_1(x) = y(x)$, $y_2(x) = y'(x)$, we can write the Vanderpool system can
be written in the form:

$$
\begin{aligned}
y_1'(x) &= y_2(x), y_1(0) = 1 \\
y_2'(x) &= \sigma(1 - y_1^2(x))y_2(x) - y_1(x), y_2(0) = 2
\end{aligned}
\tag{7.9}
$$

The MATLAB ode45 function will fail to solve the Vanderpool system (7.9).
We can use the MATLAB function, ode23s instead to solve the system (7.9).
We will declare $\sigma$ as a global variable and then solve the system as follows:

```
1   %vdpsig.m
2   function z = vdpsig(t, y)
3   global sigma
4   z = zeros(2, 1) ;
5   z(1) = y(2);
6   z(2) = sigma*(1-y(1)^2)*y(2)-y(1) ;
```

```
1   % SolveVDPSystem.m
2   clear ; clc ;
3   global sigma ;
4   tspan = [0 200] ; y0 = [2; 1] ;
5   figure(1) ;
6   sigma = 1 ;
7   [t,y] = ode23s(@vdpsig, tspan, y0);
8   subplot(3, 2, 1) ;
9   plot(t,y(:,1),'-b', 'LineWidth', 2) ;
10  xlabel('x') ; ylabel('y(x)') ;
11  title(['\sigma = ' num2str(sigma)]) ;
12  sigma = 5 ;
13  [t,y] = ode23s(@vdpsig, tspan, y0);
14  subplot(3, 2, 2) ;
15  plot(t,y(:,1),'--b', 'LineWidth', 2) ;
16  xlabel('x') ; ylabel('y(x)') ;
17  title(['\sigma = ' num2str(sigma)]) ;
18  sigma = 10 ;
19  [t,y] = ode23s(@vdpsig, tspan, y0);
20  subplot(3, 2, 3) ;
21  plot(t,y(:,1),'--b', 'LineWidth', 2) ;
22  xlabel('x') ; ylabel('y(x)') ;
23  title(['\sigma = ' num2str(sigma)]) ;
24  sigma = 20 ;
25  [t,y] = ode23s(@vdpsig, tspan, y0);
26  subplot(3, 2, 4) ;
27  plot(t,y(:,1),'--b', 'LineWidth', 2) ;
28  xlabel('x') ; ylabel('y(x)') ;
29  title(['\sigma = ' num2str(sigma)]) ;
30  sigma = 40 ;
31  [t,y] = ode23s(@vdpsig, tspan, y0);
32  subplot(3, 2, 5) ;
33  plot(t,y(:,1),'--b', 'LineWidth', 2) ;
34  xlabel('x') ; ylabel('y(x)') ;
35  title(['\sigma = ' num2str(sigma)]) ;
36  sigma = 80 ;
37  [t,y] = ode23s(@vdpsig, tspan, y0);
38  subplot(3, 2, 6) ;
39  plot(t,y(:,1),'--b', 'LineWidth', 2) ;
40  xlabel('x') ; ylabel('y(x)') ;
41  title(['\sigma = ' num2str(sigma)]) ;
```

By executing the above MATLAB code, we get the solutions for different values of $\sigma$ as shown in Figure 7.8.

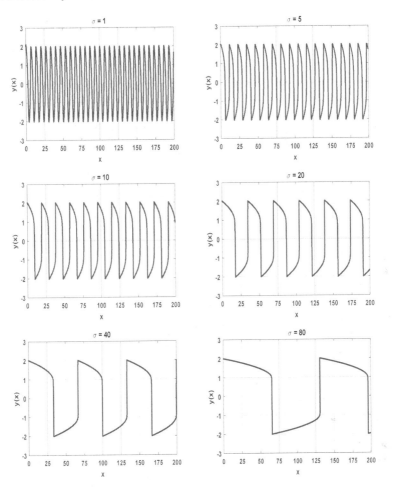

FIGURE 7.8: Solution of the Vanderpool system for $\sigma = 1, 5, 10, 20, 40$ and $80$.

## 7.5 Python Solvers for IVPs

Python has many means to solve IVPs of ODEs. Among the available tools are the `scipy.integrate.odeint` function and the `gekko` package.

### 7.5.1 Solving ODEs with odeint

The `odeint` function receives a vector containing the RHS of the ODEs system, the vector of initial conditions and the time span. It returns the solution of the ODEs system.

**Example 7.8** In this example, a python code `LogmodWithPergrow.py` will be used to solve the differential equation:

$$\dot{x}(t) = 4\sin(2t)x(t) - 3x^2(t), \ x(0) = 0.1, \ t \in [0, 20]$$

The Python code is:

```
1   import numpy as np
2   from scipy.integrate import odeint
3
4   f = lambda x, t: (4*(np.sin(2*t))*x-3*x**2)
5   t = np.linspace(0.0, 20.0, 401)
6   x0 = 0.1
7
8   x = odeint(f, x0, t)
9
10  import matplotlib.pyplot as plt
11  plt.figure(1, figsize=(10, 10))
12  plt.plot(t, x, color='orangered', lw = 2)
13  plt.grid(True, ls = ':')
14  plt.xlabel('t', fontweight='bold')
15  plt.ylabel('x(t)', fontweight='bold')
16  plt.xticks(np.arange(0, 22, 2), fontweight='bold')
17  plt.yticks(np.arange(0.0, 1.0, 0.1), fontweight='bold')
```

Executing the code, shows in Figure 7.9:

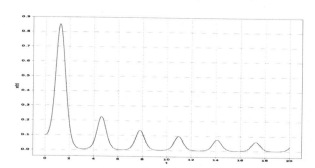

FIGURE 7.9: Solution of the initial value problem using the `odeint` function.

A Python code for solving the Vanderpole oscillator problem is as follows:

```
1   import numpy as np
2   from scipy.integrate import odeint
3   import matplotlib.pyplot as plt
4   global sigma ;
5   tspan = [0, 200] ; y0 = [2, 1] ;
6   t = np.linspace(0.0, 200.0, 10001)
7   vdpsig = lambda z, t: [z[1], sigma*(1-z[0]**2)*z[1]-z[0]]
8   plt.figure(1, figsize = (20, 10)) ;
```

```
 9  sigma = 1 ;
10  y = odeint(vdpsig, y0, t)
11  y1= y[:, 0]
12  plt.subplot(3, 2, 1) ;
13  plt.plot(t,y1,color = 'crimson', lw = 2, label = r'$\sigma = ...
        $'+str(sigma))
14  plt.xlabel('x', fontweight='bold') ; plt.ylabel('y(x)', ...
        fontweight='bold')
15  plt.legend()
16  plt.grid(True, ls=':')
17
18  plt.xticks(np.arange(0.0, 220.0, 20), fontweight='bold')
19  plt.yticks(np.arange(-2, 3), fontweight='bold')
20
21  sigma = 5 ;
22  y = odeint(vdpsig, y0, t)
23  y1= y[:, 0]
24  plt.subplot(3, 2, 2) ;
25  plt.plot(t,y1,color = 'crimson', lw = 2, label = r'$\sigma = ...
        $'+str(sigma))
26  plt.xlabel('x', fontweight='bold') ; plt.ylabel('y(x)', ...
        fontweight='bold')
27  plt.legend()
28  plt.grid(True, ls=':')
29  plt.xticks(np.arange(0.0, 220.0, 20), fontweight='bold')
30  plt.yticks(np.arange(-2, 3), fontweight='bold')
31
32  sigma = 10 ;
33  y = odeint(vdpsig, y0, t)
34  y1= y[:, 0]
35  plt.subplot(3, 2, 3) ;
36  plt.plot(t,y1,color = 'crimson', lw = 2, label = r'$\sigma = ...
        $'+str(sigma))
37  plt.xlabel('x', fontweight='bold') ; plt.ylabel('y(x)', ...
        fontweight='bold')
38  plt.legend()
39  plt.grid(True, ls=':')
40  plt.xticks(np.arange(0.0, 220.0, 20), fontweight='bold')
41  plt.yticks(np.arange(-2, 3), fontweight='bold')
42
43  sigma = 20 ;
44  y = odeint(vdpsig, y0, t)
45  y1= y[:, 0]
46  plt.subplot(3, 2, 4) ;
47  plt.plot(t,y1,color = 'crimson', lw = 2, label = r'$\sigma = ...
        $'+str(sigma))
48  plt.xlabel('x', fontweight='bold') ; plt.ylabel('y(x)', ...
        fontweight='bold')
49  plt.legend()
50  plt.grid(True, ls=':')
51  plt.xticks(np.arange(0.0, 220.0, 20), fontweight='bold')
52  plt.yticks(np.arange(-2, 3), fontweight='bold')
53
54  sigma = 40 ;
55  y = odeint(vdpsig, y0, t)
56  y1= y[:, 0]
57  plt.subplot(3, 2, 5) ;
```

```
58  plt.plot(t,y1,color = 'crimson', lw = 2, label = r'$\sigma = ...
        $'+str(sigma))
59  plt.xlabel('x', fontweight='bold') ; plt.ylabel('y(x)', ...
        fontweight='bold')
60  plt.legend()
61  plt.grid(True, ls=':')
62  plt.xticks(np.arange(0.0, 220.0, 20), fontweight='bold')
63  plt.yticks(np.arange(-2, 3), fontweight='bold')
64
65  sigma = 80 ;
66  y = odeint(vdpsig, y0, t)
67  y1= y[:, 0]
68  plt.subplot(3, 2, 6) ;
69  plt.plot(t,y1,color = 'crimson', lw = 2, label = r'$\sigma = ...
        $'+str(sigma))
70  plt.xlabel('x', fontweight='bold') ; plt.ylabel('y(x)', ...
        fontweight='bold')
71  plt.legend()
72  plt.grid(True, ls=':')
73  plt.xticks(np.arange(0.0, 220.0, 20), fontweight='bold')
74  plt.yticks(np.arange(-2, 3), fontweight='bold')
```

The solutions of the system for different values of $\sigma$ are explained by Figure 7.10.

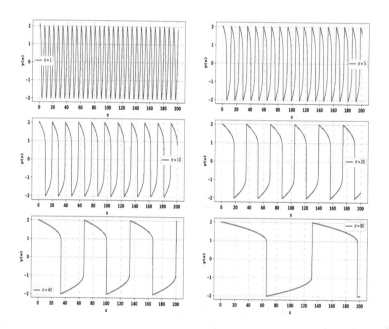

FIGURE 7.10: Solution of the Vanderpool system for $\sigma = 1, 5, 10, 20, 40$ and 80.

## 7.5.2 Solving ODEs with Gekko

GEKKO is a python optimization suite that is originally designed for solving dynamic optimization problems for mixed-integer, nonlinear, and differential algebraic equations (DAE) problems [2]. The default solver of Gekko is the IPOPT solver which is based on the interior point method. It includes other solvers such as APOPT, BPOPT, SNOPT, and MINOS [2].

The discretization of state equations is performed by using orthogonal collocation on finite elements, where in each finite element orthogonal collocation approximates control and state variables with polynomials [2].

To initialize a gekko model m in python, the following two instructions are used

```
1  from gekko import GEKKO
2  m = GEKKO(remote = False)
```

Assigning the value False to the parameter remote informs gekko to set the server to be the local machine.

If the model m contains a variable $x$ that is bounded from below by $x_{min}$, from above by $x_{max}$ and has an initial value $x_0$, then it can be declared as follows:

$$x = m.\mathrm{Var}(\mathrm{value} = x_0, \mathrm{lb} = x_{min}, \mathrm{ub} = x_{max})$$

**Example 7.9** This example uses gekko to solve the SIR model:

$$\dot{S}(t) = \Lambda - \alpha S(t)I(t) - \mu S(t), S(0) = S_0, t \in [0,T]$$
$$\dot{I}(t) = \alpha S(t)I(t) - (\mu + \beta)I(t), I(0) = I_0, t \in [0,T]$$
$$\dot{R}(t) = \beta I(t) - \mu R(t), R(0) = R_0, t \in [0,T]$$

with $\Lambda = 0.05, \mu = 0.05, \alpha = 2, \beta = 0.6$ and $T = 25$.

The gekko Python code is:

```
1  #SolveSIRgekko.py
2  from gekko import GEKKO
3  import numpy as np
4  import matplotlib.pyplot as plt
5
6  Lambda = 0.05 ; mu = 0.05 ; alpha = 2 ; beta = 0.6
7  t0 = 0
8  T = 25
9  N = 1000
10 m = GEKKO(remote=False)
11 m.time = np.linspace(t0, T, N+1)
12 S, I, R = m.Var(value=0.9), m.Var(value=0.1), m.Var(value=0.0)
13 # Equations
14 m.Equation(S.dt() == Lambda-alpha*S*I-mu*S)
15 m.Equation(I.dt() == alpha*S*I-(mu+beta)*I)
16 m.Equation(R.dt() == beta*I-mu*R)
17
18 m.options.IMODE = 4 # simulation mode
19 m.solve()
```

```
20  t = m.time
21
22  plt.figure(1, figsize=(8, 8))
23
24  plt.plot(t, S, color='darkblue', ls = '-', lw = 3, ...
            label='Suceptible Population (S(t))')
25  plt.plot(t, I, color='crimson', ls = '--', lw = 3, ...
            label='Infected Population (I(t))')
26  plt.plot(t, R, color='darkmagenta', ls = ':', lw = 3, ...
            label='Recovered Population (R(t))')
27  plt.xlabel('Time (t)', fontweight='bold')
28  plt.ylabel('Population', fontweight='bold')
29  plt.xticks(np.arange(0, T+T/10, T/10), fontweight='bold')
30  plt.yticks(np.arange(0, 1.1, 0.1), fontweight='bold')
31  plt.grid(True, ls='--')
32  plt.axis([0.0, T, 0.0, 1.1])
```

Executing the code SolveSIRgekko.py

```
runfile('/media/WindowsData1/PyFiles/SolveSIRjekko.py',
    wdir='/media/WindowsData1/PyFiles')
----------------------------------------------------------------
APMonitor, Version 0.8.9
APMonitor Optimization Suite
----------------------------------------------------------------


--------- APM Model Size ------------
Each time step contains
Objects      :              0
Constants    :              0
Variables    :              3
Intermediates:              0
Connections  :              0
Equations    :              3
Residuals    :              3

Number of state variables:          1200
Number of total equations: -        1200
Number of slack variables: -           0
-----------------------------------------
Degrees of freedom       :             0

solver            3  not supported
using default solver: APOPT
------------------------------------------------
Dynamic Simulation with APOPT Solver
------------------------------------------------


Iter   Objective  Convergence
0  2.90508E-13  1.15000E-01
```

```
1  1.50866E-13  1.14988E-01
2  4.75681E-15  3.14821E-01
3  1.74925E-15  3.13396E-01
4  1.39990E-13  2.80934E+01
5  6.11632E-16  2.59675E+00
6  1.54265E-16  2.30466E+00
7  3.91236E-17  1.83967E+00
8  9.97549E-18  9.28526E-01
9  2.46346E-18  6.43451E-01

Iter    Objective  Convergence
10  5.36666E-19  3.23197E-01
11  1.04749E-19  8.50434E-02
12  2.01098E-20  1.72751E-02
13  1.59276E-21  3.67618E-03
14  6.84198E-24  5.41524E-04
15  6.43007E-29  5.76856E-06
16  6.43007E-29  5.76856E-06
Successful solution


-----------------------------------------------------
Solver         :   IPOPT (v3.12)
Solution time  :   6.480000000010477E-002 sec
Objective      :   0.000000000000000E+000
Successful solution
-----------------------------------------------------
```

The solution graph is shown in Figure 7.11.

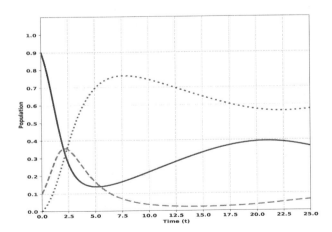

FIGURE 7.11: Solution of the SIR model in $[0, 10]$ using the gekko package.

**Example 7.10** In this example, gekko will be used to solve the stiff system of differential equations:

$$\frac{dy_1(t)}{dt} = -1002 * y_1(t) + 1000 * y_2^2(t), \ y_1(0) = 1$$

$$\frac{dy_1(t)}{dt} = y_1(t) + y_2 * (1 + y_2(t)), \ y_2(0) = 1$$

for $0 \leq t \leq 5$.

The code is:

```
1   from gekko import GEKKO
2   import numpy as np
3   import matplotlib.pyplot as plt
4
5   t0 = 0
6   T = 5
7   N = 2000
8   m = GEKKO(remote=False)
9   m.time = np.linspace(t0, T, N+1)
10  y1, y2 = m.Var(value=1.0), m.Var(value=2.0)
11  t = m.time
12  # Equations
13  m.Equation(y1.dt() == -1002*y1+1000*y2**2)
14  m.Equation(y2.dt() == y1-y2*(1+y2))
15
16  m.options.IMODE = 4 # simulation mode
17  m.solve()
18
```

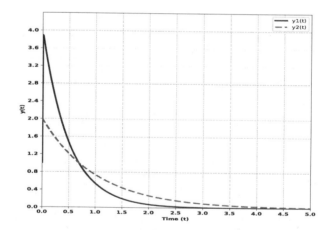

FIGURE 7.12: Solution of the stiff system of ODEs in $[0,5]$ using the gekko package.

```
19  plt.figure(1, figsize=(8, 8))
20
21  plt.plot(t, y1, color='darkblue', ls = '-', lw = 3, label='y1(t)')
22  plt.plot(t, y2, color='crimson', ls = '--', lw = 3, label='y2(t)')
23  plt.xlabel('Time (t)', fontweight='bold')
24  plt.ylabel('y(t)', fontweight='bold')
25  plt.xticks(np.arange(0, T+T/10, T/10), fontweight='bold')
26  mxS = np.ceil(max(max(y1), max(y2)))
27  plt.yticks(np.arange(0, mxS+mxS/10, mxS/10), fontweight='bold')
28  plt.grid(True, ls='--')
29  plt.axis([0.0, T, 0.0, mxS+mxS/10])
30  plt.legend()
```

Executing the code `SolveSIRgekko.py` shows the solution graph in Figure 7.12.

# 8

# Nonstandard Finite Difference Methods for Solving ODEs

## Abstract

Standard numerical methods are initially designed to solve a class of general problems without considering the structure of any individual problems. Hence, they seldom produce reliable numerical solutions to problems with complex structures such as nonlinearity, stiffness, singular perturbations and high oscillations. While the explicit difference schemes can solve such problems with low computational cost, they suffer the problems of small stability regions and hence they suffer severe restrictions on step sizes to achieve convenient results. On the other-hand, the implicit finite difference schemes enjoy wide stability regions but suffer high associated computational costs and their convergence orders cannot exceed one order above the explicit methods for the same number of stages [1].

This chapter highlights some of the cases in which the standard finite difference schemes fail to find reliable solutions, then it discusses the rules upon which the nonstandard schemes stand with several examples to explain the idea. MATLAB® and Python are used to implement the solution algorithms in all the sections.

The chapter is organized as follows. The first section discusses some numerical cases in which the standard finite difference methods give inappropriate solutions. In the second section, the construction rules of nonstandard finite difference methods are introduced. Exact finite difference schemes based on nonstandard methods are presented in Section 3, for solving some given initial value problems. Finally, in the fourth section, design of nonstandard finite difference schemes -for the case when exact finite differences are hard to find-is presented.

## 8.1 Deficiencies with Standard Finite Difference Schemes

The major problem that encounters the standard finite difference methods is the numerical instability [38, 40]. A discrete model of a differential equation is said to have numerical instability if there exist solutions to the finite difference equation that do not correspond to any of the possible solutions of the differential equations. Mickens pointed to some causes of numerical instability[37]. These causes are:

(i) Representing a derivative in the differential equation with a discrete derivative of different order. An example to this situation is when representing the logistic model

$$\frac{dP}{dt} = P(1-P), P(0) = P_0$$

using the centeral difference scheme:

$$\frac{P^{j+1} - P^{j-1}}{2h} = P^j(1-P^j) \Rightarrow P^{j+1} = P^{j-1} + 2hP^j(1-P^j).$$

Figure 8.1 shows the solution of the logistic model $\dot{P}(t) = P(t)(1 - P(t)), P(0) = 0.5$ using the centeral, forward and backward difference schemes. The figure show that the solutions obtained by both the forward and backward difference scheme have the same behavior as the original model, where as the centeral finite difference scheme is irrelative to the exact solution.

(ii) Using the standard denominator function $h$ [40]. The first and second derivatives at a point $t_j$ are approximated by:

$$\frac{du}{dt}(t_i) \approx \frac{u(t_{i+1}) - u(t_i)}{h} \quad \text{or} \quad \frac{du}{dt}(t_i) \approx \frac{u(t_i) - u(t_{i-1})}{h}$$

$$\frac{d^2u}{dt^2}(t_i) \approx \frac{u(t_{i+1}) - 2u(t_i) + u(t_{i+1})}{h^2}$$

In many cases, selection of the classical denominator function is not suitable and can lead to numerical instability. To see such an example, consider the first order IVP:

$$\frac{dy(t)}{dt} = Ay(t) + t, y(0) = 0.$$

The exact solution of this IVP is

$$y(t) = \frac{1}{A^2}\left(e^{At} - (At+1)\right).$$

The interval $[0,1]$ will be divided into 1000 subintervals, by points $0 = t_0 < t_1 < \ldots < t_{1000} = 1$ and each subinterval has a length $h = 0.001$. The forward Euler's method, Implicit trapezoidal rule and classical fourth-order Runge-Kutta method will be used for solving the given problem at the discrete points for different values of parameter $A$. Theoretically, the error of Euler's method shall be of $\mathcal{O}(h) = \mathcal{O}(10^{-3})$, the error of the

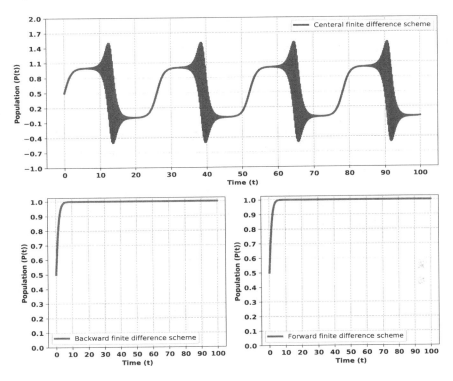

FIGURE 8.1: Solution of the logistic model $\dot{P}(t) = P(t)(1 - P(t)), P(0) = 0.5$ using the forward, backward and central finite difference schemes.

implicit trapezoidal rule shall be of $\mathcal{O}(h^2) = \mathcal{O}(10^{-6})$ and the error of the fourth-order Runge-Kutta method shall be of $\mathcal{O}(h^4) = \mathcal{O}(10^{-12})$.

The Python code to compute the errors of the three methods for different values of $A$ is:

```python
import numpy as np
import matplotlib.pyplot as plt

f = lambda s, v, B: B*v+s
N = 1000
h, t = 1.0/N, np.linspace(0, 1, N+1)
y0 = 0
MaxRows = 12
Eul = np.zeros((MaxRows, len(t)), float)
rk4 = np.zeros((MaxRows, len(t)), float)
itr = np.zeros((MaxRows, len(t)), float)
EulError = np.zeros((MaxRows, ), float)
rk4Error = np.zeros((MaxRows, ), float)
itrError = np.zeros((MaxRows, ), float)
A = np.zeros((MaxRows,), float)
```

```
16   row = 0
17   A[row] = 5.0
18   i = 0
19   while row < MaxRows:
20       # y is the exact solution of the problem
21       y = 1/A[row]**2*(np.exp(A[row]*t)-(A[row]*t+1))
22       plt.plot(t, y, lw=2, label='A = '+str(A[row]))
23       plt.legend()
24       Eul[i, 0] = y0
25       rk4[i, 0] = y0
26       itr[i, 0] = y0
27       for j in range(N):
28       # Solving with Euler's method
29           Eul[i, j+1] = Eul[i, j] + h*f(t[j], Eul[i, j], A[row])
30       # Solving with implicit trapezoidal rule
31           k1 = f(t[j], rk4[i, j], A[row])
32           k2 = f(t[j]+h, itr[i, j]+h*k1, A[row])
33           itr[i, j+1] = itr[i, j] + h/2*(k1+k2)
34       # Solving with the classical fourth-order Runge-Kutta ...
               method
35           k1 = f(t[j], rk4[i, j], A[row])
36           k2 = f(t[j]+h/2, rk4[i, j]+h/2*k1, A[row])
37           k3 = f(t[j]+h/2, rk4[i, j]+h/2*k2, A[row])
38           k4 = f(t[j]+h, rk4[i, j]+h*k3, A[row])
39           rk4[i, j+1] = rk4[i, j] +h/6*(k1+2*k2+2*k3+k4)
40       # computing the norm-infinity error for the three methods
41       EulError[i] = np.linalg.norm(y-Eul[i, :], np.inf)
42       itrError[i] = np.linalg.norm(y-itr[i, :], np.inf)
43       rk4Error[i] = np.linalg.norm(y-rk4[i, :], np.inf)
44       i += 1
45       row += 1
46       if row >= MaxRows:
47           break
48       else:
49           A[row] = A[row-1] + 2.0
50   print('--------------------------------------------------------')
51   print('  A\t Euler Error\t\t Imp. Trapz Err\t\t RK4 Error')
52   print('--------------------------------------------------------')
53   for row in range(MaxRows):
54       print('{0:4.0f}'.format(A[row]), ...
           '\t','{0:1.8e}'.format(EulError[row]), \
55               '\t', '{0:1.8e}'.format(itrError[row]), '\t', ...
               '{0:1.8e}'.format(rk4Error[row]))
56   print('--------------------------------------------------------')
```

Running the code gives the following results:

```
runfile('/media/WindowsData1/PyFiles/
numinstduetotrivstepfun.py',
wdir='/media/WindowsData1/PyFiles')
```

| A | Euler Error | Imp. Trapz Err | RK4 Error |
|---|---|---|---|
| 1 | 1.35789622e-03 | 3.56321584e-07 | 2.30926389e-14 |
| 3 | 1.00002555e-02 | 5.19127795e-06 | 4.50417481e-12 |

| 5  | 7.35013397e-02 | 4.52531609e-05 | 1.53952406e-10 |
| 7  | 5.39170234e-01 | 3.52721966e-04 | 3.11633030e-09 |
| 9  | 3.94740396e+00 | 2.65343074e-03 | 4.88584107e-08 |
| 11 | 2.88444591e+01 | 1.97153658e-02 | 6.58044428e-07 |
| 13 | 2.10373074e+02 | 1.45836606e-01 | 8.01259330e-06 |
| 15 | 1.53146894e+03 | 1.07702617e+00 | 9.07992144e-05 |
| 17 | 1.11283670e+04 | 7.94936660e+00 | 9.75035611e-04 |
| 19 | 8.07191089e+04 | 5.86610000e+01 | 1.00415309e-02 |
| 21 | 5.84468245e+05 | 4.32849955e+02 | 1.00014394e-01 |
| 23 | 4.22478467e+06 | 3.19388027e+03 | 9.69289817e-01 |
| 25 | 3.04879507e+07 | 2.35668600e+04 | 9.18238944e+00 |
| 27 | 2.19661624e+08 | 1.73896191e+05 | 8.53281952e+01 |
| 29 | 1.58017839e+09 | 1.28317310e+06 | 7.79939406e+02 |

It could be seen that the errors of the three methods are increasing as $A$ increases, although the step-size $h$ is kept constant.

(iii) using local approximations of nonlinear terms [40]: The standard finite difference schemes represent nonlinear terms such as $x^2(t_j)$ using a local approximations such as $x^{j^2}$ or $x^{j+1^2}$. Mickens showed that such local approximations of nonlinear terms could lead to numerical instability. An example given by Mickens is the solution of the exponential decay model

$$\frac{dP(t)}{dt} = -P(t), \ P(0) = 0.5.$$

The forward Euler's discretization of the exponential decay model is given by:

$$P^{n+1} = P^n - hP^n = (1.0 - h)P^n = (1 - h)^n P^0$$

For different values of the step size $h$, the resulting solutions are shown in Figure 8.2.

The solutions obtained for the exponential decay model include solution profiles with asymptotic stable, periodic and unstable dynamics, where the periodic and unstable solutions do not correspond any true solution of the exponential decay model. The first graph alone, where the population size decays to zero agrees with the true solution of the differential equation.

Another example is the use of the Heun's method to solve the logistic model

$$\frac{dx(t)}{dt} = x(t)(1 - x(t)), \ x(0) = 0.5$$

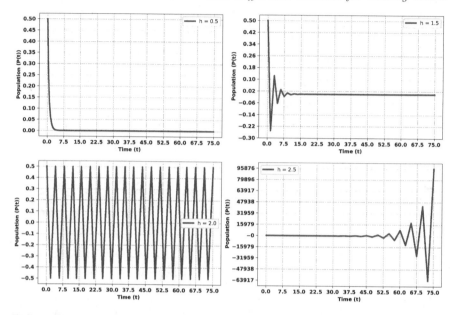

FIGURE 8.2: Numerical solution of the exponential decay model based on forward Euler's method with different step-sizes.

The Python code is:

```
1   import numpy as np
2   import matplotlib.pyplot as plt
3   steps = [0.5, 2.75, 3.0, 3.35]
4   plt.figure(1, figsize=(12, 12))
5   T = 100
6   for k in range(4):
7   h = steps[k]
8   N = round(T/h)
9   t = np.linspace(0, T, N+1)
10  P = np.zeros_like(t)
11  P[0] = 0.5
12  for j in range(N):
13          k1 = P[j] * (1.0-P[j])
14          Pest = P[j]+h*k1
15          k2 = Pest*(1.0-Pest)
16          P[j+1] = P[j] + h/2*(k1+k2)
17      plt.subplot(2, 2, k+1)
18      plt.plot(t, P, color = 'blue', lw=3, label = 'h = ...
            '+str(h))
19      plt.xlabel('Time (t)', fontweight='bold')
20      plt.ylabel('Population (P(t))', fontweight = 'bold')
21      plt.legend()
22      plt.grid(True, ls = '--')
23      plt.xticks(np.arange(0, 105, 10.0), fontweight = 'bold')
```

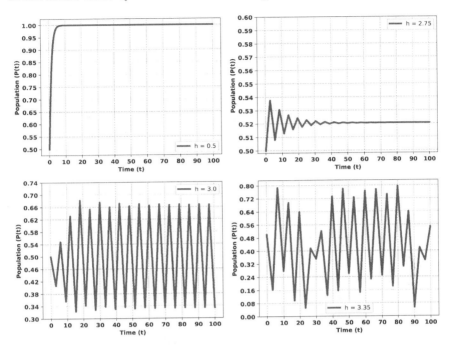

FIGURE 8.3: Solution of the logistic model using Heun's method for different values of step-size.

```
24    mnp, mxp = np.floor(10*min(P))/10, np.ceil(10*max(P))/10
25    plt.yticks(np.arange(mnp, mxp+(mxp-mnp)/10, ...
              (mxp-mnp)/10), fontweight = 'bold')
```

Figure 8.3 shows the solution of the logistic model with the Heun's method for different values of the step-size $h$.

From Figure 8.3, corresponding to some values of the step-size $h$, the numerical solution of the logistic model could have periodic and chaotic behaviors, which are not corresponding to any solution of the logistic model.

## 8.2 Construction Rules of Nonstandard Finite Difference Schemes [38]

The numerical stability that accompany the design of the standard finite difference schemes, show that due to some inherent errors in the standard models, it becomes very expensive to retrieve the true information, causing these standard schemes to present unreliable results.

Considering the causes leading to numerical instabilities in the standard finite difference schemes, four nonstandard construction rules have been established by Micken's for the design of such numerical scheme. These rules are stated in [37]:

(I) Because using a discrete derivative of order that differs than the order of the differential equation can lead to numerical instability, the order of the discrete model shall be as same as the order of the differential equation. Under this rule the central finite difference scheme cannot be used as an approximation of the first derivative in the discrete model of a first-order differential equation. Either the forward or backward difference schemes can be used.

(II) Nonstandard denominator functions have to be used for the discrete representation of the continuous derivative. A nonstandard discrete representation of $\frac{dy}{dt}$ at $t = t_j$ is of the form:

$$\frac{dy}{dt} \approx \frac{y(t_{j+1}) - \psi(\boldsymbol{\lambda}, h) y(t_j)}{\phi(\boldsymbol{\lambda}, h)}$$

where $\boldsymbol{\lambda}$ is a vector of the model's parameters, $h = t_{j+1} - t_j$. The numerator and denominator functions $\psi(\boldsymbol{\lambda}, h)$ and $\phi(\boldsymbol{\lambda}, h)$ shall fulfill the properties:

$$\psi(\boldsymbol{\lambda}, h) \to 1 \text{ and } \phi(\boldsymbol{\lambda}, h) \to h \text{ as } h \to 0$$

An example to show how the use of nontrivial denominator can lead to numerical instability, consider the exponentially decay model

$$\frac{dP(t)}{dt} = -P(t), P(0) = 0.5.$$

In Euler's method, suppose that the trivial denominator function $h$ is replaced by the denominator function

$$\phi(h) = \frac{1 - e^{-h}}{h},$$

hence the resulting discrete model is

$$P^{j+1} = (1 - \phi) P^j, \quad P^0 = 0.5$$

The Python code `expdecnsden.py` is used to solve the exponential decay model for different values of the step-size $h$:

```
1  #expdecnsden.py
2  import numpy as np
3  import matplotlib.pyplot as plt
4  steps = [0.5, 1.05, 2.5, 5.0]
5  plt.figure(1, figsize=(12, 12))
```

```
 6   T = 105
 7   for k in range(4):
 8       h = steps[k]
 9       phi = (1-np.exp(-h))/h
10       N = round(T/h)
11       t = np.linspace(0, T, N+1)
12       P = np.zeros_like(t)
13       P[0] = 0.5
14       for j in range(N):
15           P[j+1] = (1.0-phi)*P[j]
16       plt.subplot(2, 2, k+1)
17       plt.plot(t, P, color = 'blue', lw=2, label = 'h = ...
                 '+str(h))
18       plt.xlabel('Time (t)', fontweight='bold')
19       plt.ylabel('Population (P(t))', fontweight = 'bold')
20       plt.legend()
21       plt.grid(True, ls = '--')
22       plt.xticks(np.arange(0, 110, 15), fontweight = 'bold')
23       mnp, mxp = np.floor(10*min(P))/10, np.ceil(10*max(P))/10
24       plt.yticks(np.arange(mnp, mxp+(mxp-mnp)/10, ...
                 (mxp-mnp)/10), fontweight = 'bold')
```

The result of executing the code is Figure 8.4, in which the exponential decay model is solved with a nonstandard denominator function for different values of the step size $h$.

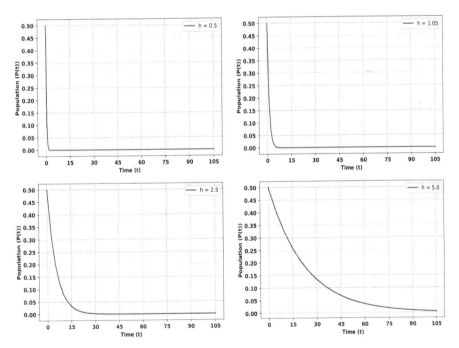

FIGURE 8.4: Solution of the decay model using a nonstandard denominator function, for different values of the step-size.

(III) Nonlinear terms shall be approximated by using non-local approximations: terms like $y^2(t)$ shall be represented by $y^j y^{j+1}$ or $y^{j-1} y^{j+1}$ instead of $(y^j)^2$, where $y^j \approx y(t_j)$. For example, the discrete model corresponding to the logistic model

$$\frac{dy}{dt} = y(t)(1 - y(t)), y(0) = 0.5$$

could be

$$\frac{y^{j+1} - y^j}{h} = y^j(1 - y^{j+1}) \Rightarrow y^{j+1} = \frac{y^j(1+h)}{1 + hy^j}, y^0 = 0.5$$

A Python code `nonlocapplog.py` is used to solve the logistic model using nonlocal approximations. The code is

```
1   # nonlocapplog.py
2   import numpy as np
3   import matplotlib.pyplot as plt
4   steps = [0.75, 3.75, 7.5, 15.0]
5   plt.figure(1, figsize=(12, 12))
6   T = 75
7   for k in range(4):
8       h = steps[k]
9       N = round(T/h)
10      t = np.linspace(0, T, N+1)
11      P = np.zeros_like(t)
12      P[0] = 0.5
13      for j in range(N):
14          P[j+1] = (1.0+h)*P[j]/(1+h*P[j])
15      plt.subplot(2, 2, k+1)
16      plt.plot(t, P, color = 'blue', lw=3, label = 'h = ...
            '+str(h))
17      plt.xlabel('Time (t)', fontweight='bold')
18      plt.ylabel('Population (P(t))', fontweight = 'bold')
19      plt.legend()
20      plt.grid(True, ls = '--')
21      plt.xticks(np.arange(0, 80, 7.5), fontweight = 'bold')
22      mnp, mxp = np.floor(10*min(P))/10, np.ceil(10*max(P))/10
23      plt.yticks(np.arange(mnp, mxp+(mxp-mnp)/10, ...
            (mxp-mnp)/10), fontweight = 'bold')
```

Figure 8.5 is obtained by executing the python code. It shows the solutions of the logistic model using different values of step-sizes. All these solutions correspond the true solution of the differential equation model. Hence, for all values of step-sizes, the discrete model does not suffer numerically instability.

(IV) Dynamical consistency between the solutions of the discrete model and differential equation. That means all the properties possessed by the solution of the differential equation must be possed by the solution of the discrete model. Particular properties include positiveness, monotonicity, boundness, limit cycles and other periodic solutions, etc.

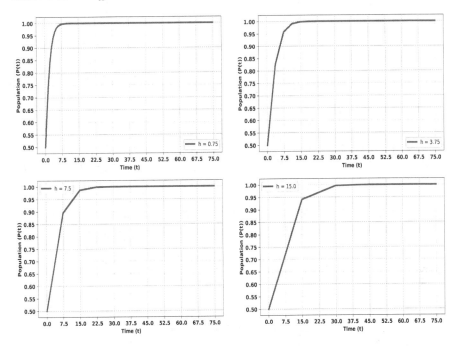

FIGURE 8.5: Solution of the logistic model using a non-local approximation of $y^2(t)$ with different values of step-sizes.

A nonstandard finite difference scheme is a discrete representation of the differential equation which is constructed based on the nonstandard rules.

## 8.3 Exact Finite Difference Schemes

Given an initial value problem

$$\frac{dy}{dt} = f(t, y(t), \boldsymbol{\lambda}), \ y(a) = y^0, t \in [a, b] \tag{8.1}$$

where $\boldsymbol{\lambda}$ is the set of model parameters. Suppose that the solution $y(t)$ of the given initial value problem is

$$y(t) = F(t, y(t), y^0, \boldsymbol{\lambda}). \tag{8.2}$$

Let $N$ be a positive integer, $h = \frac{b-a}{N}$ and $t_j = a + jh, j = 0, \ldots, N$. Consider a finite difference scheme

$$y^{j+1} = g(h, t_j, y^j, \boldsymbol{\lambda}) \tag{8.3}$$

and $y^j \approx y(t_j)$. Suppose that the solution of the discrete finite difference scheme is

$$y^j = G(t_j, h, y^j, y^0, \boldsymbol{\lambda}). \tag{8.4}$$

According to Mickens [37] Equation (8.3) is an `exact finite difference scheme` of the differential equation (8.1) if its solution (8.4) is as same as the solution of associated differential equation (8.2). That means

$$y^j = G(t_j, h, y^j, y^0, \boldsymbol{\lambda}) = F(t_j, y(t_j), y^0, \boldsymbol{\lambda}) = y(t_j).$$

Mickens [39] described a mechanism to construct exact finite difference schemes for many kinds of models. The idea is to incorporate some qualitative features of the solution of the problem into the discrete model. Qualitative features can be obtained from the linear part in the model. The nonlinear parts are dealt with using the nonlocal approximation techniques.

## 8.3.1 Exact Finite Difference Schemes for Homogeneous Linear ODEs

In this section, exact finite difference schemes will be derived for three kinds of linear and homogeneous ODEs. They include linear first order, linear second order equation and linear system of first order ODEs.

### 8.3.1.1 Exact Finite Difference Schemes for a Linear Homogeneous First-Order ODE

Given the initial value problem:

$$\frac{dy(t)}{dt} = -\alpha y(t), y(0) = y^0$$

The exact solution of the given IVP is $y(t) = y^0 e^{-\alpha t}$. The bases solution function is $y_1(t) = e^{-\alpha t}$. To determine the exact finite difference scheme of the given IVP, consider the determinant:

$$\begin{vmatrix} y^j & e^{-\alpha t_j} \\ y^{j+1} & e^{-\alpha t_{j+1}} \end{vmatrix} = e^{-\alpha t_j} \begin{vmatrix} y^j & 1 \\ y^{j+1} & e^{-\alpha h} \end{vmatrix} = 0 \Rightarrow y^{j+1} = e^{-\alpha h} y^j$$

Subtracting $y^j$ from the two sides and multiplying the right-hand side by $\alpha$ and dividing by $\alpha$ gives the discrete scheme:

$$y^{j+1} - y^j = -\alpha \left( \frac{1 - e^{-\alpha h}}{\alpha} \right) y^j$$

from which,

$$\frac{y^{j+1} - y^j}{\frac{1 - e^{-\alpha h}}{\alpha}} = -\alpha y^j$$

Hence, instead of using the standard denominator function $h$, using the denominator function $\frac{1-e^{-\alpha h}}{\alpha}$ will result in an exact finite difference scheme.

**Example 8.1 (First order linear ODE [39])** In this example, an exact finite difference scheme will be derived for the exponential decay model

$$\frac{dy(t)}{dt} = -\frac{\pi}{4}y(t), y(0) = \frac{1}{2}$$

The following Python code solves the exponential decay model using the exact finite difference scheme with different values of the step size $h$ and shows the corresponding infinity norm of the difference between the exact solution and the solution of the discrete model.

```
1   import numpy as np
2   import matplotlib.pyplot as plt
3   plt.figure(1, figsize=(12, 12))
4   alpha = np.pi/4
5   T = 100
6   steps = np.array([0.1, 0.5, 1.0, 2.0, 4.0, 5.0, 6.25, 10.0, ...
        12.5, 20.0, 25.0, 50.0, 100.0])
7   Errors = []
8   for k in range(1, len(steps)):
9       h = steps[k]
10      phi = (1-np.exp(-alpha*h))/alpha
11      N = int(round(T/h))
12      t = np.linspace(0, T, N+1)
13      P, PEx = np.zeros_like(t), np.zeros_like(t)
14      P[0] = 0.5
15      PEx[0] = 0.5
16      for j in range(N):
17          P[j+1] = (1.0-phi*alpha)*P[j]
18          PEx[j+1] = P[0]*np.exp(-alpha*t[j+1])
19      MaxError = np.linalg.norm(PEx-P, np.inf)
20      Errors.append([h, MaxError])
21  print('----------------------')
22  print('   h\t\t Error')
23  print('----------------------')
24  for j in range(len(Errors)):
25  print('{0:3.2f}'.format(Errors[j][0]) + '\t' + ...
        '{0:1.3e}'.format(Errors[j][1]))
26  print('----------------------')
```

Executing the code gives the following results:

```
--------------------------------
    h           Error
--------------------------------
   0.50     2.77555756e-17
   1.00     4.16333634e-17
   2.00     8.67361738e-19
   4.00     7.63278329e-17
```

| 5.00 | 5.20417043e-18 |
| 6.25 | 1.30104261e-17 |
| 10.00 | 4.24465151e-17 |
| 12.50 | 3.25836634e-17 |
| 20.00 | 1.24790452e-18 |
| 25.00 | 2.25704533e-17 |
| 50.00 | 4.40824356e-18 |
| 100.00 | 3.88652225e-35 |

----------------------------

### 8.3.1.2  Exact Finite Difference Scheme for Linear Homogeneous Second Order ODE

**Example 8.2 (second order linear ODE [39])** In this example, the exact finite difference scheme will be established for solving the second order BVP:

$$\frac{d^2u}{dt^2}(t) + \omega^2 u(t) = 0, u(0) = u_0, \ u(2) = u_1$$

The characteristic equation corresponding to the given BVP is

$$r^2 + \omega^2 = 0,$$

whose solutions are

$$r_{1,2} = \pm\omega i.$$

Hence, the bases solution functions of the BVP are

$$u_1(t) = \sin(\omega t) \text{ and } u_2(t) = \cos(\omega t).$$

Consider the determinant:

$$\begin{vmatrix} u^{j-1} & \cos(\omega t_{j-1}) & \sin(\omega t_{j-1}) \\ u^j & \cos(\omega t_j) & \sin(\omega t_j) \\ u^{j+1} & \cos(\omega t_{j+1}) & \sin(\omega t_{j+1}) \end{vmatrix} = 0.$$

Using the trigonometric identity $\sin(a)\cos(b) - \cos(a)\sin(b) = \sin(a-b)$ and remembering that $t_{j+1} - t_j = t_j - t_{j-1} = h, tj+1 - t_{j-1} = 2h$ the determinant is expanded into:

$$\sin(\omega h)u^{j-1} - \sin(2\omega h)u^j + \sin(\omega h)u^{j+1} = 0 \Rightarrow u^{j-1} - 2\cos(\omega h)u^j + u^{j+1} = 0$$

Subtracting $2u^j$ from and adding $2cos(\omega h)$ to the two sides and using the trigonometric identity $\cos(\omega h) = 1 - 2\sin^2\left(\frac{\omega h}{2}\right)$ gives:

$$u^{j-1} - 2u^j + u^{j+1} = -4\sin^2\frac{\omega h}{2}u^j \Rightarrow u^{j-1} - 2u^j + u^{j+1} = -\frac{4}{\omega^2}\sin^2\frac{\omega h}{2}\omega^2 u^j$$

Finally, dividing the two sides of equation by $\frac{4}{\omega^2} \sin^2 \frac{\omega h}{2} \omega^2$ and adding $\omega^2 u^j$ to the two sides gives the exact finite difference scheme

$$\frac{u^{j-1} - 2u^j + u^{j+1}}{\left(\frac{2}{\omega} \sin \frac{\omega h}{2}\right)^2} + \omega^2 u^j = 0$$

Hence, the exact finite difference scheme is obtained by selecting the denominator function

$$\phi(\omega, h) = \frac{2}{\omega} \sin \frac{\omega h}{2}.$$

As $h \to 0$, $\sin\left(\frac{\omega h}{2}\right) \to \frac{\omega h}{2}$ and hence, $\phi(\omega, h) \to h$.

The exact solution of the problem is:

$$u_{exact}(t) = \frac{3}{2} \cos(\omega t) + \frac{\frac{1}{2} - \frac{3}{2} \cos(2\omega)}{\sin(2\omega)} \sin(\omega t)$$

To solve the given BVP using MATLAB for $u_0 = 1.5$ and $u_1 = 0.5$, the time space $[0, 2]$ is divided into $N$-subintervals, where $N$ is a positive integer. The solution is obtained by solving the linear system

$$\begin{bmatrix} 1 & 0 & 0 & 0 & \cdots & 0 & 0 & 0 & 0 \\ \frac{1}{\phi^2} & \omega^2 - \frac{2}{\phi^2} & \frac{1}{\phi^2} & 0 & \cdots & 0 & 0 & 0 & 0 \\ 0 & \frac{1}{\phi^2} & \omega^2 - \frac{2}{\phi^2} & \frac{1}{\phi^2} & \cdots & 0 & 0 & 0 & 0 \\ \vdots & \vdots & \vdots & \vdots & \ddots & \vdots & \vdots & \vdots & \vdots \\ 0 & 0 & 0 & 0 & \cdots & \frac{1}{\phi^2} & \omega^2 - \frac{2}{\phi^2} & \frac{1}{\phi^2} & 0 \\ 0 & 0 & 0 & 0 & \cdots & 0 & \frac{1}{\phi^2} & \omega^2 - \frac{2}{\phi^2} & \frac{1}{\phi^2} \\ 0 & 0 & 0 & 0 & \cdots & 0 & 0 & 0 & 1 \end{bmatrix} \begin{bmatrix} u^0 \\ u^1 \\ u^2 \\ \vdots \\ u^{N-2} \\ u^{N-1} \\ u^N \end{bmatrix} = \begin{bmatrix} u_0 \\ 0 \\ 0 \\ \vdots \\ 0 \\ 0 \\ u_1 \end{bmatrix}$$

The MATLAB script `SolveHarmOsc.m` solves the BVP for $\omega = 16$ and $N = 10, 20, 40$ and $80$:

```
1   clear ; clc ; clf ;
2   a = 0.0 ; b = 2.0 ;
3   w = 16 ; u0 = 1.5; u1 = 0.5 ;
4   C1 = 1.5 ; C2 = (0.5-1.5*cos(2*w))/sin(2*w) ;
5   figure(1) ; Error = [] ;
6   N = 10 ; k = 1 ;
7   while k ≤ 4
8       t = linspace(a, b, N+1) ; h = (b-a)/N ;
9       phi = 2/w*sin(w*h/2.0) ;
10      A = ...
            full(gallery('tridiag',N+1,1/phi^2,w^2-2/phi^2,1/phi^2)) ...
            ;
11      A(1, 1) = 1.0 ; A(end, end) = 1.0 ;
12      A(1, 2) = 0.0 ; A(end, end-1) = 0.0 ;
13      g = zeros(N+1, 1) ;
14      g(1) = u0 ; g(end) = u1 ;
15      u = A\g ;
16      ux = C1*cos(w*t)' + C2*sin(w*t)' ;
```

```
17      Error = [Error; [N, norm(ux-u, inf)]] ;
18      subplot(2, 2, k) ;
19      plot(t, u, '-b', 'LineWidth', 2) ;
20      xlabel('t') ;
21      ylabel('u(t)') ;
22      legend(['N = ' num2str(N)])
23      set(gca, 'fontweight', 'bold') ;
24      set(gca, 'fontsize', 12) ;
25      set(gca, 'XTick', a:(b-a)/10:b) ;
26      mnu = floor(10*min(u))/10 ; mxu = ceil(10*max(u))/10 ;
27      set(gca, 'YTick', mnu:(mxu-mnu)/10:mxu) ;
28      axis([a, b, mnu, mxu]) ;
29      grid on ;
30      N = 2*N ;
31      k = k + 1 ;
32 end
33 fprintf('--------------------------------\n') ;
34 fprintf(' N\t\t Error\n') ;
35 fprintf('--------------------------------\n') ;
36 N = 100 ;
37 for j = 1 : 4
38      fprintf('%3i\t\t%1.12e\n', Error(j, 1), Error(j, 2)) ;
39 end
40 fprintf('--------------------------------\n')
```

By executing the MATLAB code, the solutions of the BVP for the different values of $N$ are shown in Figure 8.6 and the infinity norm errors are also shown.

| N | Error |
|---|---|
| 10 | 3.330669073875e-15 |
| 20 | 6.883382752676e-15 |
| 40 | 4.996003610813e-15 |
| 80 | 1.088018564133e-14 |

If the standard denominator function $h$ is used instead of $\phi(\omega, h)$, then the table of errors will be:

| N | Error |
|---|---|
| 10 | 1.308207805131e+00 |
| 20 | 3.921038479674e+00 |
| 40 | 1.694674604840e+00 |
| 80 | 5.894499384984e-01 |

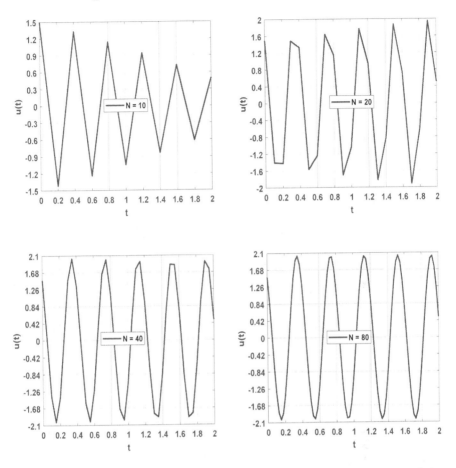

FIGURE 8.6: Solution of the harmonic oscillator system using an exact finite difference scheme for different numbers of subintervals.

### 8.3.1.3 Exact Finite Difference Scheme for a System of Two Linear ODEs

Given a system of two ODEs:

$$\frac{dx}{dt} = ax(t) + by(t)$$
$$\frac{dy}{dt} = cx(t) + dy(t)$$

which can be written in the matrix form:

$$\frac{d\mathbf{z}}{dt} = A\mathbf{z}(t) = \begin{bmatrix} a & b \\ c & d \end{bmatrix} \begin{bmatrix} x(t) \\ y(t) \end{bmatrix} \tag{8.5}$$

where

$$z(t) = \left[ \begin{array}{c} x(t) \\ y(t) \end{array} \right]$$

and $a, b, c$ and $d$ are real numbers.

Matrix $A$ is similar to its Jordan form $J \in \mathbb{R}^{2 \times 2}$, hence there exists an invertible matrix $P \in \mathbb{R}^{2 \times 2}$ such that $J = P^{-1}AP$. The diagonal elements of $J$ are the eigenvalues of $A$ and the columns of $P$ are the eigenvectors or generalized eigenvectors of matrix $A$. The Jordan form $J$ of $A$ has one of three forms:

$$J = \left[ \begin{array}{cc} \lambda_1 & 0 \\ 0 & \lambda_2 \end{array} \right], \left[ \begin{array}{cc} \lambda & 0 \\ 0 & \lambda \end{array} \right] \text{ or } \left[ \begin{array}{cc} \lambda & 1 \\ 0 & \lambda \end{array} \right]$$

with $\lambda_1 \neq \lambda_2 \neq 0$ and $\lambda \neq 0$.

In MATLAB, the Jordan form of a matrix $A$ can be found by using the MATLAB function jordan, which returns the matrices $P$ and $J$.

```
>> A = [2, 0, 0; 1, 1, -3; -1, -1, -1] ;
>> [P, J] = jordan(A) ;
>> disp('P = ') ; disp(P) ;
P =

                 0                0    1.0000e+00
        1.2500e-01    1.5000e+00   -1.2500e-01
        1.2500e-01   -5.0000e-01   -1.2500e-01
>> disp('J = ') ; disp(J) ;
J =

        -2      0      0
         0      2      1
         0      0      2
```

In Python, the Jordan form of a matrix $A$ can be found by using the method jordan_form of a symbolic matrix.

```
In [1]: import numpy as np
In [2]: import sympy as smp
In [3]: A = np.array([[2, 0, 0], [1, 1, -3], [-1, -1, -1]])
In [4]: P, J = (smp.Matrix(A)).jordan_form()
In [5]: P, J = np.array(P), np.array(J)
In [6]: print('P = \n', P)
P =
[[0 0 -2]
 [1 -3 1]
 [1 1 0]]
In [7]: print('J = \n', J)
J =
[[-2 0 0]
 [0 2 1]
 [0 0 2]]
```

Now, equation (8.5) can be written as

$$\frac{d\boldsymbol{z}}{dt} = PJP^{-1}\boldsymbol{z}.$$

Multiplying the two sides of equation (8.5) by $P^{-1}$ and using the linear transformation

$$\boldsymbol{u}(t) = P^{-1}\boldsymbol{z}(t)$$

Equation (8.5) can be written as

$$\frac{d\boldsymbol{u}}{dt} = J\boldsymbol{u} \tag{8.6}$$

The solution of the linear system (8.5) is obtained by solving the linear system (8.6) and using the linear transformation

$$\boldsymbol{z}(t) = P\boldsymbol{u}(t).$$

In [48], the exact finite difference schemes are derived for the three kinds of Jordan forms. The author proved that the linear system (8.6) has an exact finite difference scheme of the form:

$$\frac{\boldsymbol{u}^{j+1} - \boldsymbol{u}^j}{\phi} = J\left(\theta\boldsymbol{u}^{j+1} + (1-\theta)\boldsymbol{u}^j\right) \tag{8.7}$$

where $\phi$ and $\theta$ are to be determined in terms of the step-size $h$ and the eigenvalues of $A$. The functions $\phi$ and $\theta$ are as follows:

I. In the case that matrix $A$ has two distinct roots $\lambda_1$ and $\lambda_2$, that is

$$J = \begin{bmatrix} \lambda_1 & 0 \\ 0 & \lambda_2 \end{bmatrix},$$

$\phi$ and $\theta$ are

$$\phi = \frac{(\lambda_1 - \lambda_2)\left(e^{\lambda_1 h} - 1\right)\left(e^{\lambda_2 h} - 1\right)}{\lambda_1 \lambda_2 \left(e^{\lambda_1 h} - e^{\lambda_2 h}\right)}$$

and

$$\theta = \frac{\lambda_2\left(e^{\lambda_1 h} - 1\right) - \lambda_1\left(e^{\lambda_2 h} - 1\right)}{(\lambda_1 - \lambda_2)\left(e^{\lambda_1 h} - 1\right)\left(e^{\lambda_2 h} - 1\right)}$$

II. In the case that $A$ has a repeated eigenvalue $\lambda$ and $\mathrm{Dim}(A - \lambda I) = 2$, that is

$$J = \begin{bmatrix} \lambda & 0 \\ 0 & \lambda \end{bmatrix},$$

then,

$$\phi = \frac{e^{\lambda h} - 1}{\lambda\theta\left(e^{\lambda h} - 1\right) + \lambda}$$

Hence, if $\theta$ is chosen to be $\frac{1}{2}$, then

$$\phi = \frac{2\left(e^{\lambda h} - 1\right)}{\lambda\left(e^{\lambda h} + 1\right)},$$

and if

$$\theta = \frac{1 - e^{\lambda h} + \lambda h e^{\lambda h}}{\left(e^{\lambda h} - 1\right)^2},$$

then

$$\phi = \frac{\left(e^{\lambda h} - 1\right)^2}{\lambda^2 h e^{\lambda h}}$$

III. In the case that $A$ has a repeated eigenvalue $\lambda$ and $\text{Dim}(A - \lambda I) = 1$, that is

$$J = \begin{bmatrix} \lambda & 1 \\ 0 & \lambda \end{bmatrix},$$

then $\phi$ and $\theta$ will have the forms:

$$\theta = \frac{1 - e^{\lambda h} + \lambda h e^{\lambda h}}{\left(e^{\lambda h} - 1\right)^2}$$

and

$$\phi = \frac{\left(e^{\lambda h} - 1\right)^2}{\lambda^2 h e^{\lambda h}}.$$

**Example 8.3** In this example, an exact finite difference scheme will be established to solve the linear system:

$$\begin{aligned} \dot{x}(t) &= -2x(t) + y(t), x(0) = 1, \\ \dot{y}(t) &= x(t) - y(t), y(0) = 0.5 \end{aligned}$$

To solve this linear system of equations, let $z(t) = [x(t), y(t)]^T$, then the matrix form of the system is:

$$\dot{z}(t) = \begin{bmatrix} -2 & 1 \\ 1 & -1 \end{bmatrix} z(t), z(0) = \begin{bmatrix} 1 \\ 0.5 \end{bmatrix}.$$

The eigenvalues of the coefficient matrix are:

$$\lambda_{1,2} = -\frac{3}{2} \mp \frac{\sqrt{5}}{2}$$

and the corresponding eigenvectors are:

$$v_{1,2} = \begin{bmatrix} -\frac{1}{\frac{1}{2} \pm \frac{\sqrt{5}}{2}} \\ 1 \end{bmatrix}$$

The matrices $J$ (Jordan form) and $P$ are:

$$J = \begin{bmatrix} -\frac{3}{2} - \frac{\sqrt{5}}{2} & 0 \\ 0 & -\frac{3}{2} + \frac{\sqrt{5}}{2} \end{bmatrix} \text{ and } P = \begin{bmatrix} -\frac{1}{\frac{1}{2} + \frac{\sqrt{5}}{2}} & -\frac{1}{\frac{1}{2} - \frac{\sqrt{5}}{2}} \\ 1 & 1 \end{bmatrix}$$

The exact solution of the linear system is:

$$z(t) = P \begin{bmatrix} e^{\lambda_1 t} & 0 \\ 0 & e^{lambda_2 t} P^{-1} z(0) \end{bmatrix}$$

The MATLAB script `SolveLinSysExact.m` implements the exact finite difference scheme for solving the linear system, and shows the approximation errors at the discrete points:

```
%SolveLinSysExact.m
clear ; clc ; clf ;
a = 0.0 ; b = 10.0 ;
A = [-2, 1; 1, -1] ;
N = 40 ;
t = linspace(a, b, N+1) ;
h = (b-a)/N ;
[phi, tht, P, J] = CompPhiThet(A, h) ;
z0 = [2; 1] ;
z = zeros(2, N+1) ; u = zeros(2, N+1) ;
z(:, 1) = z0 ;
u(:, 1) = P\z(:, 1) ;
B = eye(2)-tht*phi*J ; C = eye(2)+(1-tht)*phi*J ;
for j = 1 : N
    u(:, j+1) = B\(C*u(:, j)) ;
    z(:, j+1) = P*u(:, j+1) ;
end

uex = zeros(size(u)) ; %Exact transformed solution
zex = zeros(size(z)) ; %Exact solution
uex(:, 1) = u(:, 1) ; zex(:, 1) = z(:, 1) ;
iP = inv(P) ;
for j = 1 : N
    uex(:, j+1) = expm(J*t(j+1))*uex(:, 1) ;
    zex(:, j+1) = P*uex(:, j+1) ;
end
Errors = abs(z-zex)' ;
fprintf('---------------------------------------------------\n') ;
fprintf('    t\t |x(t)-xexact(t)|\t |y(t)-yexact(t)|\n') ;
fprintf('---------------------------------------------------\n') ;
for j = 1 : length(t)
    fprintf('  %2i\t %8.7e\t\t  %8.7e\n', t(j), Errors(j, ...
        1), Errors(j, 2)) ;
end
fprintf('---------------------------------------------------\n') ;

plot(t, z(1, :), '-b', t, z(2, :), '--m', 'LineWidth', 2)
xlabel('t') ;
set(gca, 'XTick', linspace(a, b, 11)) ;
```

```
39  set(gca, 'YTick', linspace(0, 1, 11)) ;
40  grid on ;
41  legend('x(t)', 'y(t)') ;
42  set(gca, 'fontweight', 'bold') ;
43
44  function [phi, tht, P, J] = CompPhiThet(A, h)
45      [P, J] = jordan(A) ;
46      L1 = J(1, 1) ; L2 = J(2, 2) ;
47      if isreal(L1) && isreal(L2)
48          if abs(L1) > eps && abs(L2) > eps
49              if abs(L1-L2) > eps
50                  phi = (L1-L2)*(exp(L1*h)-1)*(exp(L2*h)-1)/...
51                      (L1*L2*(exp(L1*h)-exp(L2*h))) ;
52                  tht = (L2*(exp(L1*h)-1)-L1*(exp(L2*h)-1))/...
53                      ((L1-L2)*(exp(L1*h)-1)*(exp(L2*h)-1)) ;
54              else
55                  phi = (exp(L1*h)-1)^2/(L1^2*h*exp(L1*h)) ;
56                  tht = (1-exp(L1*h)+L1*h*exp(L1*h))...
57                      /(exp(L1*h)-1)^2 ;
58              end
59          end
60          if abs(L1) >= eps && abs(L2) < eps
61              phi = h ;
62              tht = (exp(L1*h)-1-L1*h)/(L1*h*(exp(L1*h)-1)) ;
63          end
64          if abs(L1) < eps && abs(L2) >= eps
65              phi = h ;
66              tht = (exp(L2*h)-1-L2*h)/(L2*h*(exp(L2*h)-1)) ;
67          end
68          if abs(L1) < eps && abs(L2) < eps
69              phi = h ;
70              tht = 0.5 ;
71          end
72      end
73  end
```

By executing the script, the approximation errors at $t_j, j = 0, \ldots, N$ are as follows:

| t | \|x(t)-xexact(t)\| | \|y(t)-yexact(t)\| |
|---|---|---|
| 0 | 0.0000000e+00 | 0.0000000e+00 |
| 1 | 5.5511151e-17 | 1.1102230e-16 |
| 2 | 5.5511151e-17 | 5.5511151e-17 |
| 3 | 8.3266727e-17 | 1.1102230e-16 |
| 4 | 5.5511151e-17 | 1.1102230e-16 |
| 5 | 5.5511151e-17 | 8.3266727e-17 |
| 6 | 4.1633363e-17 | 5.5511151e-17 |
| 7 | 2.7755576e-17 | 4.1633363e-17 |
| 8 | 4.1633363e-17 | 5.5511151e-17 |
| 9 | 3.4694470e-17 | 4.8572257e-17 |
| 10 | 3.4694470e-17 | 4.8572257e-17 |

The Python code of the script `SolveLinSysExact.py` is:

```python
1   import numpy as np
2   from sympy import Matrix
3   from scipy.linalg import expm
4   import matplotlib.pyplot as plt
5
6   def CompPhiThet(A, h):
7       eps = np.spacing(1.0)
8       P, J = Matrix(A).jordan_form()
9       P, J = np.array(P, float), np.array(J, float)
10      L1, L2 = J[0, 0], J[1, 1]
11      if isinstance(L1, float) and isinstance(L2, float):
12          if np.abs(L1) > eps and np.abs(L2) > eps:
13              if np.abs(L1-L2) > eps:
14                  phi = (L1-L2)*(np.exp(L1*h)-1)*(np.exp(L2*h)-1)\
15                      /(L1*L2*(np.exp(L1*h)-np.exp(L2*h)))
16                  tht = (L2*(np.exp(L1*h)-1)-L1*(np.exp(L2*h)-1))\
17                      /((L1-L2)*(np.exp(L1*h)-1)*(np.exp(L2*h)-1))
18              else:
19                  phi = (np.exp(L1*h)-1)^2/(L1^2*h*np.exp(L1*h))
20                  tht = (1-np.exp(L1*h)+L1*h*np.exp(L1*h))\
21                      /(np.exp(L1*h)-1)**2
22          if np.abs(L1) >= eps and np.abs(L2) < eps:
23              phi = h
24              tht = (np.exp(L1*h)-1-L1*h)/(L1*h*(np.exp(L1*h)-1))
25          if np.abs(L1) < eps and np.abs(L2) >= eps:
26              phi = h
27              tht = (np.exp(L2*h)-1-L2*h)/(L2*h*(np.exp(L2*h)-1))
28          if np.abs(L1) < eps and np.abs(L2) < eps:
29              phi = h
30              tht = 0.5
31      return phi, tht, P, J
32
33  a, b = 0.0, 10.0
34  A = np.array([[-3, 2], [1, -1]])
35  N = 10
36  t = np.linspace(a, b, N+1)
37  h = (b-a)/N
38  phi, tht, P, J = CompPhiThet(A, h)
39  z0 = np.array([1, 0.5])
40  z, u = np.zeros((2, N+1), float), np.zeros((2, N+1), float)
41  z[:, 0] = z0
42  u[:, 0] = np.linalg.solve(P, z[:, 0])
43  B = np.eye(2)-tht*phi*J ; C = np.eye(2)+(1-tht)*phi*J
44  for j in range(N):
45      u[:, j+1] = np.linalg.solve(B,C@u[:, j])
46      z[:, j+1] = P@u[:, j+1]
47
48  uex, zex = np.zeros((2, N+1), float), np.zeros((2, N+1), float)
49  zex[:, 0] = z0 # Exact solution of the linear system
50  uex[:, 0] = u[:, 0]
51  for j in range(N):
52      uex[:, j+1] = expm(J*t[j+1])@uex[:, 0]
53      zex[:, j+1] = P@uex[:, j+1]
54  Errors = np.abs(z-zex)
```

```
55  print('-----------------------------------------------------------') ;
56  print('      t\t  |x(t)-xexact(t)|\t  |y(t)-yexact(t)|') ;
57  print('-----------------------------------------------------------') ;
58  for j in range(len(t)):
59      print('{0:2.0f}'.format(t[j]), '\t',  ...
          '{0:8.7e}'.format(Errors[0, j]), '\t\t' \ ...
          '{0:8.7e}'.format(Errors[1, j]))
60  print('-----------------------------------------------------------') ;
61
62  plt.plot(t, z[0, :], '-b', label='x(t)', lw = 3)
63  plt.plot(t, z[1, :], '--m', label='y(t)', lw= 3)
64  plt.xlabel('t', fontweight='bold') ;
65  plt.xticks(np.linspace(a, b, 11), fontweight='bold')
66  plt.yticks(np.linspace(0, 1, 11), fontweight='bold')
67  plt.grid(True, ls=':')
68  plt.legend()
```

## 8.3.2   Exact Difference Schemes for Nonlinear Equations

In this section, exact finite difference schemes will be established to differential equations whose right-hand sides are of the form:

$$\frac{du}{dt} = -\alpha u^n(t), u(t_0) = u_0 \tag{8.8}$$

where $n \geq 2$ is a positive integer.

The exact solution of equation (8.8) is given by

$$u(t) = \sqrt[n-1]{\frac{1}{(n-1)(\alpha t + C)}}$$

Substituting the initial condition, gives

$$C = \frac{1 - (n-1)\alpha t_0 u_0^{n-1}}{(n-1)u_0^{n-1}}$$

Hence the exact solution is given by:

$$u(t) = \sqrt[n-1]{\frac{(n-1)u_0^{n-1}}{(n-1)\left(1 + \alpha(n-1)(t-t_0)u_0^{n-1}\right)}} \tag{8.9}$$

To derive the exact finite difference scheme of Equation (8.8), the substitutions:

$$t_0 \to t_k, t \to t_{k+1}, u_0 \to u_k \text{ and } u \to u_{k+1}$$

are used, with the notice that $t_{k+1} - t_k = h$. Then,

$$u_{k+1} = \sqrt[n-1]{\frac{(n-1)u_k}{(n-1)\left(1 + \alpha(n-1)hu_k^{n-1}\right)}}$$

Raising the two sides to power $n-1$ and dividing the numerator and denominator of the RHS by $n-1$ and doing few manipulations give the form:

$$u_{k+1}^{n-1} - u_k^{n-1} = -\alpha(n-1)hu_k^{n-1}u_{k+1}^{n-1}$$

or

$$(u_{k+1} - u_k)\left(u_{k+1}^{n-2} + u_{k+1}^{n-3}u_k + \cdots + u_{k+1}u_k^{n-3} + u_k^{n-2}\right) = -\alpha(n-1)hu_k^{n-1}u_{k+1}^{n-1},$$

from which the exact finite difference scheme

$$\frac{u_{k+1} - u_k}{h} = \frac{-\alpha(n-1)u_k^{n-1}u_{k+1}^{n-1}}{u_{k+1}^{n-2} + u_{k+1}^{n-3}u_k + \cdots + u_{k+1}u_k^{n-3} + u_k^{n-2}} \qquad (8.10)$$

For $n = 2$, the exact finite difference scheme for

$$\frac{du}{dt} = -\alpha u^2(t)$$

is

$$\frac{u_{k+1} - u_k}{h} = -\alpha u_k u_{k+1}$$

For $n = 3$, the exact finite difference scheme for

$$\frac{du}{dt} = -\alpha u^3(t)$$

is

$$\frac{u_{k+1} - u_k}{h} = \frac{-2\alpha u_k^2 u_{k+1}^2}{u_k + u_{k+1}}$$

For $n = 4$, the exact finite difference scheme for

$$\frac{du}{dt} = -\alpha u^4(t)$$

is

$$\frac{u_{k+1} - u_k}{h} = \frac{-3\alpha u_k^3 u_{k+1}^3}{u_k^2 + u_k u_{k+1} + u_{k+1}^2}$$

The Python script ExactFDMupown.py implements the exact finite difference scheme for the differential equation (8.8), with $u_0 = 0.5$ and shows the errors at the discrete points $t_j, j = 0, \ldots, 10$:

```
1   # ExactFDMupown.py
2   import numpy as np
3
4   def ExactSol(t, n, u0):
5       U = (n-1)*u0**(n-1)/((n-1)*(1+(n-1)*t*u0**(n-1)))
6       u = np.array([v**(1/(n-1)) for v in U])
7       return u
```

```
8
9   def ExactFDM(t, n, h, u0):
10      u = np.zeros_like(t)
11      u[0] = u0
12      for k in range(len(u)-1):
13          u[k+1] = ...
                ((n-1)*u[k]**(n-1)/((n-1)*(1+(n-1)*h*u[k]**(n-1)))) ...
                **(1/(n-1))
14      return u
15
16  a, b, N = 0.0, 10.0, 5
17  t = np.linspace(a, b, N+1)
18  h, u0 = (b-a)/N, 0.5
19  ns = [2, 4, 8, 10]
20  for n in ns:
21      uex = ExactSol(t, n, u0)
22      ufdm = ExactFDM(t, n, h, u0)
23      Errors = np.zeros((2, N+1, float)
24      Errors[0, :] = t
25      Errors[1, :] = np.abs(uex-ufdm)
26      Errors = Errors.T
27
28      print('Errors corresponding to n = '+str(n))
29      print('------------------------------')
30      print(' h  \t       Error')
31      print('------------------------------')
32      for j in range(len(Errors)):
33          print('{0:3.2f}'.format(Errors[j][0]) + '\t' + ...
                '{0:1.8e}'.format(Errors[j][1]))
34      print('------------------------------\n\n')
```

Executing the code gives the following results:

```
runfile('D:/PyFiles/ExactFDMupown.py', wdir='D:/PyFiles')
Errors corresponding to n = 2
------------------------------

t              Error
------------------------------

0.00    0.00000000e+00
2.00    0.00000000e+00
4.00    0.00000000e+00
6.00    0.00000000e+00
8.00    0.00000000e+00
10.00   1.38777878e-17
------------------------------


Errors corresponding to n = 5
------------------------------

t              Error
------------------------------

0.00    0.00000000e+00
2.00    0.00000000e+00
```

```
4.00    0.00000000e+00
6.00    0.00000000e+00
8.00    0.00000000e+00
10.00   0.00000000e+00
------------------------------
```

```
Errors corresponding to n = 8
------------------------------
   t            Error
------------------------------
0.00    0.00000000e+00
2.00    0.00000000e+00
4.00    5.55111512e-17
6.00    5.55111512e-17
8.00    5.55111512e-17
10.00   1.11022302e-16

------------------------------
```

```
Errors corresponding to n = 10
------------------------------
   t            Error
------------------------------
0.00    0.00000000e+00
2.00    0.00000000e+00
4.00    0.00000000e+00
6.00    5.55111512e-17
8.00    5.55111512e-17
10.00   5.55111512e-17

------------------------------
```

The code of the MATLAB script `ExactFDMupown.m` is:

```
1  a = 0.0 ; b = 10.0 ;
2  N = 5 ;
3  t = linspace(a, b, N+1) ;
4  h = (b-a)/N;
5  u0 = 0.5 ;
6  ns = [2, 4, 8, 10] ;
7  for n = ns
8      uex = ExactSol(t, n, u0) ;
9      ufdm = ExactFDM(t, n, h, u0) ;
10     Errors = zeros(2, N+1) ;
11     Errors(1, :) = t ;
12     Errors(2, :) = abs(uex-ufdm) ;
13     Errors = Errors' ;
14     fprintf('Errors corresponding to n = %i\n', n) ;
15     fprintf('-------------------------------\n') ;
16     fprintf(' t \t      Error\n') ;
```

```
17          fprintf('-----------------------------\n') ;
18          for j = 1 : N+1
19              fprintf('%2i\t%8.7e\n', Errors(j, 1), Errors(j, 2)) ;
20          end
21          fprintf('-----------------------------\n\n') ;
22  end
23
24  function u = ExactSol(t, n, u0)
25      U = (n-1)*u0^(n-1)./((n-1)*(1+(n-1)*t*u0^(n-1))) ;
26      u = U.^(1/(n-1)) ;
27  end
28
29  function u = ExactFDM(t, n, h, u0)
30      u = zeros(size(t)) ;
31      u(1) = u0 ;
32      for k = 1 : length(t)-1
33          u(k+1) = ...
                    ((n-1)*u(k)^(n-1)/((n-1)*(1+(n-1)*h*u(k)^(n-1)))) ...
                    ^(1/(n-1)) ;
34      end
35  end
```

### 8.3.3 Exact Finite Difference Schemes for Differential Equations with Linear and Power Terms

Given a differential equation:

$$\frac{du}{dt} = \alpha u(t) - \beta u^n(t), u(t_0) = u_0, \; n \geq 2 \tag{8.11}$$

To construct the exact finite difference scheme, the denominator function $\phi$ is derived from the linear term $\alpha u(t)$ as

$$\phi(h, \alpha) = \frac{1 - e^{-\alpha h}}{\alpha}$$

and the nonlocal approximation is used for the nonlinear term $-\beta u^n(t)$ as

$$\frac{-\beta(n-1)u_k^{n-1}u_{k+1}^{n-1}}{u_{k+1}^{n-2} + u_{k+1}^{n-3}u_k + \cdots + u_{k+1}u_k^{n-3} + u_k^{n-2}}$$

Then, the exact finite difference scheme for the differential equation (8.11) will be:

$$\frac{u_{k+1} - u_k}{\frac{1 - e^{-\alpha h}}{\alpha}} = \alpha u_k - \frac{-\beta(n-1)u_k^{n-1}u_{k+1}^{n-1}}{u_{k+1}^{n-2} + u_{k+1}^{n-3}u_k + \cdots + u_{k+1}u_k^{n-3} + u_k^{n-2}} \tag{8.12}$$

A special case is when $n = 2$ which yields the logistic equation

$$\frac{du}{dt} = \alpha u(t) - \beta u^2(t),$$

whose exact finite difference scheme is

$$\frac{u_{k+1} - u_k}{\frac{1-e^{-\alpha h}}{\alpha}} = \alpha u_k - \beta u_k u_{k+1}$$

from which:

$$u_{k+1} = \frac{(1+\phi\alpha)u_k}{1+\phi\beta u_k}$$

The Python script `ExactFDMLogistic.py` solves the logistic model, using an exact finite difference scheme:

```
import numpy as np
import matplotlib.pyplot as plt
a, b, N = 0.0, 50.0, 100
t = np.linspace(a, b, N+1)
h = (b-a)/N
u0 = 0.1

u = np.zeros_like(t)
u[0] = u0
alpha, beta = 0.25, 0.25 ;
phi = (1-np.exp(-alpha*h))/alpha ;
for k in range(N):
    u[k+1] = (1+phi*alpha)*u[k]/(1+phi*beta*u[k]) ;
MaxError = np.linalg.norm(u-uex, np.inf)
print('Maximum error = ', MaxError)
plt.plot(t, u, color='brown', lw=3)
plt.xlabel('Time (t)', fontweight = 'bold')
plt.ylabel('Population u(t)', fontweight = 'bold')
plt.grid(True, ls=':')
s = (b-a)/10.
plt.xticks(np.arange(a, b+s, s), fontweight = 'bold')
plt.yticks(np.arange(0., 1.1, 0.1), fontweight = 'bold')
```

By executing the code, the maximum error is shown and the solution of the model is explained in Figure 8.7.

```
runfile('D:/PyFiles/ExactFDMLogistic.py', wdir='D:/PyFiles')
Maximum error =  3.3306690738754696e-16
```

The MATLAB script `ExactFDMLogistic.m` applies an exact finite difference scheme to solve the logistic model:

```
%ExactFDMLogistic.m
a = 0.0 ; b = 50.0 ; N = 100 ;
t = linspace(a, b, N+1) ; h = (b-a)/N ;
u0 = 0.1 ;
u = zeros(1, N+1) ;
alpha = 0.25 ; beta = 0.25 ;
phi = (exp(alpha*h)-1)/alpha ;
u(1) = u0 ;
uex = zeros(1, N+1) ; uex(1) = u0 ;
```

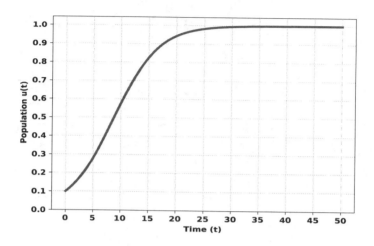

FIGURE 8.7: Solution of the logistic model, using exact finite difference scheme.

```
10   for k = 1 : N
11       u(k+1) = (1+phi*alpha)*u(k)/(1+phi*beta*u(k)) ;
12       uex(k+1) = ...
             alpha*u0/((alpha-beta*u0)*exp(-alpha*t(k+1))+beta*u0) ;
13   end
14
15   MaxError = norm(u-uex, inf) ;
16   disp(['Max Error = ' num2str(MaxError)]) ;
17
18   plot(t, u, 'r', 'LineWidth', 3) ;
19   xlabel('Time (t)') ;
20   ylabel('Population u(t)') ;
21   set(gca, 'fontweight', 'bold') ;
22   grid on ;
23   s = (b-a)/10 ;
24   set(gca, 'XTick', a:s:b) ;
25   set(gca, 'YTick', linspace(0, 1.1, 12)) ;
```

## 8.4   Other Nonstandard Finite Difference Schemes

In the case that an exact finite difference scheme cannot be obtained, then the purpose in constructing a nonstandard finite difference scheme is to establish the best possible finite difference scheme that solves the problem.

In [37] Mickens put the rules for selecting the denominator function and dealing with the nonlinear terms. Given an initial value problem,

$$\frac{du}{dt} = f(u), \ u(0) = u_0$$

To find the denominator function, the equilibria points of the differential equations are calculated by solving: $f(u) = 0$ for $u(t)$. Let $u_1, u_2, \ldots, u_n$ be the equilibria points of the differential equation. Let

$$r_k = \left[\frac{df}{du}\right]_{u=u_k} , k = 1, \ldots, n$$

and $r = \max\{|r_k|, k = 1, \ldots, n\}$. Then the denominator function is given by:

$$\phi(r, h) = \frac{1 - e^{-rh}}{r}$$

The linear and nonlinear terms at the right-hand side can be approximated using nonlocal approximations. For example a term $u(t)$ in the differential equation can be approximated at $t_k$ by $2u_k - u_{k+1}$, a term $u^2(t)$ in the differential equation can be approximated at $t_k$ by $2u_k^2 - u_k u_{k+1}$, etc.

**Example 8.4** This example is taken from [37] and constructs a nonstandard finite difference scheme for the differential equation:

$$\frac{du}{dt} = u^2(t) - u^3(t), u(0) = u_0$$

The equilibria points of the model are $u_1 = u_2 = 0$ and $u_3 = 1$. The denominator function is $\phi(h) = 1 - e^{-h}$. The term $u^2(t)$ is approximated at $t_k$ by $2u_k^2 - u_k u_{k+1}$ and the nonlinear term $-u^3(t)$ is approximated by $-u_k^2 u_{k+1}$. The nonstandard finite difference scheme is:

$$\frac{u_{k+1} - u_k}{\phi} = 2u_k^2 - u_k u_{k+1} - u_k^2 u_{k+1}$$

which after few manipulations gives:

$$u_{k+1} = \frac{(1 + 2\phi u_k) u_k}{1 + \phi(u_k + u_k^2)}$$

The Python script `MickPopModel.py` implements the nonstandard finite difference scheme of Mickens and plots the solution:

```
1   import numpy as np
2   import matplotlib.pyplot as plt
3   a, b, N = 0.0, 25.0, 100
```

```
4   t = np.linspace(a, b, N+1)
5   h = (b-a)/N
6   u0 = 0.1
7
8   u = np.zeros_like(t)
9   u[0] = u0
10  phi = 1-np.exp(-h)
11  for k in range(N):
12      u[k+1] = (1+2*phi*u[k])*u[k]/(1+phi*(u[k]+u[k]**2))
13  plt.plot(t, u, color='orangered', lw=3)
14  plt.xlabel('t', fontweight = 'bold')
15  plt.ylabel('u(t)', fontweight = 'bold')
16  plt.grid(True, ls=':')
17  s = (b-a)/10.
18  plt.xticks(np.arange(a, b+s, s), fontweight = 'bold')
19  plt.yticks(np.arange(0., 1.1, 0.1), fontweight = 'bold')
```

By executing the code Figure 8.8 is obtained.

The MATLAB code is

```
1   a = 0.0 ; b = 25.0 ; N = 100 ;
2   t = linspace(a, b, N+1) ; h = (b-a)/N ;
3   u0 = 0.1 ;
4   u = zeros(1, N+1) ;
5   phi = 1-exp(-h) ;
6   u(1) = u0 ;
```

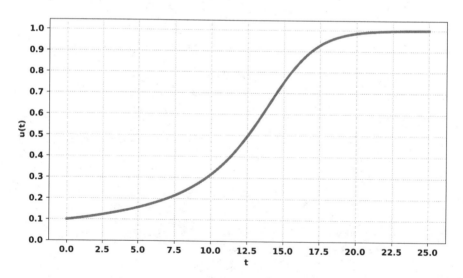

FIGURE 8.8: Solution of the model $u'(t) = u^2(t) - u^3(t)$ using Micken's nonstandard finite difference scheme.

```
7   for k = 1 : N
8       u(k+1) = (1+2*phi*u(k))*u(k)/(1+phi*(u(k)+u(k)^2)) ;
9   end
10
11  plot(t, u, 'r', 'LineWidth', 3) ;
12  xlabel('Time (t)') ;
13  ylabel('Population u(t)') ;
14  set(gca, 'fontweight', 'bold') ;
15  grid on ;
16  s = (b-a)/10 ;
17  set(gca, 'XTick', a:s:b) ;
18  set(gca, 'YTick', linspace(0, 1.1, 12)) ;
```

# Part III

# Solving Linear, Nonlinear and Dynamic Optimization Problems

# 9

## Solving Optimization Problems: Linear and Quadratic Programming

### Abstract

Linear programming is one of the most important methods of mathematical programming and most applied in practical life to ensure optimal use of limited resources. Such examples include the optimal mix of the products produced by a particular plant to achieve the maximum profit according to the available labors and raw materials. Also, moving certain products from production areas to consumption centers so that each consumer center satisfies its demand at the lowest possible cost.

This chapter is organized as follows. The first section discusses the form of a linear programming problem. The second and third sections discuss the solutions of linear programming problems using the built-in functions in MATLAB® and Python. Section 4 uses the Python pulp package for solving linear programming problems. The package Pyomo is another Python package for solving linear programming problems, and is discussed in Section 5. Section 6 uses the gekko Python package for solving linear programming problems. Section 7 is devoted to solving quadratic programming problems using many MATLAB and Python packages.

## 9.1 Form of a Linear Programming Problem

The general form of the linear programming problem is:

$$\min \ \alpha_1 x_1 + \alpha_2 x_2 + \cdots + \alpha_n x_n \tag{9.1}$$

subject to equality constraints:

$$
\begin{aligned}
a_{11}x_1 + a_{12}x_2 + \ldots + a_{1n}x_n &= b_1 \\
a_{21}x_1 + a_{22}x_2 + \ldots + a_{2n}x_n &= b_2 \\
a_{31}x_1 + a_{32}x_2 + \ldots + a_{3n}x_n &= b_3 \\
&\vdots \qquad \vdots \\
a_{m1}x_1 + a_{m2}x_2 + \ldots + a_{mn}x_n &= b_m
\end{aligned}
\tag{9.2}
$$

inequality constraints

$$
\begin{aligned}
c_{11}x_1 + c_{12}x_2 + \ldots + c_{1n}x_n &= d_1 \\
c_{21}x_1 + c_{22}x_2 + \ldots + c_{2n}x_n &= d_2 \\
c_{31}x_1 + c_{32}x_2 + \ldots + c_{3n}x_n &= d_3 \\
&\vdots \qquad \vdots \\
c_{l1}x_1 + c_{l2}x_2 + \ldots + c_{ln}x_n &= d_l
\end{aligned}
\tag{9.3}
$$

and box constraints:

$$
x_1 \in (X_1^{min}, X_1^{max}), \ldots, x_n \in (X_n^{min}, X_n^{max})
\tag{9.4}
$$

It is worthy to notice that an inequality constraint of the form:

$$
\sum_{j=1}^{n} \beta_j x_j \geq \gamma_j
$$

is equivalent to the inequality constraint:

$$
-\sum_{j=1}^{n} \beta_j x_j \leq -\gamma_j,
$$

therefore, it is convenient to always write any inequality constraint in the form

$$
LHS \leq RHS.
$$

The linear programming problem described by equations (9.1)-(9.4) can be written in the matrix form as:

$$
\min \boldsymbol{\alpha}^T \boldsymbol{x}
\tag{9.5}
$$

subject to constraints

$$
\begin{aligned}
A\boldsymbol{x} &= \boldsymbol{b} & \text{(9.6)} \\
C\boldsymbol{x} &\leq \boldsymbol{d} & \text{(9.7)} \\
\boldsymbol{x} &\in \boldsymbol{X} & \text{(9.8)}
\end{aligned}
$$

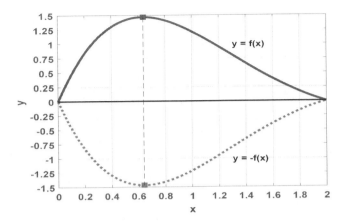

FIGURE 9.1: Graphs of the functions $f(x) = e^{-(x-0.5)} \sin\left(\frac{\pi x}{2}\right)$ and $-f(x)$, which show that the value of $x$ that maximizes $f(x)$ minimizes $-f(x)$.

where,

$$A = \begin{bmatrix} a_{11} & a_{12} & \cdots & a_{1n} \\ a_{21} & a_{22} & \cdots & a_{2n} \\ \vdots & \vdots & \ddots & \vdots \\ a_{m1} & a_{ms} & \cdots & a_{mn} \end{bmatrix}, x = \begin{bmatrix} x_1 \\ x_2 \\ \vdots \\ x_n \end{bmatrix}, b = \begin{bmatrix} b_1 \\ b_2 \\ \vdots \\ b_m \end{bmatrix},$$

$$C = \begin{bmatrix} c_{11} & c_{12} & \cdots & c_{1n} \\ c_{21} & c_{22} & \cdots & c_{2n} \\ \vdots & \vdots & \ddots & \vdots \\ c_{l1} & c_{ls} & \cdots & c_{ln} \end{bmatrix}, d = \begin{bmatrix} d_1 \\ d_2 \\ \vdots \\ d_l \end{bmatrix}, X = \begin{bmatrix} (X_1^{min}, X_1^{max}) \\ (X_2^{min}, X_2^{max}) \\ \vdots \\ (X_n^{min}, X_n^{max}) \end{bmatrix}$$

The problem of maximizing some objective function $f(x)$ is equivalent to the problem of minimizing $-f(x)$, as the maximization problem of $f(x)$ and the minimization problem of $-f(x)$ have the same optimal solution $x^{opt}$. In Figure 9.1 the graphs of the functions $f(x) = e^{-(x-0.5)} \sin\left(\frac{\pi x}{2}\right)$ and $-f(x)$ are plotted. From the graphs of the two functions, it is clear that the maximum value of $f(x)$ is obtained at the same point that minimizes $-f(x)$.

Then, the problem:

$$\max \; \alpha^T x$$

subject to constraints

$$Ax = b$$
$$Cx \leq d$$
$$x \in X$$

is equivalent to the problem:

$$\min \; \beta^T x$$

subject to constraints

$$Ax = b$$
$$Cx \leq d$$
$$x \in X$$

$(\beta = -\alpha)$ without further change in the equality or inequality constraints.

## 9.2   Solving Linear Programming Problems with `linprog`

Considering the problem:

$$\min_{x \in \mathbb{R}^n} \alpha^T x,$$

subject to constraints

$$Ax = b,$$
$$Cx \leq d,$$
$$\text{lb} \leq x \leq \text{ub},$$

the MATLAB function `linprog` can be used for solving linear programming problems. It is of the form:

`[xopt, fval] = linprog(objfun, C, d, A, b, lb, ub)`

where x0 is an initial starting point. If the problem does not contain lower and upper bounds, then the form is:

`[xopt, fval] = linprog(objfun, C, d, A, b)`

If the problem does not contain equality constraints, then A and b are replaced by empty squared brackets []. The form will be:

`[xopt, fval] = linprog(objfun, C, d, [], [], lb, ub)`

To show how to use the function `linprog`, the following example is considered:

**Example 9.1** In this example, MATLAB's function `linprog` will be used to solve the linear programming problem:

$$\max \quad 3x_1 + x_2 + 2x_3$$

subject to:

$$3x_1 + x_2 \leq 40$$
$$x_1 + 2x_3 \leq 60$$
$$x_2 + 2x_3 \leq 60$$
$$x_1 \geq 0, x_2 \geq 0, x_3 \geq 0$$

The following MATLAB instructions are used to find the optimal solution:

```
>> objfun = [-3; -1; -2] ;
>> C = [3, 1, 0; 1, 0, 2; 0, 1, 2] ;
>> d = [40; 60; 60] ;
>> lb = [0; 0; 0] ;
>> [xopt, fval] = linprog(objfun, C, d, [], [], lb, [])

Optimal solution found.


xopt =

10
10
25


fval =

-90
```

The MATLAB `linprog` function uses by default the *dual-simplex* algorithm. The choice of the algorithm can be changed through the `optimoptions` struct. For example to switch to the `interior-point` solver, and solve Example 9.1 the following instructions can be used [17].

```
>> Options = optimoptions('linprog', 'Algorith', 'interior-point');
>> [xopt, fval] = linprog(objfun, C, d, [], [], lb, [], Options)

Minimum found that satisfies the constraints.

Optimization completed because the objective function is non-
decreasing in feasible directions, to within the selected value
of the function tolerance, and constraints are satisfied to
within the selected value of the constraint tolerance.

xopt =
    10.0000
    10.0000
    25.0000

fval =
    -90.0000
>> Options = optimoptions('linprog', 'Algorith',
    'interior-point-legacy', 'Disp', 'Iter') ;
>> [xopt, fval] = linprog(objfun, C, d, [], [], lb, [], Options)
```

```
Optimization terminated.

xopt =
    10.0000
    10.0000
    25.0000

fval =
    -90.0000
```

Python has also a function `linprog` located in the `scipy.optimize` library. To use the `linprog` function, it shall imported from the `optimize` library of `scipy` [18]. This can be done as follows:

In [1]: from scipy.optimize import linprog

The Python function `linprog` form is close to the MATLAB's `linporg` form. Its syntax is:

OptSol = linprog(objfun, A_ub = C, b_ub = d, A_eq = A, b_eq = b, bounds = bnds,\ method='optmethod', options = optoptions)

To solve Example 9.1 with Python, the following Python instructions can be used:

```
In [1]: import numpy as np
In [2]: import scipy.optimize as opt
In [3]: objfun = np.array([-3, -1, -2])
In [4]: C = np.array([[3, 1, 0], [1, 0, 2], [0, 1, 2]])
In [5]: d = np.array([40, 60, 60])
In [6]: bnds = [(0., None), (0., None), (0., None)]
In [7]: OptSol = opt.linprog(objfun, A_ub = C, b_ub = d, bounds = bnds, \
options = ({"disp":True}))
```

| Primal Feasibility | Dual Feasibility | Duality Gap | Step | Path Parameter | Objective |
|---|---|---|---|---|---|
| 1.0 | 1.0 | 1.0 | – | 1.0 | -6.0 |
| 0.1157362093423 | 0.1157362093423 | 0.1157362093423 | 0.8915842403063 | 0.1157362093423 | -31.98924052603 |
| 0.01711151690354 | 0.0171115169036 | 0.0171115169036 | 0.8667218148033 | 0.01711151690367 | -76.59885574796 |
| 0.0001832497415752 | 0.0001832497416004 | 0.0001832497416007 | 0.9929871273485 | 0.0001832497416554 | -89.85504503568 |
| 9.462525147506e-09 | 9.462525023246e-09 | 9.462524985793e-09 | 0.9999484444422 | 9.462524789912e-09 | -89.99999255556 |

```
Optimization terminated successfully.
Current function value: -89.999993
Iterations: 4

In [8]: print(OptSol)
    con: array([], dtype=float64)
    fun: -89.99999255555932
    message: 'Optimization terminated successfully.'
    nit: 4
    slack: array([2.26564158e-06, 5.24875590e-06, 7.23457025e-06])
    status: 0
    success: True
    x: array([ 9.99999993,  9.99999794, 24.99999741])
```

The default solver of the `linprog` function is the *interior-point* method. But there are other two options of the method, which are the **simplex** and **revised simplex** methods. They are used as follows:

```
In [9]: OptSol = opt.linprog(objfun, A_ub = C, b_ub = d, bounds = bnds,\
    method='simplex', options = ({"disp":True}))
    Optimization terminated successfully.
    Current function value: -90.000000
    Iterations: 3

In [10]: print(OptSol)
    con: array([], dtype=float64)
    fun: -90.0
    message: 'Optimization terminated successfully.'
    nit: 3
    slack: array([0., 0., 0.])
    status: 0
    success: True
    x: array([10., 10., 25.])
```

```
In [11]: OptSol = opt.linprog(objfun, A_ub = C, b_ub = d, bounds = bnds,\
    method='revised simplex', options = ({"disp":True}))
```

| Phase | Iteration | Minimum Slack | Constraint Residual | Objective |
|-------|-----------|---------------|---------------------|-----------|
| 1 | 0 | 40.0 | 0.0 | 0.0 |

| Phase | Iteration | Minimum Slack | Constraint Residual | Objective |
|-------|-----------|---------------|---------------------|-----------|
| 2 | 0 | 40.0 | 0.0 | 0.0 |
| 2 | 1 | 0.0 | 0.0 | -40.0 |
| 2 | 2 | 0.0 | 0.0 | -86.66666666667 |
| 2 | 3 | 0.0 | 0.0 | -90.0 |

```
    Optimization terminated successfully.
    Current function value: -90.000000
    Iterations: 3

In [12]: print(OptSol)
    con: array([], dtype=float64)
    fun: -90.0
    message: 'Optimization terminated successfully.'
    nit: 3
    slack: array([0., 0., 0.])
    status: 0
    success: True
    x: array([10., 10., 25.])
```

## 9.3 Solving Linear Programming Problems with fmincon MATLAB's Functions

The MATLAB function fmincon can be used to solve a linear programming problem. To solve Example 9.1 with the *fmincon* function, a function handle to the objective function (or a user-defined objective function) shall be initiated.

Also, an initial guess x0, the linear inequality and equality constraints and the variables bounds are passed to the fmincon function.

The following MATLAB instructions can be used to solve Example 9.1 using the fmincon function:

```
>> f = @(x) objfun'*x ;
>> x0 = [0.; 0.; 0.] ;
>> C = [3, 1, 0; 1, 0, 2; 0, 1, 2] ;
>> d = [40; 60; 60] ;
>> lb = [0; 0; 0] ;
>> [xopt, fval] = fmincon(f, [0.;0; 0.], C, d, [], [], lb, [])

Local minimum found that satisfies the constraints.

Optimization completed because the objective function is non-
decreasing in feasible directions, to within the default value
of the optimality tolerance, and constraints are satisfied to
within the default value of the constraint tolerance.

<stopping criteria details>

xopt =
10.0000
10.0000
25.0000


fval =
-90.0000
```

## 9.4    Solving Linear Programming Problems with pulp Python

Python possesses a library pulp to solve linear programming problems [41]. Installing pulp is easy and can be done by typing:

```
>>> pip install pulp
```

In anaconda it can be installed by typing:

```
In [13]: conda install -c conda-forge pulp
```

To use pulp, the pulp library must be imported first:

```
In [14]: import pulp as plp
```

Then, some variable (for example LPP) shall be used to define the problem and solve it. pulp enables the user to name his/her problem and defining its type (*maximization, minimization*) through the function LpProblem. To define a maximization problem the parameter LpMaximize shall be passed as a second argument to LpProblem, and to define a minimization problem the parameter LpMinimize shall be passed. For example:

```
In [15]: LPP = plp.LpProblem("Problem of maximizing profit",
    LpMaximize)
```

The variables of the problem shall be defined by using the **pulp** function LpVariable. It receives a string and bounds of the variable. For example if the problem contains a variable $0 \leq x \leq 5$:

```
x = plp.LpVariable("x", 0., 5.)
```

Those variables can be used with problem instance LPP.

Next, the equality and inequality constraints of the problem are added to the problem instance. If the problem contains an inequality constraint $2x+y \leq 10$ then this constraint is added to the problem using an instruction:

```
LPP += 2*x + y <= 10
```

In pulp the user does not have to change the forms of the inequality constraints. if the problem contains a constraint $x+y \geq 2$, it can be added to the problem instance directly without transforming it to other form:

```
LPP += x + y >= 2
```

Finally, after defining the whole problem, the solve() method of the LPP instance can be used to solve the problem and displaying the results.

The Python script SolveExLinProg.py is used to solve the problem of Example 9.1 and show the results:

```
 1  # SolveExLinProg.py
 2  import pulp as plp
 3  LPP = plp.LpProblem('Problem statement: \n', plp.LpMaximize)
 4  x, y, z = plp.LpVariable("x", lowBound=0.), ...
        plp.LpVariable("y", lowBound=0.), plp.LpVariable("z", ...
        lowBound=0.)
 5  LPP += 3*x + y + 2*z
 6  LPP += 3*x + y <= 40
 7  LPP += x + 2*z <= 60
 8  LPP += y + 2*z <= 60
 9  LPP += x >= 0.
10  LPP += y >= 0.
11  LPP += z >= 0.
12  print(LPP)
13  status = LPP.solve()
14  print('Status of the optimization process: ' + ...
        plp.LpStatus[status])
```

```
15  print('Optimal maximizer found at: \n', plp.value(x), '\n', ...
        plp.value(y), '\n', plp.value(z))
16  print('Optimal value of the objective function: ', ...
        plp.value(LPP.objective))
```

Executing the code gives the following results:

```
runfile('D:/PyFiles/SolveExLinProg.py', wdir='D:/PyFiles')
Problem statement:
:
MAXIMIZE
3*x + 1*y + 2*z + 0
SUBJECT TO
_C1: 3 x + y <= 40

_C2: x + 2 z <= 60

_C3: y + 2 z <= 60

_C4: x >= 0

_C5: y >= 0

_C6: z >= 0

VARIABLES
x Continuous
y Continuous
z Continuous

Status of the optimization process: Optimal
Optimal maximizer found at:
10.0
10.0
25.0
Optimal value of the objective function:  90.0
```

## 9.5  Solving Linear Programming Problems with pyomo

Pyomo is a software based on Python to model and solve optimization problems
[23]. There are two kinds of modelling presented by pyomo: *concrete models*
where data are specified first before model construction and *abstract models*
where the model is constructed first before data specification and the model

data are passed to the model during the runtime of the optimization process. This section presents the creation of concrete models for solving linear programming problems.

In anaconda, *pyomo* can be installed by using the command:

```
conda install -c conda-forge pyomo
```

It is important to be sure that the targeted optimization solver is installed before proceeding to the use of pyomo. For example, if the 'glpk' solver is not installed, it can be installed by typing the command:

```
conda install -c conda-forge glpk
```

The first step for solving a linear programming problem is to prepare the pyomo environment and solver, through importing both. This can be done by:

```
import pyomo as pym
from pyomo.opt import SolverFactory
```

Then a variable of either type *ConcreteModel* or *AbstractModel* is constructed by using:

```
Model = pym.ConcreteModel()
```

or

```
Model = pym.AbstractModel()
```

If for example $1 \leq x \leq 5$ and $0 \leq y < \infty$ are two variables of the optimization problem, they can be declared by using:

```
Model.x = pym.Var(bounds=(1., 5.))
Model.y = pym.Var(within=NonNegativeReals)
```

The objective function is defined by using the **pym.Objective** method which receives the mathematical expression of the objective function (for example $10\ x + y/100$):

```
Model.objfun = pym.Objective(expr = 10.*Model.x+Model.y/100)
```

If the problem contains a constraint $10x + y \leq 100$, this constraint can be added to the model by typing:

```
Model.con1 = pym.Constraint(expr=10*Model.x + Model.y <= 100)
```

After completely defining the model, an optimization solver is used to solve the problem defined by the model.

To solve Example 9.1 using pyomo, a Python script `SolveEx0WithPyomo.py` is used. Its code is:

```
1  import pyomo.environ as pym
2  from pyomo.opt import SolverFactory
3  LPP_model = pym.ConcreteModel()
4  LPP_model.x, LPP_model.y, LPP_model.z = pym.Var(bounds=(0., ...
       None)), pym.Var(bounds=(0., None)), pym.Var(bounds=(0., None))
```

```
5  LPP_model.objfun = ...
       pym.Objective(expr=3*LPP_model.x+LPP_model.y+2*LPP_model.z, ...
       sense=maximize)
6  LPP_model.con1 = pym.Constraint(expr=3*LPP_model.x + ...
       LPP_model.y <= 40)
7  LPP_model.con2 = pym.Constraint(expr=LPP_model.x + ...
       2*LPP_model.z <= 60)
8  LPP_model.con3 = pym.Constraint(expr=LPP_model.y + ...
       2*LPP_model.z <= 60)
9  opt = SolverFactory('glpk')
10 OptResults = opt.solve(LPP_model)
11 print('\nStatus of optimization process: ', ...
       OptResults.solver.Status, '\n')
12 print('Status of termination condition: ', ...
       OptResults.solver.termination_condition, '\n')
13 print('Optimal solution obtained at: \n x\t', ...
       LPP_model.x.value, '\n y   = ', LPP_model.y.value, '\n ...
       z\t', LPP_model.z.value, '\n')
14 print('Value of the objective function at optimal solution: ', ...
       LPP_model.objfun())
```

By executing the code, the following results are obtained:

```
Status of optimization process:   ok

Status of termination condition:  optimal

Optimal solution obtained at:
x        10.0
y   =    10.0
z        25.0

Value of the objective function at optimal solution:  90.0
```

## 9.6   Solving Linear Programming Problems with gekko

Optimization problems are solved in gekko at mode 3 [2]. To solve Example 9.1 with gekko, the Python script SolveEx2LPWithGekko.py is used:

```
1  from gekko import GEKKO
2  m = GEKKO()
3  x = m.Var(1., 0, None)
4  y = m.Var(1., 0, None)
5  z = m.Var(1., 0, None)
6  m.Obj(-3*x-y-2*z)
7  m.Equation(3*x + y <= 40)
8  m.Equation(x + 2*z <= 60)
9  m.Equation(y + 2*z <= 60)
```

```
10   m.options.IMODE = 3
11   xopt=m.solve()
12   print('Solution found at \n')
13   print('{0:10.7f}'.format(x[0]))
14   print('{0:10.7f}'.format(y[0]))
15   print('{0:10.7f}'.format(z[0]))
16   print('Value of the objective function: ', ...
            '{0:10.7f}'.format(3*x[0]+y[0]+2*z[0]))
```

Executing the code gives:

```
EXIT: Optimal Solution Found.

The solution was found.

The final value of the objective function is    -89.9999999911064


 ----------------------------------------------------
 Solver          :  IPOPT (v3.12)
 Solution time   :   9.100000001126318E-003 sec
 Objective       :   -89.9999999911064
 Successful solution
 ----------------------------------------------------


Solution found at

10.0000000
10.0000000
25.0000000
Value of the objective function:  90.0000000
```

## 9.7   Solving Quadratic Programming Problems

A quadratic programming problem is an optimization problem whose objective function is quadratic and constraints are linear. A general form of a quadratic programming problem is:

$$\min \frac{1}{2}\boldsymbol{x}^T \boldsymbol{H} \boldsymbol{x} + \boldsymbol{\alpha}^T \boldsymbol{x}, \ \boldsymbol{H} \in \mathbb{R}^{n \times n}, \ \boldsymbol{\alpha} \text{ and } \boldsymbol{x} \in \mathbb{R}^n \qquad (9.9)$$

together with the linear equality constraints described by (9.2) and linear inequality constraints described by Equation (9.3).

Through this section, solution of quadratic programming problem described in the following example will be considered:

## Example 9.2

$$\min \ x_1^2 + 4x_2^2 + 4x_3^2 + x_4^2 + 4x_1x_3 + 4x_2x_4 + 2x_1 + x_2 + x_3 + 2x_4$$

subject to the inequality constraints:

$$
\begin{aligned}
x_1 \qquad\qquad + x_4 &\geq 10 \\
2x_1 + x_2 + x_3 + 2x_4 &\geq 24 \\
x_2 + x_3 \qquad &\geq 20 \\
x_1 + x_2 + x_3 + x_4 &\leq 30 \\
0 \leq x_1, x_2, x_3, x_4 &\leq 20
\end{aligned}
$$

MATLAB has a function quadprog for solving quadratic programming problems. The function fmincon can also be used for solving quadratic problems.

## Solution with quadprog MATLAB's function

The MATLAB's function quadprog receives the matrix $H$, the vector $\alpha$, matrix of inequality constraints $C$ and its right-hand side vector $d$, matrix of equality constraints $A$ and its right-hand side $b$, and vectors of lower bounds $l_b$ and upper bounds $u_b$. It is of the form:

[xopt, fval] = quadprog(H, alpha, C, d, A, b, lb, ub) ;

If any of the inputs is missing, it can be replaced by blank squared brackets []. The MATLAB script SolveQuadProg.m is used to solve Example 9.2.

```
1  % SolveQuadProg.m
2  clear ; clc ;
3  H = [2, 0, 4, 0; 0, 8, 0, 4; 4, 0, 8, 0; 0, 4, 0, 2] ;
4  alp = [2; 1; 1; 2] ;
5  C = [-1, -0, -0, -1; -2, -1, -1, -2; 0, -1, -1, 0; 1, 1, 1, 1] ;
6  d = [-10; -24; -20; 30] ;
7  lb = [0; 0; 0; 0] ;
8  ub = [20; 20; 20; 20] ;
9  [xopt, fval] = quadprog(H, alp, C, d, [], [], lb, ub) ;
10 fprintf('Optimal solution found at: \n') ; disp(xopt) ;
11 fprintf('Objective function at optimal point: %10.7f\n', fval) ;
```

Executing the codes gives:

Minimum found that satisfies the constraints.

Optimization completed because the objective function is non-decreasing in feasible directions, to within the default value of the optimality tolerance, and constraints are satisfied to within the default value of the constraint tolerance.

Optimal solution found at:
5.0000
10.0000
10.0000
5.0000

Objective function at optimal point: 1290.0000000

## Solution with the MATLAB's function fmincon

The MATLAB's script SolveQPWithfmincon.m is used to solve the Example 9.2.

```
1  clear ; clc ;
2  H = [2, 0, 4, 0; 0, 8, 0, 4; 4, 0, 8, 0; 0, 4, 0, 2] ;
3  alp = [2; 1; 1; 2] ;
4  objfun = @(x) 0.5*x'*H*x + alp'*x ;
5  C = [-1, -0, -0, -1; -2, -1, -1, -2; 0, -1, -1, 0; 1, 1, 1, 1] ;
6  d = [-10; -24; -20; 30] ;
7  lb = [0; 0; 0; 0] ;
8  ub = [20; 20; 20; 20] ;
9  x0 = [1; 1; 1; 1] ;
10 [xopt, fval] = fmincon(objfun, x0, C, d, [], [], lb, ub) ;
11 fprintf('Optimal solution found at: \n') ; disp(xopt) ;
12 fprintf('Objective function at optimal point: %10.7f\n', fval) ;
```

Executing the code gives:

Local minimum found that satisfies the constraints.

Optimization completed because the objective function is non-decreasing in feasible directions, to within the default value of the optimality tolerance, and constraints are satisfied to within the default value of the constraint tolerance.

<stopping criteria details>

Optimal solution found at:
5.0000
10.0000
10.0000
5.0000

Objective function at optimal point: 1290.0000000

## Solution with gekko Python

The Python script `SolveQPWithGekko.py` solves the problem of Example 9.2:

```
1  from gekko import GEKKO
2  m = GEKKO()
3  x1 = m.Var(1., 0, 20)
4  x2 = m.Var(1., 0, 20)
5  x3 = m.Var(1., 0, 20)
6  x4 = m.Var(1., 0, 20)
7
8  m.Obj(x1**2 + 4*x2**2+ 4*x3**2 + x4**2 + 4*x1*x3 + 4*x2*x4 + ...
        2*x1 + x2   + x3  + 2*x4)
9  m.Equation(x1+ x4 >= 10)
10 m.Equation(2*x1 + x2 + x3 + 2*x4 >= 24)
11 m.Equation(x2 + x3 >= 20)
12 m.Equation(x1 + x2 + x3 + x4 <= 30)
13
14 m.options.IMODE = 3
15 xopt=m.solve()
16 print('Solution found at \n')
17 print('{0:10.7f}'.format(x1[0]))
18 print('{0:10.7f}'.format(x2[0]))
19 print('{0:10.7f}'.format(x3[0]))
20 print('{0:10.7f}'.format(x4[0]))
21 xopt = x1[0]**2 + 4*x2[0]**2+ 4*x3[0]**2 + x4[0]**2 + ...
        4*x1[0]*x3[0] + 4*x2[0]*x4[0] + 2*x1[0] + x2[0] + x3[0]  + ...
        2*x4[0]
22 print('Value of the objective function: ', ...
        '{0:10.7f}'.format(xopt))
```

Results of execution:

```
EXIT: Optimal Solution Found.

The solution was found.

The final value of the objective function is    1289.99999909229

----------------------------------------------------
Solver         :  IPOPT (v3.12)
Solution time  :   1.630000000295695E-002 sec
Objective      :   1289.99999909229
Successful solution
----------------------------------------------------

Solution found at

4.9999999
10.0000000
```

```
10.0000000
5.0000000
Value of the objective function:   1289.9999991
```

## Solution with pyomo Python

The GNU Linear Programming Kit (glpk) solver cannot be used for solving a quadratic programming problem in pyomo. Since pyomo does not include any solvers, it is important that the user installs a suitable solver that is compatible with pyomo. Such an example is the IPOPT solver. To install the IPOPT solver in anaconda, the following command can be used:

```
conda install -c conda-forge ipopt
```

The Python script SolveQPWithpyomo.py solves Example 9.2:

```
 1  import pyomo.environ as pym
 2  from pyomo.opt import SolverFactory
 3  m = pym.ConcreteModel()
 4  m.x1, m.x2, m.x3, m.x4 = pym.Var(bounds=(0., 20)), ...
        pym.Var(bounds=(0., 20)), pym.Var(bounds=(0., 20)), ...
        pym.Var(bounds=(0., 20))
 5  m.objfun = pym.Objective(expr=m.x1**2 + 4*m.x2**2+ 4*m.x3**2 + ...
        m.x4**2 + 4*m.x1*m.x3 + 4*m.x2*m.x4 + 2*m.x1 + m.x2   + ...
        m.x3   + 2*m.x4)
 6  m.con1 = pym.Constraint(expr=m.x1+ m.x4 ≥ 10)
 7  m.con2 = pym.Constraint(expr=2*m.x1 + m.x2 + m.x3 + 2*m.x4 ≥ 24)
 8  m.con3 = pym.Constraint(expr=m.x2 + m.x3 ≥ 20)
 9  m.con4 = pym.Constraint(expr=m.x1 + m.x2 + m.x3 + m.x4 ≤ 30)
10  opt = SolverFactory('ipopt')
11  OptResults = opt.solve(m)
12  print('\nStatus of optimization process:  ', ...
        OptResults.solver.Status, '\n')
13  print('Status of termination condition: ', ...
        OptResults.solver.termination_condition, '\n')
14  print('Optimal solution obtained at: \n x\t', \
15  '{0:8.6f}'.format(m.x1.value), '\n y   = ', ...
        '{0:8.6f}'.format(m.x2.value), '\n z\t', ...
        '{0:8.6f}'.format(m.x3.value),\
16  '\n z\t', '{0:8.6f}'.format(m.x4.value), '\n')
17  print('Value of the objective function at optimal solution: ', ...
        '{0:8.6f}'.format(m.objfun()))
```

Executing the code gives the results:

```
Status of optimization process:   ok

Status of termination condition:  optimal
```

```
Optimal solution obtained at:
x          5.000000
y    =    10.000000
z         10.000000
z          5.000000
```

Value of the objective function at optimal solution: 1289.999975

# 10

## Solving Optimization Problems: Nonlinear Programming

### Abstract

Nonlinear programming problems are divided into two kinds of problems: unconstrained and constrained [19, 20]. An unconstrained optimization problem is a problem of the form

$$\min f(\boldsymbol{x}), \quad \boldsymbol{x} = [x_1, x_2, \ldots, x_n]^T \in \mathbb{R}^n \qquad (10.1)$$

A general constrained optimization problem consists of the objective function in Equation (10.1) with equality constraints:

$$\boldsymbol{E}(\boldsymbol{x}) = 0, \qquad (10.2)$$

and inequality constraints:

$$\boldsymbol{I}(\boldsymbol{x}) \leq 0 \qquad (10.3)$$

where $\boldsymbol{x} \in \mathbb{R}^n$, $\boldsymbol{E}(\boldsymbol{x}) \in \mathbb{R}^p$ and $\boldsymbol{I}(\boldsymbol{x}) \in \mathbb{R}^q$.

In this chapter, methods for solving both unconstrained and constrained optimization problems using MATLAB® and Python will be implemented.

## 10.1 Solving Unconstrained Problems

In this section, problem (10.1) is considered. If $\boldsymbol{x}^* = [x_1^*, x_2^*, \ldots, x_n^*]^T \in \mathbb{R}^n$ is the optimal solution of the problem, then

$$\boldsymbol{g}(\boldsymbol{x}^*) \;=\; \boldsymbol{0} \qquad (10.4)$$
$$\boldsymbol{y}^T \boldsymbol{H}(\boldsymbol{x}^*)\boldsymbol{y} \;\geq\; 0, \; \forall \, \boldsymbol{y} \in \mathbb{R}^n, \qquad (10.5)$$

where $g(x) = \nabla f(x)$ is the gradient vector and $H(x)$ is the Hessian matrix defined by:

$$g(x) \;=\; \nabla f(x) = \begin{bmatrix} \frac{\partial f(x)}{\partial x_1} \\ \frac{\partial f(x)}{\partial x_2} \\ \vdots \\ \frac{\partial f(x)}{\partial x_n} \end{bmatrix}$$

$$H(x) \;=\; \begin{bmatrix} \frac{\partial^2 f(x)}{\partial x_1^2} & \frac{\partial^2 f(x)}{\partial x_1 \partial x_2} & \cdots & \frac{\partial^2 f(x)}{\partial x_1 \partial x_n} \\ \frac{\partial^2 f(x)}{\partial x_2 \partial x_1} & \frac{\partial^2 f(x)}{\partial x_2^2} & \cdots & \frac{\partial^2 f(x)}{\partial x_2 \partial x_n} \\ \vdots & \vdots & \ddots & \vdots \\ \frac{\partial^2 f(x)}{\partial x_n \partial x_1} & \frac{\partial^2 f(x)}{\partial x_n \partial x_2} & \cdots & \frac{\partial^2 f(x)}{\partial x_n^2} \end{bmatrix}$$

Equation (10.5) is to say that the Hessian matrix is a positive semi-definite. Equations (10.4)-(10.5) are *necessary* optimality conditions of $x^*$. If condition (10.5) is replaced by the condition:

$$y^T H(x^*) y > 0, \ \forall \ y \in \mathbb{R}^n \tag{10.6}$$

(that is the Hessian matrix is positive definite), then the conditions (10.4)-(10.6) are *sufficient* optimality conditions.

The numerical methods for solving (10.1) are iterative. They agree in starting from an initial guess $x^{(0)}$ of the solution $x^*$ and then construct a sequence of solutions $x^{(0)}, x^{(1)}, x^{(2)}, \ldots$, where $f(x^{(0)}) > f(x^{(1)}) > \ldots$. At iteration $k$, moving from the point $x^{(k)}$ to a point $x^{(k+1)}$ in numerical optimization generally passes through two steps: in the first step a *search direction* $p^{(k)}$ is determined, then in the second step a *line search* in the direction of $p^{(k)}$ (where $\|p^{(k+1)}\| = 1$) to locate the minimum point $x^{(k+1)} = x^{(k)} + \alpha p^{(k)}$ in the line is carried-out, such that $f(x^{(k+1)}) < f(x^{(k)})$. The iterative process stops at a solution $x^{(k)}$ if

$$\|g(x^{(k)})\| < \varepsilon,$$

where $\varepsilon > 0$ is an arbitrary small positive real and $H(x^{(k)})$ is positive semi-definite [3, 20].

The numerical optimization techniques for unconstrained optimization care about two problems:

(i) **Line search problem:** given a function $f(x)$, its gradient $g(x) = \nabla f(x)$, a point $x^{(k)}$ at iteration $k$ a descent direction $p^{(k)}$, find $\alpha^{(k)} > 0$ such that for $\alpha = \alpha^{(k)}$ the function $f$ is minimized along the ray $x^{(k)} + \alpha p^{(k)}$, that is

$$\alpha^{(k)} = \arg\min_{\alpha > 0} \{ f(x^{(k)} + \alpha p^{(k)}) \}.$$

(ii) **Search direction problem:** given a function $f(x)$ and point $x^{(k)}$ at iteration $k$, find a unit vector $p^{(k)}$, such that $f(x^{(k)}+\alpha p^{(k)})$ is a decreasing function for $0 < \alpha < \alpha_{max}$. That is $p^{(k)}$ is a `descent direction`.

The numerical optimization methods differ from each other by the methods through which the search directions are determined and the gradient vector and Hessian matrices are approximated and (or) updated.

## 10.1.1 Line Search Algorithm

There are two kinds of line search algorithms: *exact line search* and *inexact line search* algorithm. The exact line search algorithm looks for the exact step size $\alpha$ ($\alpha = \alpha_{Exact}$) such that

$$\alpha_{Exact} = \arg\min_{\alpha > 0}\{f(x^{(k)}+\alpha p^{(k)})\},$$

which is numerically impractical. The inexact line search algorithms look for approximations to the exact step size $\alpha_{Exact}$. This is done iteratively, starting from an initial guess $\alpha^0$ and constructing a sequence of step sizes $\alpha^0, \alpha^1, \alpha^2, \ldots, \alpha^k$ such that $\alpha^k \approx \alpha_{Exact}$ [49].

The most famous inexact line search algorithm is the *backtracking line search* algorithm, which employs two parameters $a$ and $b$ and is based on *Armijo condition*:

$$f(x^{(k)}+\alpha^k p^{(k)}) < f(x) + a\alpha^k g(x^{(k)})^T p^{(k)}.$$

The parameter $a$ is associated with the termination condition of Armijo. The parameter $0 < b < 1$ is used to update the step size in each iteration, where $\alpha^j = b\alpha^{j-1}$, starting from a large step size $\alpha^0$(usually $= 1$).

The MATLAB function `LineSearch.m` receives a function $f$, a vector $g$ representing the gradient of $f$ the starting vector $x$ and a unit (direction) vector $p$ and returns the optimal step size $\alpha$:

```
1  function alpha = LineSearch(f, g, x, p)
2  a = 0.3 ; b = 0.9 ;
3  alpha = 1.0 ;
4  while f(x+alpha*p) > f(x) + a*alpha*g(x)'*p
5  alpha = b*alpha ;
6  end
7  end
```

The Python code of the function `LineSearch.py` is:

```
1  import numpy as np
2  def LineSearch(f, g, x, p):
3  a, b = 0.3, 0.9
4  alpha = 1.0
5  while f(x+alpha*p) > f(x) + a*alpha*np.dot(g(x), p):
```

```
6  alpha *= b
7  return alpha
```

## 10.1.2    The Steepest Descent Method

From the elementary calculus, the gradient vector $g(x) = \nabla f(x)$ points to the direction of greatest increase in a function of several variables $f(x)$. Hence, $-g(x)$ points to the direction of greatest decrease in $f(x)$. Any descent direction $p(x)$ makes an acute angle with $-g(x)$. The steepest descent method chooses the search direction to be $p(x) = -\frac{g(x)}{\|g(x)\|}$. Hence, in the steepest descent method the iterative process to progress from a point $x^{(k)}$ at iteration $k$ to a point $x^{(k+1)}$ is given by the formula:

$$x^{(k+1)} = x^{(k)} - \alpha^{(k)} \frac{g(x^{(k)})}{\|g(x^{(k)})\|}$$

where $\alpha^{(k)}$ is found from a line search algorithm [20, 49].

For example, to minimize the function

$$f(x_1, x_2) = x_1^2 - x_1 x_2 + 4x_2^2 + 1$$

the gradient at any point $x = [x_1, x_2]^T$ is given by:

$$g(x) = \begin{bmatrix} 2x_1 - x_2 \\ -x_1 + 8x_2 \end{bmatrix}$$

The MATLAB script `SolveEx1withSteepDesc.m` finds the minimum of the given objective function based on the steepest descent method.

```
1   % SolveEx1withSteepDesc.m
2   clear ; clc ; clf ;
3   f = @(x) x(1)^2 -x(1)*x(2)+ 4*x(2)^2 + 1 ;
4   g = @(x) [2*x(1)-x(2); -x(1)+8*x(2)] ;
5   x0 = [1; 1] ; Eps = 1e-8 ;
6   [x, Iterations] = GradDec(f, g, x0, Eps) ;
7   disp('Optimum solution = ') ; disp(x) ;
8   disp(['Iterations = ' num2str(Iterations)]) ;
9
10  function [x, Iterations] = GradDec(f, g, x0, Eps)
11  x = x0 ;
12  Iterations = 0 ;
13  while norm(g(x), 2) ≥ Eps
14  p = -g(x)/norm(g(x), 2) ;
15  alpha = LineSearch(f, g, x, p) ;
16  x = x + alpha * p ;
17  fprintf('%3i\t\t%14.9f\t\t%12.10e\n', Iterations, f(x), ...
        norm(g(x), 2)) ;
18  Iterations = Iterations + 1 ;
19  end
```

```
20   end
21
22   function alpha = LineSearch(f, g, x, p)
23   a = 1-2/(1+sqrt(5)) ;
24   b = 2/(1+sqrt(5)) ;
25   alpha = 1.0 ;
26   while f(x+alpha*p) > f(x) + a*alpha*g(x)'*p
27   alpha = b*alpha ;
28   end
29   end
```

By executing the code, the optimal solution is obtained in 30 iterations.

```
-------------------------------------------------------
Iteration      f(x)                ||g(x)||
-------------------------------------------------------
    0        1.728932188        1.8761048104e+00
    1        1.288589321        1.1953530385e+00
    2        1.116213769        7.5167383791e-01

   29        1.000000000        3.1130274879e-08
   30        1.000000000        4.5572657815e-09
-------------------------------------------------------


Optimum solution =
1.0e-09 *

-0.306552140415857
0.513653144168223


Iterations = 31
```

The corresponding Python script `SolveEx1withSteepDesc.py` is:

```python
1    # SolveEx1withSteepDesc.py
2    import numpy as np
3    def LineSearch(f, g, x, p):
4        a, b = 1-2/(1+np.sqrt(5)), 2/(1+np.sqrt(5))
5        alpha = 1.0
6        while f(x+alpha*p) > f(x) + a*alpha*np.dot(g(x), p):
7            alpha *= b
8        return alpha
9
10   def GradDec(f, g, x0, Eps):
11       x = x0 ;
12       Iterations = 0 ;
13       print('-------------------------------------------------------')
14       print('Iteration\t f(x)\t ||g(x)||')
15       print('-------------------------------------------------------')
16       while np.linalg.norm(g(x), 2) >= Eps:
17           p = -g(x)/np.linalg.norm(g(x), 2)
```

```
18            alpha = LineSearch(f, g, x, p)
19            x = x + alpha * p
20            print('{0:5.0f}'.format(Iterations), \
21             '{0:12.10f}'.format(f(x)),\
22             '{0:10.8e}'.format(np.linalg.norm(g(x))))
23            Iterations += 1
24        print('-------------------------------------------------')
25        return x, Iterations
26
27   f = lambda x: x[0]**2-x[0]*x[1]+4*x[1]**2 + 1
28   g = lambda x: np.array([2*x[0]-x[1], -x[0]+8*x[1]])
29   x0 = np.array([1, 1])
30   Eps = 1e-8
31   x, Iterations = GradDec(f, g, x0, Eps)
32   print('x = ', x)
33   print('Iterations = ', Iterations)
```

The disadvantage with the steepest descent method is that at the beginning it converges quickly to the optimum solution, but as it comes closer to the solution its convergance rate drops rapidly and its progress towards the optimum solution becomes too slow [3].

## 10.1.3    Newton's Method

At iteration $k$, the function $f$ is approximated near $\boldsymbol{x}^{(k)}$ by a quadratic formula from Taylor series as:

$$f(\boldsymbol{x}) \approx f(\boldsymbol{x}^{(k)}) + g(\boldsymbol{x}^{(k)})^T(\boldsymbol{x} - \boldsymbol{x}^{(k)}) + (\boldsymbol{x} - \boldsymbol{x}^{(k)})^T \boldsymbol{H}(\boldsymbol{x}^{(k)})(\boldsymbol{x} - \boldsymbol{x}^{(k)}) \quad (10.7)$$

The quadratic function at the right-hand side has a unique minimizer, obtained by differentiating the RHS of Equation (10.7) with respect to $\boldsymbol{x}$ and equating the resulting linear equation to $\boldsymbol{0}$:

$$g(\boldsymbol{x}^{(k)}) + \boldsymbol{H}(\boldsymbol{x}^{(k)})(\boldsymbol{x} - \boldsymbol{x}^{(k)}) = \boldsymbol{0}$$

Setting $\boldsymbol{x} = \boldsymbol{x}^{(k+1)}$ and solving for $\boldsymbol{x}^{(k+1)}$ gives:

$$\boldsymbol{x}^{(k+1)} = \boldsymbol{x}^{(k)} - \boldsymbol{H}^{-1}(\boldsymbol{x}^{(k)})g(\boldsymbol{x}^{(k)}).$$

Hence, in Newton's method the search direction at iteration $k$ is given by:

$$\boldsymbol{p}^{(k)} = -\boldsymbol{H}^{-1}(\boldsymbol{x}^{(k)})g(\boldsymbol{x}^{(k)}),$$

and

$$\boldsymbol{x}^{(k+1)} = \boldsymbol{x}^{(k)} + \alpha^{(k)}\boldsymbol{p}^{(k)} = \boldsymbol{x}^{(k)} - \alpha^{(k)}\boldsymbol{H}^{-1}(\boldsymbol{x}^{(k)})g(\boldsymbol{x}^{(k)})$$

**Example 10.1** In this example, Newoton's method will be used to find the optimum solution of the unconstrained minimization problem:

$$\min_{\boldsymbol{x} \in \mathbb{R}^2} 5x_1^2 + \frac{x_2^2}{2} + 5 \log e^{-x_1 - x_2}$$

The MATLAB script `MinEx2WithPureNewton.m` implements a function `PureNewton` to apply the Newton's iterations for solving the minimization problem:

```
1  % MiWithPureNewton.m
2  clear ; clc ; clf ;
3  f = @(x) 5 *x(1)^2 + x(2)^2/2 + 5*log(1+exp(-x(1)-x(2))) ;
4  g = @(x) [10*x(1) - 5*exp(-x(1) - x(2))/(exp(-x(1) - x(2)) + 1);
5  x(2) - 5*exp(-x(1) - x(2))/(exp(-x(1) - x(2)) + 1)] ;
6  H = @(x) [5*(5*exp(x(1) + x(2)) + 2*exp(2*x(1) + 2*x(2)) + ...
      2)/(2*exp(x(1) + x(2)) + exp(2*x(1) + 2*x(2)) + 1) ...
7  5*exp(x(1) + x(2))/(2*exp(x(1) + x(2)) + exp(2*x(1) + 2*x(2)) ...
      + 1);
8  5*exp(x(1) + x(2))/(2*exp(x(1) + x(2)) + exp(2*x(1) + 2*x(2)) ...
      + 1) ...
9  (7*exp(x(1) + x(2)) + exp(2*x(1) + 2*x(2)) + 1)/(2*exp(x(1) + ...
      x(2)) + exp(2*x(1) + 2*x(2)) + 1)] ;
10 x0 = [10; 10] ; Eps = 1e-8 ;
11 [x, Iterations] = PureNewton(f, g, H, x0, Eps) ;
12 disp('Optimum solution = ') ; fprintf('%15.14e\n%15.14e\n', ...
      x(1), x(2)) ;
13 disp(['Iterations = ' num2str(Iterations)]) ;
14
15 function [x, Iterations] = PureNewton(f, g, H, x0, Eps)
16     x = x0 ;
17     Iterations = 0 ;
18     fprintf('-------------------------------------------------\n') ;
19     fprintf('Iteration\t\t f(x)\t\t\t ||g(x)||\n') ;
20     fprintf('-------------------------------------------------\n') ;
21     fprintf('%5i\t\t%14.9f\t\t%12.10e\n', Iterations, f(x0), ...
          norm(g(x0), 2)) ;
22     while norm(g(x), 2) >= Eps
23         p = -H(x)\g(x) ;
24         alpha = LineSearch(f, g, x, p) ;
25         x = x + alpha*p ;
26         Iterations = Iterations + 1 ;
27         fprintf('%5i\t\t%14.9f\t\t%12.10e\n', Iterations, f(x), ...
              norm(g(x), 2)) ;
28     end
29     fprintf('-------------------------------------------------\n') ;
30 end
31
32 function alpha = LineSearch(f, g, x, p)
33     a = 1-2/(1+sqrt(5)) ;
34     b = 2/(1+sqrt(5)) ;
35     alpha = 1.0 ;
36     while f(x+alpha*p) > f(x) + a*alpha*g(x)'*p
37         alpha = b*alpha ;
38     end
39 end
```

By running the code, the minimization problem is solved in only 5 iterations:

```
--------------------------------------------------
Iteration          f(x)                  ||g(x)||
--------------------------------------------------
0              6.134640055           9.4126587853e+00
1              1.976896063           2.2780206611e-01
2              1.969726543           2.6269970505e-03
3              1.969725575           3.6873326702e-07
4              1.969725575           7.2224253099e-15
--------------------------------------------------

Optimum solution =
0.1125
1.1247


Iterations = 4
```

The code of the Python's script `MinEx2WithPureNewton.py` is:

```python
 1  import numpy as np
 2
 3  def LineSearch(f, g, x, p):
 4      a, b = 1-2/(1+np.sqrt(5)), 2/(1+np.sqrt(5))
 5      alpha = 1.0
 6      while f(x+alpha*p) > f(x) + a*alpha*np.dot(g(x), p):
 7          alpha *= b
 8  return alpha
 9
10  def PureNewton(f, g, x0, Eps):
11      x = x0 ;
12      Iterations = 0 ;
13      print('--------------------------------------------------')
14      print('Iteration\t f(x)\t ||g(x)||')
15      print('--------------------------------------------------')
16      while np.linalg.norm(g(x), 2) ≥ Eps:
17          p = -np.linalg.solve(H(x), g(x))
18          alpha = LineSearch(f, g, x, p)
19          x = x + alpha * p
20          print('{0:5.0f}'.format(Iterations), '\t\t', ...
                  '{0:12.10f}'.format(f(x)),\
21              '\t', '{0:10.8e}'.format(np.linalg.norm(g(x))))
22          Iterations += 1
23      print('--------------------------------------------------')
24      return x, Iterations
25
26  f = lambda x: (10*x[0]**2 + x[1]**2)/2 + ...
        5*np.log(1+np.exp(-x[0]-x[1]))
27  g = lambda x: np.array([10*x[0] - 5*np.exp(-x[0] - ...
        x[1])/(np.exp(-x[0] - x[1]) + 1),
28      x[1] - 5*np.exp(-x[0] - x[1])/(np.exp(-x[0] - x[1]) + 1)])
```

```
29  Den = lambda x: 1/(2*np.exp(x[0] + x[1]) + np.exp(2*x[0] + ...
         2*x[1]) + 1)
30  H = lambda x: Den(x) * np.array([[5*(5*np.exp(x[0] + x[1]) + ...
         2*np.exp(2*x[0] +\
31  2*x[1]) + 2), 5*np.exp(x[0] + x[1])], [5*np.exp(x[0] + x[1]), \
32  7*np.exp(x[0] + x[1]) + np.exp(2*x[0] + 2*x[1]) + 1]])
33  x0 = np.array([1, 1])
34  Eps = 1e-8
35  x, Iterations = PureNewton(f, g, x0, Eps)
36  print('x = ', x)
37  print('Iterations = ', Iterations)
```

## 10.1.4    Quasi Newton's Methods

The pure Newton's method for solving unconstrained minimization problems suffer the problem that the inverse of the Hessian matrix must be evaluated at iteration $k$ to find the search direction $p^{(k)}$. When the Hessian matrix is ill-conditioned the resulting search direction will be inaccurate and the iteration process could be unsuccessful.

Instead of directly inverting the Hessian matrix, the quasi-Newton's methods look for approximations to the Hessian matrix in the different iterations. Therefore, $\left(H^{(k)}\right)^{-1}$ is replaced by a positive definite matrix $B^{(k)}$ and the search direction at iteration $k$ is $p^{(k)} = -B^{(k)}g^{(k)}$. This iterative process usually starts from $B^{(0)} = I$ where $I \in \mathbb{R}^{n \times n}$ is the identity matrix of type $n \times n$. The quasi Newton methods differ in the way by which matrix $B^{(k)}$ is updated at iteration $k$, and following from that the research direction is computed. The most famous quasi Newton's methods are the Davidon-Fletcher-Powell (DFP) and the Broyden-Flethcher-Goldfarb-Shanno (BFGS) methods [19, 3, 20].

### 10.1.4.1    The Broyden-Fletcher-Goldfarb-Shanno (BFGS) Method

The BFGS statrs from initial approximation of the Hessian matrix $H^{(0)} \approx I$. At iteration $k$, the approximate Hessian matrix $H^{(k)}$ is updated to obtain $H^{(k+1)}$ which will be used in the next iteration [20]. This is done as follows. At iteration $k$ given $x^{(k)}$, $g^{(k)} = g(x^{(k)})$ and $H^{(k)}$. The procedure starts with finding the search direction $p^{(k)}$ by solving the linear system $H^{(k)}p^{(k)} = -g^{(k)}$ and the step size $\alpha^{(k)}$ from the line search algorithm.Then two vectors are computed: $s^{(k)} = \alpha^{(k)}p^{(k)}$ and $y^{(k)} = g(x^{(k)} + s^{(k)}) - g(x^{(k)})$. Finally, the Hessian matrix is updated as:

$$H^{(k+1)} = H^{(k)} + \frac{y^{(k)}\left(y^{(k)}\right)^T}{\left(y^{(k)}\right)^T s^{(k)}} - \frac{H^{(k)}\left(s^{(k)}\left(s^{(k)}\right)^T\right)\left(H^{(k)}\right)^T}{\left(s^{(k)}\right)^T H^{(k)} s^{(k)}}$$

The Python code `MinWithBFGS.py` is used to find the optimal solution of example 10.1 based on the BFGS algorithm:

```python
import numpy as np

def LineSearch(f, g, x, p):
    a, b = 1-2/(1+np.sqrt(5)), 2/(1+np.sqrt(5))
    alpha = 1.0
    while f(x+alpha*p) > f(x) + a*alpha*np.dot(g(x), p):
        alpha *= b
    return alpha

def BFGS(f, g, x0, Eps):
    x = x0 ;
    Iterations = 0 ;
    print('-----------------------------------------------')
    print('Iteration\t f(x)\t\t ||g(x)||')
    print('-----------------------------------------------')
    H = np.eye(len(x0), dtype=float)
    while np.linalg.norm(g(x), 2) >= Eps:
        p = -np.linalg.solve(H, g(x))
        alpha = LineSearch(f, g, x, p)
        s = alpha * p
        y = g(x+alpha*s) - g(x)
        x = x + s
        H = H + np.outer(y, y)/np.inner(y,s)-(H@np.outer(s, ...
            s)@H.T)/(s.T@H@s)
        print('{0:5.0f}'.format(Iterations), '\t    ', ...
            '{0:12.10f}'.format(f(x)),\
        '\t', '{0:10.8e}'.format(np.linalg.norm(g(x))))
        Iterations += 1
    print('-----------------------------------------------')
    return x, Iterations

f = lambda x: (10*x[0]**2 + x[1]**2)/2 + ...
    5*np.log(1+np.exp(-x[0]-x[1]))
g = lambda x: np.array([10*x[0] - 5*np.exp(-x[0] - ...
    x[1])/(np.exp(-x[0] - x[1]) + 1),
x[1] - 5*np.exp(-x[0] - x[1])/(np.exp(-x[0] - x[1]) + 1)])
x0 = np.array([1, 1])
Eps = 1e-8
x, Iterations = BFGS(f, g, x0, Eps)
print('x = ', x)
print('Iterations = ', Iterations)
```

By executing the code, the optimal solution is obtained in 13 iterations:

| Iteration | f(x) | \|\|g(x)\|\| |
|---|---|---|
| 0 | 1.9971128385 | 3.93916248e-01 |
| 1 | 1.9842525656 | 2.84906636e-01 |
| 2 | 1.9703729526 | 8.38711419e-02 |

| 10 | 1.9697255747 | 4.32879304e-08 |
| 11 | 1.9697255747 | 3.93521430e-08 |
| 12 | 1.9697255747 | 1.62355225e-10 |

```
------------------------------------------
x =  [0.11246719 1.12467185]
Iterations = 13
```

The corresponding MATLAB code is:

```matlab
clear ; clc ;
f = @(x) 5 *x(1)^2 + x(2)^2/2 + 5*log(1+exp(-x(1)-x(2))) ;
g = @(x) [10*x(1) - 5*exp(-x(1) - x(2))/(exp(-x(1) - x(2)) + 1);
    x(2) - 5*exp(-x(1) - x(2))/(exp(-x(1) - x(2)) + 1)] ;
x0 = [1.0; 1.0] ; Eps = 1e-8 ;
[x, Iterations] = BFGS(f, g, x0, Eps) ;
disp('Optimum solution = ') ; disp(x) ;
disp(['Iterations = ' num2str(Iterations)]) ;

function [x, Iterations] = BFGS(f, g, x0, Eps)
    x = x0 ;
    Iterations = 0 ;
    fprintf('--------------------------------------------\n') ;
    fprintf('Iteration\t\t f(x)\t\t\t ||g(x)||\n') ;
    fprintf('--------------------------------------------\n') ;
    fprintf('%5i\t\t%14.9f\t\t%12.10e\n', Iterations, f(x0), ...
        norm(g(x0), 2)) ;
    H = eye(length(x0)) ;
    while norm(g(x), 2) >= Eps
        p = -H\g(x) ;
        alpha = LineSearch(f, g, x, p) ;
        s = alpha*p ;
        y = g(x+s)-g(x) ;
        x = x + alpha*p ;
        H = H + y*y'/(y'*s)-H*(s*s')*H'/(s'*H*s) ;
        Iterations = Iterations + 1 ;
        fprintf('%5i\t\t%14.9f\t\t%12.10e\n', Iterations, ...
            f(x), norm(g(x), 2)) ;
    end
    fprintf('--------------------------------------------\n') ;
end

function alpha = LineSearch(f, g, x, p)
    a = 1-2/(1+sqrt(5)) ;
    b = 2/(1+sqrt(5)) ;
    alpha = 1.0 ;
    while f(x+alpha*p) > f(x) + a*alpha*g(x)'*p
        alpha = b*alpha ;
    end
end
```

## 10.1.4.2    The Davidon-Fletcher-Powell (DFP) Algorithm

The Davidon-Fletcher-Powell (DFP) method has a close form as the BFGS to update the approximate Hessian matrix. At iteration $k$, the vectors $\mathbf{s}^{(k)}$ and $\mathbf{y}^{(k)}$ are computed the same way as in the BFGS algorithm. When formulating the Hessian matrix $\boldsymbol{H}^{(k+1)}$, vectors $\mathbf{y}^{(k)}$ and $\mathbf{s}^{(k)}$ of the BFGS method exchange their role in the DFP method [3, 20]. The formula of updating the Hessian matrix at iteration $k$ is:

$$\boldsymbol{H}^{(k+1)} = \boldsymbol{H}^{(k)} + \frac{\mathbf{s}^{(k)}\left(\mathbf{s}^{(k)}\right)^T}{\left(\mathbf{y}^{(k)}\right)^T \mathbf{s}^{(k)}} - \frac{\boldsymbol{H}^{(k)}\left(\mathbf{y}^{(k)}\left(\mathbf{y}^{(k)}\right)^T\right)\left(\boldsymbol{H}^{(k)}\right)^T}{\left(\mathbf{y}^{(k)}\right)^T \boldsymbol{H}^{(k)}\mathbf{y}^{(k)}}$$

**Example 10.2** In this example, the DFP method will be implemented to minimize the quadratic function

$$f(x_1, x_2, x_3) = (x_1 - 2)^2 + (1 - x_3)^2 + x_2^2 + 1$$

The Python code is:

```
1   import numpy as np
2
3   def LineSearch(f, g, x, p):
4   a, b = 1-2/(1+np.sqrt(5)), 2/(1+np.sqrt(5))
5   alpha = 1.0
6   while f(x+alpha*p) > f(x) + a*alpha*np.dot(g(x), p):
7   alpha *= b
8   return alpha
9
10  def DFP(f, g, x0, Eps):
11  x = x0 ;
12  Iterations = 0 ;
13  print('------------------------------------------------')
14  print('Iteration\t f(x)\t\t ||g(x)||')
15  print('------------------------------------------------')
16  H = np.eye(len(x0), dtype=float)
17  while np.linalg.norm(g(x), 2) ≥ Eps:
18  p = -np.linalg.solve(H, g(x))
19  alpha = LineSearch(f, g, x, p)
20  s = alpha * p
21  y = g(x+alpha*s) - g(x)
22  x = x + s
23  H = H + np.outer(s, s)/np.inner(y,s)-(H@np.outer(y, ...
        y)@H.T)/(y.T@H@y)
24  print('{0:5.0f}'.format(Iterations), '\t    ', ...
        '{0:12.10f}'.format(f(x)),\
25  '\t', '{0:10.8e}'.format(np.linalg.norm(g(x))))
26  Iterations += 1
27  print('------------------------------------------------')
28  return x, Iterations
29
30  f = lambda x: ((x[0]-2)**2+(1+x[2])**2)+(1+x[1]**2)
31  g = lambda x: np.array([2*x[0] - 4, 2*x[1], 2*x[2] + 2])
```

```
32  x0 = np.array([1, 1, 1])
33  Eps = 1e-8
34  x, Iterations = DFP(f, g, x0, Eps)
35  print('x = ', x)
36  print('Iterations = ', Iterations)
```

By executing the code, the minimization problem is solved in 8 iterations:

```
-------------------------------------------
Iteration        f(x)              ||g(x)||
-------------------------------------------
0         1.3343685400        1.15649218e+00
1         1.0010384216        6.44491002e-02
2         1.0000032249        3.59162526e-03
3         1.0000000100        2.00154416e-04
4         1.0000000000        1.11542233e-05
5         1.0000000000        6.21603559e-07
6         1.0000000000        3.46407792e-08
7         1.0000000000        1.93046439e-09
-------------------------------------------
x =  [ 2.00000000e+00   3.94054416e-10  -9.99999999e-01]
Iterations =  8
```

The MATLAB code is:

```
 1  clear ; clc ;
 2  f = @(x) (x(1)-2)^2 + (1+x(3))^2 + x(2)^2+1 ;
 3  g = @(x) [2*(x(1)-2); 2*x(2); 2*(x(3)+1)] ;
 4  x0 = [1.0; 1.0; 1] ; Eps = 1e-8 ;
 5  [x, Iterations] = DFP(f, g, x0, Eps) ;
 6  disp('Optimum solution = ') ; format long e ; disp(x) ;
 7  disp(['Iterations = ' num2str(Iterations)]) ;
 8
 9  function [x, Iterations] = DFP(f, g, x0, Eps)
10      x = x0 ;
11      Iterations = 0 ;
12      fprintf('-------------------------------------------------------\n') ;
13      fprintf('Iteration\t\t f(x)\t\t\t ||g(x)||\n') ;
14      fprintf('-------------------------------------------------------\n') ;
15      fprintf('%5i\t\t\t%14.9f\t\t%12.10e\n', Iterations, f(x0), ...
            norm(g(x0), 2)) ;
16      H = eye(length(x0)) ;
17      while norm(g(x), 2) >= Eps
18          p = -H\g(x) ;
19          alpha = LineSearch(f, g, x, p) ;
20          s = alpha*p ;
21          y = g(x+s)-g(x) ;
22          x = x + alpha*p ;
23          H = H + s*s'/(y'*s)-H*(y*y')*H'/(y'*H*y) ;
24          Iterations = Iterations + 1 ;
```

```
25          fprintf('%5i\t\t%14.9f\t\t%12.10e\n', Iterations, ...
                f(x), norm(g(x), 2)) ;
26      end
27      fprintf('----------------------------------------------------------\n') ;
28  end
29
30  function alpha = LineSearch(f, g, x, p)
31      a = 1-2/(1+sqrt(5)) ;
32      b = 2/(1+sqrt(5)) ;
33      alpha = 1.0 ;
34      while f(x+alpha*p) > f(x) + a*alpha*g(x)'*p
35          alpha = b*alpha ;
36      end
37  end
```

### 10.1.5   Solving Unconstrained Optimization Problems with MATLAB

The MATLAB function fminunc is used to solve the unconstrained minimization problem:

$$\min_{\boldsymbol{x}\in\mathbb{R}^n} f(\boldsymbol{x}), \boldsymbol{x}\in\mathbb{R}^n$$

The function fminunc receives mainly an objective function $f$ and a starting point $\boldsymbol{x}_0$ and optionally a parameter options to return mainly the optimum solution $\boldsymbol{x}$ and optionally the value of the objective function at the optimal solution and other optional outputs. Through the parameter options the user can customize the parameters of the optimization process such as the *algorithm, maximum number of iteration, maximum number of function evaluations, the tolerance in the value of the objective function, etc.*

**Example 10.3** In this example the MATLAB function fminunc will be used to solve the unconstrained minimization problem:

$$\min_{\boldsymbol{x}\in\mathbb{R}^2} \frac{10x_1^2+x_2^2}{2} + 5\log e^{-x_1-x_2}$$

starting from the point $(1,1)^T$.

```
>> f = @(x) 5 *x(1)^2 + x(2)^2/2 + 5*log(1+exp(-x(1)-x(2))) ;
>> x0 = [1; 1] ;
>> Options = optimset('Display', 'Iter', 'TolFun', 1e-7) ;
>> x = fminunc(f, [1; 1], Options)
```

|           |            |         | First-order |            |
| Iteration | Func-count | f(x)    | Step-size   | optimality |
|-----------|------------|---------|-------------|------------|
| 0         | 3          | 6.13464 |             | 9.4        |
| 1         | 6          | 2.08295 | 0.106338    | 1.39       |
| 2         | 9          | 1.98574 | 1           | 0.238      |
| 3         | 12         | 1.98063 | 1           | 0.194      |
| 4         | 15         | 1.96973 | 1           | 0.00244    |

| 5 | 18 | 1.96973 | 1 | 4.72e-05 |
|---|----|---------|---|----------|
| 6 | 21 | 1.96973 | 1 | 1.67e-06 |
| 7 | 24 | 1.96973 | 1 | 2.98e-08 |

```
Local minimum found.

Optimization completed because the size of the gradient is less than
the selected value of the optimality tolerance.

<stopping criteria details>

x =

    0.1125
    1.1247
```

## 10.1.6    Solving an Unconstrained Problem with Python

Python has many functions to solve the unconstrained minimization problems. Those solvers include the functions fmin, fmin_bfgs, fmin_ncg and minimize.

To solve the problem of example 10.3 with Python, the following Python instructions can be used:

```
In [1]: import numpy as np
In [2]: from scipy.optimize import fmin
In [3]: x0 = np.array([1., 1.])
In [4]: f = lambda x: (10*x[0]**2 + x[1]**2)/2 + 5*np.log(1+np.exp(-x[0]-x[1]))
In [5]: Eps = 1e-8
In [6]: xopt = fmin(f, x0, disp=1, xtol=Eps)
Optimization terminated successfully.
Current function value: 1.969726
Iterations: 65
Function evaluations: 126
In [7]: print(xopt)
[0.11246719 1.12467185]


In [8]: from scipy.optimize import fmin_bfgs
In [9]: g = lambda x: np.array([10*x[0] - 5*np.exp(-x[0] - x[1])/\
(np.exp(-x[0] - x[1]) + 1), x[1] - 5*np.exp(-x[0] - x[1])/\
(np.exp(-x[0] - x[1]) + 1)])
In [10]: xopt = fmin_bfgs(f, x0, fprime=g, disp=1)
    Optimization terminated successfully.
    Current function value: 1.969726
    Iterations: 7
    Function evaluations: 8
    Gradient evaluations: 8
In [11]: print(xopt)
    [0.11246715 1.1246719 ]


In [12]: from scipy.optimize import fmin_ncg
In [13]: Den = lambda x: 1/(2*np.exp(x[0] + x[1]) + np.exp(2*x[0] + 2*x[1]) + 1)
In [14]: H = lambda x: Den(x) * np.array([[5*(5*np.exp(x[0] \
+ x[1]) + 2*np.exp(2*x[0] + 2*x[1]) + 2), 5*np.exp(x[0] + x[1])],\
```

```
[5*np.exp(x[0] + x[1]), 7*np.exp(x[0] + x[1]) + np.exp(2*x[0] + 2*x[1]) + 1]])
In [15]: xopt = fmin_ncg(f, x0, fprime=g, fhess=H, disp=1)
    Optimization terminated successfully.
    Current function value: 1.969726
    Iterations: 4
    Function evaluations: 5
    Gradient evaluations: 8
    Hessian evaluations: 4
    In [16]: print(xopt)
    [0.11246719 1.12467185]

In [17]: Sol = minimize(f, x0)
In [18]: print(Sol)
    fun: 1.969725574672448
    hess_inv: array([[ 0.09705991, -0.04667767],
    [-0.04667767,  0.55809497]])
    jac: array([-2.98023224e-07,  5.96046448e-08])
    message: 'Optimization terminated successfully.'
    nfev: 32
    nit: 7
    njev: 8
    status: 0
    success: True
    x: array([0.11246715, 1.1246719 ])
```

## 10.1.7    Solving Unconstrained Optimization Problems with Gekko

In Gekko the optimization problems can be solved at mode 3. In this section two unconstrained optimization problems. The two problems are taken from the Mathworks website https://www.mathworks.com/help/optim/ug/fminunc.html#butpb7p-4. The problems are given in two following examples.

**Example 10.4** In this example Gekko will be used to solve the unconstrained minimization problem:

$$\min_{\boldsymbol{x}\in\mathbb{R}^2} f(x_1,x_2) = 3x_1^2 + 2x_1x_2 + x_2^2 - 4x_1 + 5x_2$$

The code is embedded on the Python script MinEx1uncWithGekko.py

```
1  from gekko import GEKKO
2  m = GEKKO()
3  x1 = m.Var(1)
4  x2 = m.Var(1)
5
6  m.Obj(3*x1**2 + 2*x1*x2 + x2**2 - 4*x1 + 5*x2)
7  m.options.IMODE = 3
8  m.solve()
9  print('Solution found at x = ')
```

```
10  print('{0:12.10e}'.format(x1[0]))
11  print('{0:12.10e}'.format(x2[0]))
```

By running the code, the following results are obtained:

```
EXIT: Optimal Solution Found.

The solution was found.

The final value of the objective function is   -16.3750000000000


    --------------------------------------------------

    Solver        :   IPOPT (v3.12)
    Solution time :   4.699999990407377E-003 sec
    Objective     :   -16.3750000000000
    Successful solution
    --------------------------------------------------


Solution found at x =
     2.2500000000e+00
    -4.7500000000e+00
```

**Example 10.5** The unconstrained optimization problem is to minimize

$$\min_{x \in \mathbb{R}^3} 100(x_2 - x_1^2)^2 + (1 - x_1)^2$$

The Python script MinEx2uncWithGekko.py solves the problem:

```
1   from gekko import GEKKO
2   m = GEKKO()
3   x1 = m.Var(-1.)
4   x2 = m.Var(2.)
5
6   m.Obj(100*(x2 - x1**2)**2 + (1-x1)**2)
7   m.options.IMODE = 3
8   m.solve()
9   print('Solution found at x = ')
10  print('{0:12.10e}'.format(x1[0]))
11  print('{0:12.10e}'.format(x2[0]))
```

Executing the code gives the following results:

```
EXIT: Optimal Solution Found.

The solution was found.

The final value of the objective function is
      7.912533725331136E-015
```

```
-----------------------------------------------------
Solver        :   IPOPT (v3.12)
Solution time :    1.490000000922009E-002 sec
Objective     :    7.912533725331136E-015
Successful solution
-----------------------------------------------------


Solution found at x =
    9.9999991265e-01
    9.9999982362e-01
```

## 10.2    Solving Constrained Optimization Problems

At the beginning an optimization problem with equality constraints is considered. The form of the problem is:

$$\min f(\boldsymbol{x}), \ \boldsymbol{x} \in \mathbb{R}^n \tag{10.8}$$

subject to the equality constraints:

$$\boldsymbol{E}(\boldsymbol{x}) = \boldsymbol{0} \tag{10.9}$$

where $\boldsymbol{E} \in \mathbb{R}^p$, $p \le n$.

To derive the necessary and sufficient optimality conditions, construct the Lagrangian system:

$$L(\boldsymbol{x}) = f(\boldsymbol{x}) + \boldsymbol{E}^T(\boldsymbol{x})\boldsymbol{\lambda} \tag{10.10}$$

with $\boldsymbol{\lambda} \in \mathbb{R}^p$ is the vector of Lagrange multipliers. Then, the *first-order* necessary conditions for $(\boldsymbol{x}^*, \boldsymbol{\lambda}^*)$ to be optimal are:

$$\begin{cases} \nabla_{\boldsymbol{x}} L(\boldsymbol{x}^*, \boldsymbol{\lambda}^*) = 0 = \nabla_{\boldsymbol{x}} f(\boldsymbol{x}^*) + \nabla_{\boldsymbol{x}} \boldsymbol{E}^T(\boldsymbol{x}^*)\boldsymbol{\lambda}^* \\[2mm] \nabla_{\boldsymbol{\lambda}} L(\boldsymbol{x}^*, \boldsymbol{\lambda}^*) = 0 = \boldsymbol{E}(\boldsymbol{x}^*) \end{cases} \tag{10.11}$$

where $\nabla_{\boldsymbol{x}} \boldsymbol{E}(\boldsymbol{x})$ is the Jacobian matrix evaluated at the point $\boldsymbol{x} \in \mathbb{R}^n$.

The *second-order* necessary condition is:

$$\nabla_{\boldsymbol{xx}}^2 L(\boldsymbol{x}^*, \boldsymbol{\lambda}^*) = \nabla_{\boldsymbol{xx}}^2 f(\boldsymbol{x}^*) + \sum_{j=1}^{p} \lambda_j^* \nabla_{\boldsymbol{xx}}^2 \boldsymbol{E}(\boldsymbol{x}^*) = \boldsymbol{H}(\boldsymbol{x}^*) \in \mathbb{R}^n \tag{10.12}$$

be positive semi-definite.

The second-order sufficient condition for $(\boldsymbol{x}^*, \boldsymbol{\lambda}^*)$ to be optimal is:

$$\nabla^2_{\boldsymbol{xx}} L(\boldsymbol{x}^*, \boldsymbol{\lambda}^*) = \nabla^2 f(\boldsymbol{x}^*) + \sum_{j=1}^{p} \lambda_j^* \nabla^2 \boldsymbol{E}(\boldsymbol{x}^*) = \boldsymbol{H}(\boldsymbol{x}^*) \in \mathbb{R}^n \qquad (10.13)$$

be positive-definite.

If $\nabla_{\boldsymbol{x}} \boldsymbol{E}^T(\boldsymbol{x}) \in \mathbb{R}^{n \times p}$ is a full-rank matrix, it has a $QR$ factorization of the form:

$$\nabla \boldsymbol{E}^T(\boldsymbol{x}) = \begin{bmatrix} \tilde{\boldsymbol{Q}} & \boldsymbol{Z} \end{bmatrix} \begin{bmatrix} \boldsymbol{R} \\ \boldsymbol{0} \end{bmatrix}$$

where $\tilde{\boldsymbol{Q}} \in \mathbb{R}^{n \times p}$, $\boldsymbol{Z} \in \mathbb{R}^{n \times (n-p)}$, $\boldsymbol{R} \in \mathbb{R}^{p \times p}$ and $\boldsymbol{0}$ is a matrix of zeros of type $(n-p) \times p$. The matrix $\tilde{\boldsymbol{Q}}$ is a basis for the column space of $\nabla_{\boldsymbol{x}} \boldsymbol{E}^T(\boldsymbol{x})$, whereas $\boldsymbol{Z}$ is a basis for its null space.

At the optimum solution $\boldsymbol{x}^*$, the gradient of the objective function is orthogonal to the constraints surface. That is the projection of the gradient vector onto the constraint surface is zero. This can be expressed as $\boldsymbol{Z}^{*T} \nabla f(\boldsymbol{x}^*) = 0$, which is equivalent to the first order necessary conditions (10.11). The second order necessary conditions for optimality of $(\boldsymbol{x}^*, \boldsymbol{\lambda}^*)$ that are equivalent to (10.12) is $\boldsymbol{Z}^{*T} \boldsymbol{H}^* \boldsymbol{Z}^*$ be positive semi-definite, and the second order sufficient condition of optimality of $(\boldsymbol{x}^*, \boldsymbol{\lambda}^*)$ that is equivalent to (10.13) is that $\boldsymbol{Z}^{*T} \boldsymbol{H}^* \boldsymbol{Z}^*$ be positive definite [20].

The Newoton's optimization methods iterate to find a couple $(\boldsymbol{x}^*, \boldsymbol{\lambda}^*) \in \mathbb{R}^{(n-p) \times (n-p)}$ such that the necessary conditions (10.11) and (10.12) are fulfilled. At iteration $k$, the Karush-Kuhn-Tucker *(KKT)* system is composed from a previously computed couple $(\boldsymbol{x}^{(k)}, \boldsymbol{\lambda}^{(k)})$ to find $(\boldsymbol{x}^{(k+1)}, \boldsymbol{\lambda}^{(k+1)})$ by solving the KKT system:

$$\begin{bmatrix} \boldsymbol{H}(\boldsymbol{x}^{(k)}) & \left(\nabla \boldsymbol{E}^T(\boldsymbol{x}^{(k)})\right)^T \\ \nabla \boldsymbol{E}^T(\boldsymbol{x}^{(k)}) & \boldsymbol{0} \end{bmatrix} \begin{bmatrix} \boldsymbol{s}^{(k)} \\ \boldsymbol{\lambda}^{(k+1)} \end{bmatrix} = \begin{bmatrix} -\nabla f(\boldsymbol{x}^{(k)}) \\ -\boldsymbol{E}(\boldsymbol{x}^{(k)}) \end{bmatrix} \qquad (10.14)$$

where $\boldsymbol{s}^{(k)} = \boldsymbol{x}^{(k+1)} - \boldsymbol{x}^{(k)}$.

Now, if considered a minimization problem with inequality constraints of the form:

$$\begin{aligned} \min \ & f(\boldsymbol{x}), \qquad f : \mathbb{R}^n \to \mathbb{R}, \ \boldsymbol{x} \in \mathbb{R}^n \\ & \boldsymbol{E}(\boldsymbol{x}) = \boldsymbol{0}, \qquad \boldsymbol{E} : \mathbb{R}^n \to \mathbb{R}^p \\ & \boldsymbol{I}(\boldsymbol{x}) \leq \boldsymbol{0}, \qquad \boldsymbol{I} : \mathbb{R}^n \to \mathbb{R}^q. \end{aligned}$$

Now, if $\tilde{\boldsymbol{x}} \in \mathbb{R}^n$ is a feasible point, it is called *regular* if the columns of $\nabla \boldsymbol{I}^T(\tilde{\boldsymbol{x}})$ are linearly independent in $\mathbb{R}^q$. At a regular point $\tilde{\boldsymbol{x}}$ an equality or inequality constraint $\boldsymbol{c}(\tilde{\boldsymbol{x}})$ is called *active* if $\boldsymbol{c}(\tilde{\boldsymbol{x}}) = 0$, otherwise it is *inactive* at $\tilde{\boldsymbol{x}}$. Let $\tilde{\boldsymbol{A}}$ be the set containing all the active constraints at $\tilde{\boldsymbol{x}}$, then $\tilde{\boldsymbol{A}}$ is called the *active set* at $\tilde{x}$. At the optimal solution $x^*$ all the equality constraints shall satisfy:

$$E_j(x^*) = 0, \ j = 1, \ldots, p.$$

The inequality constraints are divided between active and inactive. Let $S^*$ be the indices set of inequality constraints $\{j : I_j(x^*) = 0, \ j \in \{1, \ldots, q\}\}$ Therefore, the active set $A^*$ at the optimal solution $x^*$ consists of $E(x^*) \cup \{I_j(x^*) : j \in S^*\}$.

The Lagrangian system of the constrained minimization problem (10.1)-(10.3) is defined by:

$$L(x, \lambda, \mu) = f(x) + E^T(x)\lambda + I^T(x)\mu, \qquad (10.15)$$

where $\lambda \in \mathbb{R}^p$ and $\mu \in \mathbb{R}^q$.

Then, the first-order necessary conditions of optimality for a point $(x^*, \lambda^*, \mu^*)$ are:

$$\begin{aligned}
\nabla_x L(xb^*, \lambda^*, \mu^*) &= 0 = \nabla f(x^*) + \nabla E^T(x^*)\lambda^* + \nabla I^T(x^*)\mu^* & (10.16)\\
\nabla_\lambda L(xb^*, \lambda^*, \mu^*) &= 0 = E(x^*) & (10.17)\\
\nabla_\mu L(xb^*, \lambda^*, \mu^*) &= I(x^*) \leq 0, & (10.18)\\
\mu_j^* &\geq 0, \ j = 1, \ldots, q & (10.19)\\
\mu_j^* I_j(x^*) &= 0, \ j = 1, \ldots, q & (10.20)
\end{aligned}$$

Condition (10.20) is called the *complementary slack condition* which means that $\mu_j^* > 0$ if $j \in S^*$ ($I_j(x^*)$ is active) and $\mu_j^* = 0$ otherwise.

The second order necessary optimality condition is that:

$$y\nabla_{xx}L(x^*, \lambda^*, \mu^*)y \qquad (10.21)$$

be positive semi-definite.

The second order sufficient optimality condition is:

$$y\nabla_{xx}L(x^*, \lambda^*, \mu^*)y \qquad (10.22)$$

be positive definite.

## 10.2.1    Solving Constrained Optimization Problems with MATLAB `fmincon` Function

The MATLAB function `fmincon` is used for solving constrained minimization problems. It receives an objective function $f(x)$, a starting point $x^0$, a function handle to nonlinear constraints function, the optimization process options. If linear equality and inequality constraints present, they can be passed to the function `fmincon` as well, so are the variables bounds. It receives mainly the solution of the problem and optionally the value of the objective function at the optimal solution, the exit flag, the Lagrange multipliers, gradient and Hessian matrix at the optimum point [17].

The form of implementing the fmincon solver is:

```
>> [xopt, fmin, exFlg, Lamd, Grad, Hess] = fmincon(@f, x0, Aiq,
biq, ... Aeq, beq, lowbnd, upbnd, @nlcon, Options) ;
```

**Example 10.6** In this example, the function fmincon will be used for solving
the constrained minimization problem:

$$\min_{x \in \mathbb{R}^4} x_1 x_4 (x_1 + x_2 + x_3) + x_3$$

subject to inequality constraint:

$$x_1 \cdot x_2 \cdot x_3 \cdot x_4 \geq 25$$

and equality constraint:

$$x_1^2 + x_2^2 + x_3^2 + x_4^2 = 40$$

where

$$-1 \leq x_1, x_2, x_3, x_4 \leq 5$$

The MATLAB script MinEx1Withfmincon.m solves the optimization problem
using the fmincon function:

```
1   % MinEx1Withfmincon.m
2   clear ; clc ;
3   x0 = [1; 5; 5; 1];        % Make a starting guess at the solution
4   Algorithms = ["sqp", "active-set", "interior-point", ...
        "sqp-legacy"] ;
5   options = optimoptions(@fmincon,'Algorithm', 'sqp', 'Disp', ...
        'Iter', 'PlotFcns','optimplotfval') ;
6   [x,fval] = fmincon(@objfun,x0,[],[],[],[],[],[],@confun,options);
7   fprintf('Optimal solution found at \nx ...
        =\n%12.6e\n%12.6e\n%12.6e\n%12.6e\n\n', x(1), x(2), x(3), ...
        x(4)) ;
8   fprintf('Value of objective function = %10.7f\n\n', fval) ;
9
10  function f = objfun(x)
11      f = x(1)*x(4)*(x(1)+x(2)+x(3))+x(3) ;
12  end
13
14  function [ic, ec] = confun(x)
15      % Nonlinear inequality constraints
16      ic(1) = -x(1)*x(2)*x(3)*x(4)+25.0 ;
17      ic(2) = -x(1) + 1 ;
18      ic(3) =  x(1) - 5 ;
19      ic(4) = -x(2) + 1 ;
20      ic(5) =  x(2) - 5 ;
21      ic(6) = -x(3) + 1;
22      ic(7) =  x(3) - 5;
23      ic(8) = -x(4) + 1;
24      ic(9) =  x(4) - 5 ;
25      % Nonlinear equality constraints
26      ec = -sum(x.*x) + 40 ;
27  end
```

Running the MATLAB code gives:

```
Iter  Func-count    fval   Feasibility  Step Length Norm of   First-order
step    optimality
0     5     1.600000e+01  1.200e+01     1.000e+00   0.000e+00   1.200e+01
1    10     1.606250e+01  1.387e+00     1.000e+00   1.159e+00   2.077e+00
2    15     1.696396e+01  8.077e-02     1.000e+00   1.875e-01   1.347e-01
3    20     1.701372e+01  4.614e-04     1.000e+00   1.682e-02   1.317e-02
4    25     1.701402e+01  8.243e-08     1.000e+00   2.871e-04   5.985e-05
5    30     1.701402e+01  3.579e-11     1.000e+00   5.956e-06   2.446e-07


Local minimum found that satisfies the constraints.

Optimization completed because the objective function is non-decreasing in
feasible directions, to within the default value of the optimality tolerance,
and constraints are satisfied to within the default value of the constraint
tolerance.

<stopping criteria details>

Optimal solution found at
x =
     1.000000e+00
     4.743000e+00
     3.821150e+00
     1.379408e+00

Value of objective function = 17.0140173
```

**Example 10.7** In this example, the `fmincon` will use the algorithms `sqp`, `interior-point`, `active-set` and `sqp-legacy` to solve the constrained minimization problem:

$$\min_{\boldsymbol{x}\in\mathbb{R}^3} x_2 \left(x_1^2 - x_3^3\right)^2 + \frac{4}{x_1 x_2}$$

subject to:

$$-6 \le x_1 + x_2 + x_3 \le 6,$$

and

$$-1 \le x_2 \le 1.$$

The MATLAB script `MinEx2Withfmincon.m` solves the above minimization proplem. Its code is:

```
1   % MinEx2Withfmincon.m
2   clear ; clc ;
3   x0 = [-1;-1; -1];       % Make a starting guess at the solution
4   Algorithms = ["sqp", "active-set", "interior-point", ...
        "sqp-legacy"] ;
5
6   for n = 1 : length(Algorithms)
7       options = optimoptions(@fmincon,'Algorithm',Algorithms(n), ...
            'Display', 'off');
```

```
 8        fprintf('Solution with %s algorithm:\n', Algorithms(n)) ;
 9        [x,fval] = ...
             fmincon(@objfun,x0,[],[],[],[],[],[],@confun,options);
10        fprintf('Optimal solution found at \nx ...
             =\n%12.6e\n%12.6e\n%12.6e\n', x(1), x(2), x(3)) ;
11        fprintf('Value of objective function = %10.7f\n\n', fval) ;
12   end
13
14   function f = objfun(x)
15        f = x(2)*(1+2*x(1)^2 - x(3)^3)^2 + 4/(x(1)*x(2)) ;
16   end
17
18   function [ic, ec] = confun(x)
19        % Nonlinear inequality constraints
20        ic(1) = -x(1) -x(2) -x(3)+6 ;
21        ic(2) = x(1) + x(2) + x(3) - 6 ;
22        ic(3) = -x(2) + 1 ;
23        ic(4) = x(2) - 1 ;
24        % Nonlinear equality constraints
25        ec = [];
26   end
```

Running the code gives the following results:

```
Solution with sqp algorithm:
Optimal solution found at
x =
     2.575360e+00
     1.000000e+00
     2.424640e+00
Value of objective function =  1.5532974


Solution with active-set algorithm:
Optimal solution found at
x =
     2.575360e+00
     1.000000e+00
     2.424640e+00
Value of objective function =  1.5532974


Solution with interior-point algorithm:
> In backsolveSys
In solveKKTsystem
In computeTrialStep
In barrier
In fmincon (line 800)
In minex3withfmincon (line 8)
Warning: Matrix is singular to working precision.
Optimal solution found at
```

```
x =
     2.575362e+00
     1.000000e+00
     2.424638e+00
Value of objective function =   1.5532974


Solution with sqp-legacy algorithm:
Optimal solution found at
x =
     2.575360e+00
     1.000000e+00
     2.424640e+00
Value of objective function =   1.5532974
```

## 10.2.2   Solving Constrained Minimization Problems in Python

Constrained minimization problems are solved in Python with the function
`minimize` located in the library `scipy.optimize`. It receives the objective
function, an initial starting point and the options of the optimization process.
It returns the optimum solution and the value of the objective function at the
optimal solution [18].

In this section Examples 10.6 and 10.7 will be solved using Python scripts
`MinEx1minimize.py` and `MinEx2minimize.py`, respectively.

The Python script `MinEx1Withminimize.py` solves Example 10.6. Its
code is:

```
1   # MinEx1Withminimize
2   import numpy as np
3   from scipy.optimize import minimize
4
5   def objfun(x):
6         return x[0]*x[3]*(x[0]+x[1]+x[2])+x[2]
7
8   def ic(x):
9         return x[0]*x[1]*x[2]*x[3]-25.0
10
11  def ec(x):
12        return 40.0 - sum(x*x)
13
14  x0 = np.array([1., 5., 5., 1.])
15
16  cons = [{'type': 'ineq', 'fun': ic}, {'type': 'eq', 'fun': ec}]
17  lubs = [(1.0, 5.0), (1.0, 5.0), (1.0, 5.0), (1.0, 5.0)]
18  Sol = minimize(objfun,x0, method='SLSQP', bounds=lubs, ...
          constraints=cons)
19  x = Sol.x
20  fval = Sol.fun
21  print('Solution found at x = ')
22  print('{0:12.10e}'.format(x[0]))
23  print('{0:12.10e}'.format(x[1]))
```

```
24  print('{0:12.10e}'.format(x[2]))
25  print('{0:12.10e}'.format(x[3]))
26  print('Value of the Objective function = ' + ...
          str(('{0:10.7f}'.format(fval))))
```

Executing the code gives:

```
runfile('D:/PyFiles/MinEx1Withminimize.py', wdir='D:/PyFiles')
Solution found at x =
    1.0000000000e+00
    4.7429960656e+00
    3.8211546642e+00
    1.3794076394e+00
Value of the Objective function = 17.0140172
```

To solve Example 10.7 with Python, the script `MinEx2Withminimize.py` implements the solution:

```
1   import numpy as np
2   from scipy.optimize import minimize
3
4   def objfun(x):
5       return 4*x[1]*(1+2*x[0]**2 - x[2]**3)**2 + x[0]/x[1]
6
7   def ic(x):
8       icon = np.zeros(4)
9       icon[0] = -x[0] - x[1]**2 - x[2] + 6
10      icon[1] =  x[0] + x[1]**2 + x[2] - 6
11      icon[2] = -x[1] + 1
12      icon[3] =  x[1] - 1
13      return icon
14
15  x0 = np.array([1., 1., 1.])
16
17  cons = [{'type': 'ineq', 'fun': ic}]
18  bnds = [(-np.inf, np.inf), (-1.0, 1.0), (-np.inf, np.inf)]
19  OptSol = minimize(objfun,x0, method='SLSQP', bounds=bnds, ...
          constraints=cons)
20  print(OptSol)
21  x = OptSol.x
22  print('Solution found at x = ')
23  print('{0:12.10e}'.format(x[0]))
24  print('{0:12.10e}'.format(x[1]))
25  print('{0:12.10e}'.format(x[2]))
26  print('Value of the Objective function = ' + ...
          str(('{0:10.7f}'.format(OptSol.fun))))
```

By executing the code, the following results are obtained:

```
runfile('D:/PyFiles/MinEx2Withminimize.py', wdir='D:/PyFiles')
fun: 2.5748936729216085
jac: array([ 0.63140237, -2.57473353,  0.63150746])
message: 'Optimization terminated successfully.'
```

```
nfev: 46
nit: 8
njev: 8
status: 0
success: True
x: array([2.57481362, 1.            , 2.42518638])

Solution found at x =
    2.5748136179e+00
    1.0000000000fe+00
    2.4251863821e+00
Value of the Objective function =   2.5748937
```

## 10.2.3   Solving Constrained Optimization with Gekko Python

Gekko Python software solves constrained and unconstrained optimization
problems at mode 3 solver [2]. To solve Example 10.6 with Gekko, the the
script MinEx1WithGekko.py can be used:

```python
1  # MinEx1WithGekko.py
2  from gekko import GEKKO
3  m = GEKKO ( )
4  x1 = m.Var(1, lb=1, ub=5)
5  x2 = m.Var(5, lb=1, ub=5)
6  x3 = m.Var(5, lb=1, ub=5)
7  x4 = m.Var(1, lb=1, ub=5)
8  m.Equation (x1 * x2 * x3 * x4 >= 25)
9  m.Equation (x1**2+x2**2+x3**2+x4**2 == 40)
10 m.Obj(x1 * x4 * (x1 + x2 + x3) + x3)
11 m.options.IMODE = 3
12 m.solve()
13 print('Solution found at x = ')
14 print('{0:12.10e}'.format(x1[0]))
15 print('{0:12.10e}'.format(x2[0]))
16 print('{0:12.10e}'.format(x3[0]))
17 print('{0:12.10e}'.format(x4[0]))
```

Running the code gives the following results:

```
EXIT: Optimal Solution Found.

The solution was found.

The final value of the objective function is    17.0140171270735

-----------------------------------------------------
Solver         :  IPOPT (v3.12)
Solution time  :   1.000000000931323E-002 sec
Objective      :    17.0140171270735
Successful solution
-----------------------------------------------------
```

```
Solution found at x =
     1.0000000570e+00
     4.7429996300e+00
     3.8211500283e+00
     1.3794081795e+00
```

The Python script `MinEx2WithGekko.py` is used to solve Example 10.7.

```
 1  # MinEx2WithGekko.py
 2  from gekko import GEKKO
 3  m = GEKKO ( )
 4  x1 = m.Var(1, lb=0, ub=6.)
 5  x2 = m.Var(1, lb=-1., ub=6.)
 6  x3 = m.Var(1, lb=0, ub=6.)
 7  m.Equation(x1 + x2**2 + x3 >= -6)
 8  m.Equation(x1 + x2**2 + x3 <=  6)
 9
10  m.Obj(x2*(1+2*x1**2 - x3**3)**2 + 4./(x1*x2))
11  m.options.IMODE = 3
12  m.solve()
13  print('Solution found at x = ')
14  print('{0:12.10e}'.format(x1[0]))
15  print('{0:12.10e}'.format(x2[0]))
16  print('{0:12.10e}'.format(x3[0]))
17  fval = x2[0]*(1+2*x1[0]**2 - x3[0]**3)**2 + 4./(x1[0]*x2[0])
18  print('Value of the Objective function = ' + ...
           str(('{0:10.7f}'.format(fval))))
```

Running the program gives the following results:

```
EXIT: Optimal Solution Found.

The solution was found.

The final value of the objective function is    1.43580274955049


-----------------------------------------------------
Solver         :   IPOPT (v3.12)
Solution time  :   1.930000001448207E-002 sec
Objective      :   1.43580274955049
Successful solution
-----------------------------------------------------


Solution found at x =
     2.1288804479e+00
     1.3087777155e+00
     2.1582204319e+00
Value of the Objective function =  1.4358027
```

# 11

## Solving Optimal Control Problems

## Abstract

Optimal Control Problems (OCPs) represent a model for a wide variety of real-life phenomena. Some of these applications include the control of infectious diseases, the continuous stirred tank reactor (CSTR), biological populations, population harvesting, etc.

This chapter presents the basics of optimal control. It then discusses the uses of indirect and direct transcription methods for solving them numerically using MATLAB® and Python. It also discusses the use of the gekko Python for solving the optimal control problems.

The chapter consists of six sections, organized as follows. The statement of the problem is in the first section. The second and third sections discuss the necessary conditions for the optimal control. Ideas about some of numerical methods for solving the optimal control problems are presented in Section 4. Section 5 discusses the numerical solution of optimal control problems based on the indirect transcription methods. The numerical methods for solving optimal control problems based on the direct transcription methods are presented in Section 6.

## 11.1    Introduction

This chapter considers an optimal control problem (OCP) of the form:

$$\underset{\boldsymbol{u}\in\mathbb{U}}{\text{minimize}} \quad \varphi\big(\boldsymbol{x}(t_f)\big) + \int_{t_0}^{t_f} L_0\big(t, \boldsymbol{x}(t), \boldsymbol{u}(t)\big)\, dt \tag{11.1}$$

The dynamics is described by a system of ordinary differential equations defined on the interval $[t_0, t_f]$

$$\dot{\boldsymbol{x}}(t) = \boldsymbol{f}\left(t, \boldsymbol{x}(t), \boldsymbol{u}(t)\right), \forall t \in [t_0, t_f], \tag{11.2}$$

with initial condition

$$\boldsymbol{x}(t_0) = \boldsymbol{x}^0 \tag{11.3}$$

The system might be subject to continuous state inequality constraints

$$I(t, x(t)) \leq 0, \quad \forall t \in [t_0, t_f].$$ (11.4)

and equality constraints

$$E(t, x(t)) = 0, \quad \forall t \in [t_0, t_f].$$ (11.5)

It is subject to terminal conditions:

$$\psi(t_f, x(t_f)) = 0,$$ (11.6)

where $x(t) = [x_1(t), \cdots, x_n(t)]^T \in R^n$ is the state vector and $u(t) = [u_1(t), \cdots, u_m(t)]^T$
$\in R^m$ is the control vector.

The terminal time $t_f$ is a *fixed* positive real number. $\mathbb{U}$ is the class of all piecewise continuous functions $u(t)$ defined on $[t_0, t_f)$, with $a_i \leq u_i(t) \leq b_i, a_i, b_i \in \mathbb{R}$.

The function $f : \mathbb{R} \times \mathbb{R}^n \times \mathbb{R}^m \to \mathbb{R}^n$, is continuously differentiable with respect to each component of $x$ and $u$, and piecewise continuous with respect to $t$.

The term $\varphi(x(t_f))$ in equation (11.1) is called the `Mayer part`, and the integral term of Equation (11.1) is called the `Bolza part`. A cost function could consist of only a Mayer part, only a Bolza part or a combination of both [10].

## 11.2    The First-Order Optimality Conditions and Existence of Optimal Control

Notation: Throughout the chapter, we will use the following notations:
$L_0[t] \equiv L_0(t, x(t), u(t))$ and $f[t] \equiv f(t, x(t), u(t))$.

The Hamiltonian for problem (11.1)-(11.6) is given by

$$\mathcal{H}[t] = \mathcal{H}(t, x, u, p) = L_0[t] + p(t)^T f[t]$$

then, the augmented cost function can be written as:

$$J(u) = \varphi(t_f, x(t_f)) + \nu^T \Psi(t_f, x(t_f)) + \int_{t_0}^{t_f} \left[ \mathcal{H}[t] - p(t)^T \dot{x}(t) \right] dt$$

The first-order necessary conditions for the optimality are found by applying the variational principle to the *augmented cost function*, also referred to as *Lagrangian* [24, 11]. The augmented cost function is given by

$$J(u) = \varphi(t_f, x(t_f)) + \nu^T \Psi(t_f, x(t_f)) + \int_{t_0}^{t_f} \left[ L_0[t] + p(t)^T (f[t] - \dot{x}(t)) \right] dt$$ (11.7)

where $p(t) \in \mathbb{R}^n$ are the *adjoint* variables or the *co-state* variables, $\nu \in \mathbb{R}^l$ are Lagrange multipliers, the final time $t_f$, may be fixed or free.

The minimum of $J(u)$ is determined by computing the variation $\delta J$ with respect to all the free variables and then equating it to the zero [15]. *i.e.* $\delta J = 0$.

$$\delta J = \frac{\partial \varphi(t_f, \boldsymbol{x}(t_f))}{\partial t_f} \delta t_f + \frac{\partial \varphi(t_f, \boldsymbol{x}(t_f))}{\partial \boldsymbol{x}(t_f)} \delta \boldsymbol{x}_f + \boldsymbol{\nu}^T \left( \frac{\partial \boldsymbol{\Psi}}{\partial t_f} \delta t_f + \frac{\partial \boldsymbol{\Psi}}{\partial \boldsymbol{x}(t_f)} \delta \boldsymbol{x}_f \right)$$

$$+ (\delta \boldsymbol{\nu}^T) \boldsymbol{\Psi} + \left( \mathcal{H}[t_f] - \boldsymbol{p}^T(t_f) \dot{\boldsymbol{x}}(t_f) \right) \delta t_f + \int_{t_0}^{t_f} \left[ \frac{\partial \mathcal{H}}{\partial \boldsymbol{x}} \delta \boldsymbol{x} + \frac{\partial \mathcal{H}}{\partial \boldsymbol{u}} \delta \boldsymbol{u} + \frac{\partial \mathcal{H}}{\partial \boldsymbol{p}} \delta \boldsymbol{p} \right] dt$$

$$(11.8)$$

where $\delta \boldsymbol{x}_f = \delta \boldsymbol{x}(t_f) + \dot{\boldsymbol{x}}(t_f) \delta t_f$.

Integrating $\int_{t_0}^{t_f} -\boldsymbol{p}^T(t) \delta \dot{\boldsymbol{x}}(t) dt$ by parts, gives

$$\int_{t_0}^{t_f} -\boldsymbol{p}^T(t) \delta \dot{\boldsymbol{x}}(t) dt = -\boldsymbol{p}^T(t_f) \delta \boldsymbol{x}(t_f) + \int_{t_0}^{t_f} \dot{\boldsymbol{p}}^T(t) \delta \boldsymbol{x}(t) dt \qquad (11.9)$$

Inserting (11.9) into (11.8) and rearranging the terms we get

$$\delta J = \left( \frac{\partial \varphi(t_f, \boldsymbol{x}(t_f))}{\partial \boldsymbol{x}(t_f)} + \boldsymbol{\nu}^T \frac{\partial \boldsymbol{\Psi}(t_f, \boldsymbol{x}(t_f))}{\partial \boldsymbol{x}(t_f)} - \boldsymbol{p}^T(t_f) \right) \delta \boldsymbol{x}(t_f) + \delta \boldsymbol{\nu}^T \boldsymbol{\Psi}(t_f, \boldsymbol{x}(t_f))$$

$$+ \left( \frac{\partial \varphi}{\partial t_f} + \frac{\partial \varphi}{\partial \boldsymbol{x}(t_f)} \boldsymbol{f}[t_f] + \boldsymbol{\nu}^T \left( \frac{\partial \boldsymbol{\Psi}}{\partial t_f} + \frac{\partial \boldsymbol{\Psi}}{\partial \boldsymbol{x}(t_f)} \boldsymbol{f}[t_f] \right) + \mathcal{H}[t_f] - \boldsymbol{p}^T(t_f) \boldsymbol{f}[t_f] \right) \delta t_f$$

$$+ \int_{t_0}^{t_f} \left[ \left( \frac{\partial \mathcal{H}}{\partial \boldsymbol{x}}[t] + \dot{\boldsymbol{p}}^T(t) \right) \delta \boldsymbol{x} + \left( \frac{\partial \mathcal{H}}{\partial \boldsymbol{u}}[t] \right) \delta \boldsymbol{u} + \left( \frac{\partial \mathcal{H}}{\partial \boldsymbol{p}}[t] - \dot{\boldsymbol{x}}(t) \right) \delta \boldsymbol{p} \right] dt \quad (11.10)$$

The necessary optimality conditions define a stationary point, at which any arbitrary variations in the free variables, result in no change in the total cost [57, 54]. The first order optimality conditions are given by

(i) equating the coefficient of $\delta \boldsymbol{u}$ to zero gives

$$\frac{\partial \mathcal{H}}{\partial \boldsymbol{u}}[t] = 0, \quad t_0 \leq t \leq t_f \qquad (11.11)$$

(ii) equating the coefficient of $\delta \boldsymbol{x}$ to zero gives the *adjoint equations*

$$\dot{\boldsymbol{p}}^T(t) = -\frac{\partial \mathcal{H}}{\partial \boldsymbol{x}}[t], \quad t_0 \leq t \leq t_f \qquad (11.12)$$

(iii) equating the coefficient of $\delta \boldsymbol{p}$ to zeros gives the *state equations*

$$\dot{\boldsymbol{x}}(t) = \frac{\partial \mathcal{H}}{\partial \boldsymbol{p}}[t] = \boldsymbol{f}(t, \boldsymbol{x}(t), \boldsymbol{u}(t)), \quad t_0 \leq t \leq t_f \qquad (11.13)$$

(iv) equating the coefficient of $\delta\boldsymbol{x}(t_f)$ to zero, gives the *transversality conditions*

$$\boldsymbol{p}(t_f) = \frac{\partial\varphi(t_f,\boldsymbol{x}(t_f))}{\partial\boldsymbol{x}(t_f)} + \boldsymbol{\nu}^T\frac{\partial\boldsymbol{\Psi}(t_f,\boldsymbol{x}(t_f))}{\partial\boldsymbol{x}(t_f)} \qquad (11.14)$$

(v) equating the coefficient of $\delta\boldsymbol{\nu}^T$ to zero, gives the terminal constraints

$$\boldsymbol{\Psi}(t_f,\boldsymbol{x}(t_f)) = 0 \qquad (11.15)$$

(vi) equating the coefficient of $\delta t_f$ to zeros, gives the condition on the Hamiltonian at the terminal time,

$$\mathcal{H}[t_f] = -\frac{\partial\varphi(t_f,\boldsymbol{x}(t_f))}{\partial t_f} - \boldsymbol{\nu}^T\frac{\partial\boldsymbol{\Psi}(t_f,\boldsymbol{x}(t_f))}{\partial t_f} \qquad (11.16)$$

Equations (11.11, 11.12, 11.13) define the *Euler-Lagrange equations* [54, 15] for the optimal control problems.

## Example 11.1 (Illustrative Example:)

$$\underset{0\leq u(t)\leq u_{max}}{\text{minimize}} \quad J(u) = x(T) + \alpha\int_0^T u^2(t)dt$$

subject to the dynamics:

$$\dot{x}(t) = ax(t) + u(t), \ x(0) = x_0$$

The Hamiltonian of the optimal control problem is:

$$\mathcal{H}(x(t),u(t),p(t)) = \alpha u^2(t) + p(t)\left(ax(t) + u(t)\right),$$

and the Euler-Lagrange equations are:

$$\frac{\partial\mathcal{H}}{\partial u} = 2\alpha u(t) + p(t) = 0$$

$$\dot{p}(t) = -\frac{\partial\mathcal{H}}{\partial x} = -ap(t)$$

$$\dot{x}(t) = ax(t) + u(t), x(0) = x_0$$

in addition to the transversality condition

$$p(T) = \frac{\partial x(T)}{\partial x(T)} = 1$$

If $u^*(t)$ and $x^*(t)$ are the optimal control and the resulting optimal trajectory and $p^*(t)$ is the co-state variable corresponding to the optimal trajectory, then the tuple $(u^*(t), x^*(t), p^*(t))$ can be found by solving the Euler-Lagrange equations as follows.

By solving the second equation in $p(t)$ together with the transversality condition, we get

$$p^*(t) = e^{a(T-t)}$$

By solving the first equation in $u(t)$, we get

$$u^*(t) = -\frac{p^*(t)}{2\alpha} = -\frac{e^{a(T-t)}}{2\alpha}$$

Finally, by solving the third equation in $x(t)$, we get

$$x^*(t) = \left(x_0 - \frac{e^{aT}}{4a\alpha}\right)e^{at} + \frac{e^{a(T-t)}}{4a\alpha}$$

## 11.3   Necessary Conditions of the Discretized System

We develop in this section a numerical method for solving an OCP without transforming it into one without time delays.

Let $N$ be a positive integer and $h = \frac{T-t_0}{N}$ then $s_0 = t_0 < s_1 < \ldots < s_{KN} = t_f$ where $s_{i+1} = s_i + h$ is a partition for the interval $[t_0, t_f]$. Let $x(s_i), p(s_i)$ and $u(s_i)$ be the values of $x(t), p(t)$ and $u(t)$ at $t = s_i$, $i = 0, 1, \ldots, N$. The Hamiltonian at $s_i$ is given by

$$\mathcal{H}[s_i] = \mathcal{H}(s_i, x(s_i), x(s_i - \tau), p(s_i)) = L_0[s_i] + p^T(s_i)f[s_i] \qquad (11.17)$$

The optimality conditions for the discretized problem are

1.
$$\frac{\partial \mathcal{H}}{\partial u}[s_i] = 0, s_0 \leq s_i \leq s_{KN} \qquad (11.18)$$

2.
$$\begin{cases} p(s_i) = p(s_{i+1}) + h\left[\frac{\partial \mathcal{H}}{\partial x}[s_i] + \frac{\partial \mathcal{H}}{\partial x_\tau}[s_{i+1} + \tau]\right], & s_0 \leq s_i \leq s_{(K-1)N} \\ p(s_i) = p(s_{i+1}) + h\frac{\partial \mathcal{H}}{\partial x}[s_{i+1}], & s_{(K-1)N} \leq s_i \leq s_{KN} \end{cases} \qquad (11.19)$$

3.
$$x(s_{i+1}) = x(s_i) + h\frac{\partial \mathcal{H}}{\partial p}[s_i] = x(s_i) + hf[s_i] \qquad s_0 \leq s_i \leq s_{KN} \quad (11.20)$$

4.

$$p(s_{KN}) = \frac{\partial \varphi(s_{KN}, \boldsymbol{x}(s_{KN}))}{\partial \boldsymbol{x}(s_N)} \tag{11.21}$$

## 11.4    Numerical Solution of Optimal Control

The numerical methods for solving optimal control problems are generally divided into two classes, the *indirect methods* and the *direct methods*.

The indirect methods involve forming the optimality conditions by using the calculus of variations and the Pontryagin maximum principle [10, 54]. These methods transcribe the optimal control problem into a set of two-points boundary value problems in the state and co-state (adjoint) variables. The resulting equations are then solved using the shooting methods [10, 55].

The direct methods involve the transformation of the optimal control problem into a large scale nonlinear programing problem [11], and then solving the resulting constrained nonlinear programming problem using standard sequential quadratic programming (SQP) methods [12, 25]. are also known as discretize then optimize methods, because the problem discretization is a prior stage to the optimization process [10].

The direct methods include

1. the collocation methods, in which both the control and state variables are discretized using the collocation methods.

2. the control parameterization methods, in which the control variables are parameterized and the optimal control problem, becomes an optimal parameter selection problem [32]. and,

3. the iterative dynamic programming (IDP), which is an extension to the Bellman dynamical programming concept [35].

A classic example of Optimal Control Software based on an indirect method is MISER3, [25]. Dieter Kraft (1985) presented the software package, *Trajectory Optimization by Mathematical Programming (TOMP)*, by which numerical solutions of OCPs may be calculated by a direct shooting approach [27]. In 2001, J. Kiezenka and L.F.Shampine presented the MATLAB procedure *bvp4c* as an example of an effective application of indirect collocation methods to solve the two-point boundary value problems resulting from the Pontryagin maximum principle. In 1997, John T. Betts and W.P Hoffman presented the Sparse Optimal Control Software (SOCS) [11]. Another FORTRAN code DIRCOL by O. Von Stryk [1993] is an example of the direct collocation approach [57].

## 11.5 Solving Optimal Control Problems Using Indirect Methods

### 11.5.1 Numerical Solution Using Indirect Transcription Method

Given the Euler-Lagrange equations:

$$\frac{\partial \mathcal{H}}{\partial u} = 0 \tag{11.22}$$

$$\dot{x}(t) = \frac{\partial \mathcal{H}}{\partial p} = f(t, x(t), u(t)), \ x(t_0) = x_0 \tag{11.23}$$

$$\dot{p}(t) = -\frac{\partial \mathcal{H}}{\partial x}, \ p(T) = p_T \tag{11.24}$$

Suppose that Equation 11.22 can be solved such that

$$u^*(t) = g(x(t), p(t)),$$

then the following steps can be carried out to find the optimal control and optimal trajectory.

**Step one:** Divide the interval $[t_0, t_f]$ into $N$ equally spaced sub-intervals, by the points $s_0 < s_1 < \cdots < s_N$, where $h = s_{i+1} - s_i = \frac{t_f - t_0}{N}$.

**Step two:** Choose an initial guess $u = u^{(0)}$ for the control $u(t)$, a maximum number of iterations $MaxIters$ and a tolerence $\epsilon$.

**Step three:** set $n = 0$.

**Step four:** repeat the following steps:

**Step five:** solve the state equations (11.20) with initial condition (11.3) as follows:

**Algorithm 1 (Solving state equations)**

*5.1 Set $x^{0(n)} = x^0$.*

*5.2 for $i = 0, 1, \ldots, N-1$*

    *5.2.1 advance to $x^{i+1(n)}$ using an integration method of ODEs, such as the fourth-order Runge-Kutta method*

$$k_1 = f\left(s_i, x^{i(n)}, u^{i(n)}\right)$$

$$k_2 = f\left(s_i + \frac{h}{2}, x^{i(n)} + \frac{hk_1}{2}, \frac{u^{i(n)} + u^{i+1(n)}}{2}\right)$$

$$k_3 = f\left(s_i + \frac{h}{2}, x^{i(n)} + \frac{hk_2}{2}, \frac{u^{i(n)} + u^{i+1(n)}}{2}\right)$$

$$k_4 = f\left(s_i + h, x^{i(n)} + hk_3, u^{i+1(n)}\right)$$

$$x^{i+1(n)} = x_i^{(n)} + \frac{h}{6}(k_1 + 2k_2 + 2k_3 + k4).$$

_5.3 interpolate the points_ $(s_i, \boldsymbol{x}^{i(n)}), i = 0, \ldots, N$ _by a cubic spline_ $\boldsymbol{\varphi}^{1(n)}(t)$
_5.4_ **return** $(s_i,)$

**Step six:** using the state functions $\boldsymbol{x}^{(n)}$, solve the co-state equations as follows:

## Algorithm 2 (solving co-state equations)

_6.1 set_ $\boldsymbol{p}^{N(n)} = \dfrac{\partial \varphi(t_f, \boldsymbol{x}(t_f))}{\partial \boldsymbol{x}(t_f)}\Big|_{(s_N, \boldsymbol{x}^{N(n)})}$
_6.2_ **for** $i = N - 1$ **down to** $0$

    _6.2.1 advance to_ $\boldsymbol{p}^{i(n)}$ _using the an integration method of ODEs, such as the fourth-order Runge-Kutta method_

$$
\begin{aligned}
k_1 &= \boldsymbol{g}\left(s_{i+1}, \boldsymbol{x}^{i+1(n)}, \boldsymbol{u}^{i+1(n)}\right) \\[2mm]
k_2 &= \boldsymbol{g}\left(s_{i+1} - \frac{h}{2}, \boldsymbol{x}^{i+1(n)} - \frac{hk_1}{2}, \frac{\boldsymbol{u}^{i(n)} - \boldsymbol{u}^{i+1(n)}}{2}\right) \\[2mm]
k_3 &= \boldsymbol{g}\left(s_{i+1} - \frac{h}{2}, \boldsymbol{x}^{i+1(n)} - \frac{hk_2}{2}, \frac{\boldsymbol{u}^{i(n)} + \boldsymbol{u}^{i+1(n)}}{2}\right) \\[2mm]
k_4 &= \boldsymbol{g}\left(s_{i+1} - h, \boldsymbol{x}^{i+1(n)} - hk_3, \boldsymbol{u}^{i(n)}\right) \\[2mm]
\boldsymbol{p}^{i(n)} &= \boldsymbol{p}_{i+1}^{(n)} - \frac{h}{6}(k_1 + 2k_2 + 2k_3 + k4)
\end{aligned}
$$

_6.3_ **return** $(s_i, \boldsymbol{p}^{i(n)}), \quad i = 0, \ldots, N$

**Step seven:** update the control $\boldsymbol{u}$ by

$$
\boldsymbol{u}^{n+1} = \varphi(\boldsymbol{x}^{(n)}, \boldsymbol{p}^{(n)})
$$

**Step eight:** set $n = n + 1$.
**Step nine:** repeat the steps five, six, seven and eight until a convergence is achieved or the number of maximum iterations is passed, that is $\|\boldsymbol{u}^{(n)} - \boldsymbol{u}^{(n-1)}\| < \epsilon$ or $n > MaxIters$.
**Step ten:** in the case of success return $(\boldsymbol{u}^{(n)}, \boldsymbol{x}^{(n)})$.

## An Illustrative Example

### Example 11.2

$$
\min_{u} J(u) = I(T) + \int_0^T \left[bu^2(t) + I(t)\right] dt
$$

subject to:

$$
\dot{I}(t) = \beta(1 - I(t))I(t) - \alpha u(t)I(t), \ I(0) = I_0, \quad t \in [0, T]
$$

and control box constraints:

$$u_{min} \leq u(t) \leq u_{max}$$

where the state variable $I(t)$ fulfils the constraints:

$$0 \leq I(t) \leq 1$$

The Hamiltonian of this problem is given by

$$\mathcal{H}(u(t), I(t), \lambda(t)) = I + bu^2 + \lambda(\beta(1 - I) - (\gamma + \alpha u) * I, \quad t \in [0, T]$$

and the first order necessary conditions are given by

$$\frac{\partial \mathcal{H}}{\partial u}\bigg|_{u=u^\star} = -\alpha I(t)\lambda(t) + 2bu(t) = 0 \Rightarrow u^\star = -\frac{\alpha I(t)\lambda(t)}{2b}$$

$$\dot{I}(t) = \beta(1 - I(t))I(t) - (\gamma + \alpha u(t))I(t)$$

$$\dot{\lambda}(t) = -1 + (2\beta I + \gamma + \alpha u(t) - \beta)\lambda$$

The transversality condition is given by:

$$\lambda(T) = \frac{\partial I(T)}{\partial I(T)} = 1$$

Let $N$ be a positive integer and $h = \frac{T}{N}$ is the step size. The interval $[0, T]$ is partitioned by $N + 1S$ points $t_0, \ldots, t_N$, where $t_i = ih, i = 0, \ldots, N$. Let $v_j \approx u(t_j), z_j \approx I(t_j), l_j \approx \lambda(t_j), uh_j \approx u(t_j + h/2)$ and $zh_j \approx I(t_j + h/2)$ for $j = 1, \ldots, N$. We set

$$f(z_j, v_j) = (\beta - \beta z_j - \gamma - \alpha v_j)z_j \text{ and } g(z_j, v_j, l_j) = -1.0 + (2\beta z_j + \gamma + \alpha v_j - \beta)l_j.$$

Starting from any initial guess $u_j^{(0)}$ for the control, and setting $uh_j^{(0)}$ be the piecewise cubic Hermite interpolation of $u(t)$ at $t = t_j + h/2$ the classical fourth-order Runge-Kutta methods can be used to solve the state and co-state equations as follows:

$$k_1 = f(z_j, v_j)$$
$$k_2 = f(z_j + h * k_1/2, uh_j)$$
$$k_3 = f(z_j + h * k_2/2, uh_j)$$
$$k_4 = f(z_j + h * k_3, v_{j+1})$$
$$z_{j+1} = z_j + h/6 * (k_1 + 2k_2 + 2k_3 + k_4), j = 0, \ldots, N - 1$$
$$\kappa_1 = g(z_{j+1}, v_{j+1}, l_{j+1})$$
$$\kappa_2 = g(zh_j, uh_j, l_{j+1} - h/2\kappa_1)$$
$$\kappa_3 = g(zh_j, uh_j, l_{j+1} - h/2\kappa_2)$$
$$\kappa_4 = g(z_j, v_j, l_{j+1} - h\kappa_3)$$
$$l_j = l_{j+1} - h/6(\kappa_1 + 2\kappa_2 + 2\kappa_3 + \kappa_4), \ j = N - 1, \ldots, 0$$

where $zh_j$ is the cubic Hermite interpolation of $I(t)$ at $t = t_j + h/2$.

For $T = 20, b = 1, \alpha = 0.5, \gamma = 0.2, \beta = 1.0, I_0 = 0.01,$ $\lambda(T) = 1,$ $N = 5000$ and $\varepsilon = 10^{-15}$, the following Python code is used to solve the optimal control problem.

```python
import numpy as np
import matplotlib.pylab as plt

def UpdateControl(x, p):
    return np.array([min(u, umax) for u in [max(u, umin) for u ...
        in a*p*x/(2.0*b**2)]])
def SolveStateEquation(u):
    from scipy import interpolate
    f = lambda z, v: (bt-bt*z-gm-a*v)*z
    th = np.arange(h/2.0, T, h)
    uh = interpolate.pchip(t, u)(th)
    x[0] = x0
    for j in range(len(t)-1):
        k1 = f(x[j], u[j])
        k2 = f(x[j]+h*k1/2, uh[j])
        k3 = f(x[j]+h*k2/2, uh[j])
        k4 = f(x[j]+h*k3, u[j+1])
        x[j+1] = x[j] + h/6*(k1+2*k2+2*k3+k4)
    return x

def SolveCostateEquation(x, u):
    p[N] = pT
    th = np.arange(h/2.0, T, h)
    uh = interpolate.pchip(t, u)(th)
    xh = interpolate.pchip(t, x)(th)
    g = lambda z, lm, v: -1.0+(2*bt*z+gm+a*v-bt)*lm
    for j in [N-int(i)-1 for i in list(range(len(t)-1))]:
        k1 = g(x[j+1], p[j+1], u[j+1])
        k2 = g(xh[j], p[j+1]-h*k1/2, uh[j])
        k3 = g(xh[j], p[j+1]-h*k2/2, uh[j])
        k4 = g(x[j], p[j+1]-h*k3, u[j])
        p[j] = p[j+1] - h/6.0*(k1+2*k2+2*k3+k4)
    return p

b, a, gm, bt = 1.0, 0.5, 0.2, 1.0
T = 20.0
x0, pT = 0.01, 1.0
umin, umax = 0.0, 1.0
N = 10000
t = np.linspace(0.0, T, N+1)
h = T/N
x, p, u = np.ones_like(t), np.ones_like(t), 0.5*np.ones_like(t)

x = SolveStateEquation(u)
from scipy import interpolate
p = SolveCostateEquation(x, u)
uold = u
u = UpdateControl(x, p)
Error = np.linalg.norm(u-uold, np.inf)
Iterations = 1
while Error >= 1e-15:
```

```
51      uold = u
52      x = SolveStateEquation(u)
53      p = SolveCostateEquation(x, u)
54      u = UpdateControl(x, p)
55      Error = np.linalg.norm(u-uold, np.inf)
56      print(Iterations, Error)
57      Iterations += 1
58      print(Iterations, '\t', Error)
59      u = UpdateControl(x, p)
60
61  plt.figure(1)
62  plt.subplot(1, 2, 1)
63  plt.plot(t, u, color='crimson', lw=2)
64  plt.xlabel('Time (t)', fontweight='bold')
65  plt.ylabel('Optimal Treatment (u(t))', fontweight='bold')
66  plt.grid(True, ls = '--')
67  plt.xticks(np.arange(0, T*1.1, T/10), fontweight='bold')
68  mnu, mxu = np.floor(1000*min(u))/1000, np.ceil(1010*max(u))/1000
69  hu = (mxu-mnu)/10
70  plt.yticks(np.arange(mnu, mxu+hu, hu), fontweight='bold')
71  plt.axis([0, T, mnu, mxu])
72  plt.subplot(1, 2, 2)
73  plt.plot(t, x, color = 'purple', lw = 2)
74  plt.xlabel('Time (t)', fontweight='bold')
75  plt.ylabel('Infective Population (I(t))', fontweight='bold')
76  plt.grid(True, ls = '--')
77  plt.xticks(np.arange(0, T*1.1, T/10), fontweight='bold')
78  mnx, mxx = np.floor(10*min(x))/10, np.ceil(10*max(x))/10
79  hx = (mxx-mnx)/10
80  plt.yticks(np.arange(mnx, mxx+hx, hx), fontweight='bold')
81  plt.grid(True, ls = '--')
82  plt.axis([0, T, mnx, mxx])
```

By running the script, we obtain the output:

```
runfile('D:/PyFiles/SolveOCPSI.py', wdir='D:/PyFiles')
Iteration              Error

1               0.03923191725605521
2               0.0033203169324064197
3               0.0002443004136510607
4               1.7888301100443815e-05
5               1.3080226375083992e-06
6               9.563564312697892e-08
7               6.9922647949471894e-09
8               5.112286149966394e-10
9               3.737765652545022e-11
10              2.7327029528123603e-12
11              1.9961809982760315e-13
12              1.432187701766452e-14
13              1.1102230246251565e-15
14              1.1102230246251565e-16
```

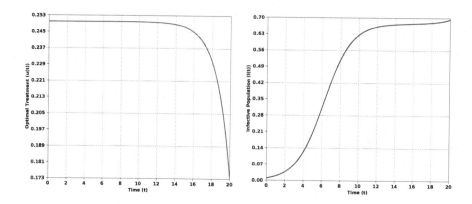

FIGURE 11.1: The optimal control and optimal trajectory of Example 11.1.

The optimal treatment and corresponding density of infected population are explained figure (11.1)

Using MATLAB, the optimal treatment and corresponding density of infected population can be computed by the following code:

```
1   clear ; clc ; clf ;
2   global a x0 gm b bt N h t T pT ;
3   a = 0.5 ; b = 1; gm = 0.2 ; bt = 1.0; N = 10000 ;
4   T = 20 ; x0 = 0.01 ; pT = 1 ;
5   h = T/N ;
6   t = linspace(0, T, N+1) ;
7   uold = 0.5*ones(1, N+1) ; %p = ones(1, N+1) ; x = ones(1, N+1) ;
8   x = SolveStateEquation(uold) ;
9   p = SolveCostateEquation(x, uold) ;
10  u = UpdateControl(x, p) ;
11
12  Iterations = 1 ;
13  Error = norm(u-uold, inf) ;
14  while Error >= 1e-15
15      uold = u ;
16      x = SolveStateEquation(u) ;
17      p = SolveCostateEquation(x, u) ;
18      u = UpdateControl(x, p) ;
19      Error = norm(u-uold, inf) ;
20      disp([num2str(Iterations) '\t\t' num2str(Error)]) ;
21      Iterations = Iterations + 1 ;
22  end
23  figure(1) ;
24  plot(t, u, '-b', 'LineWidth', 2) ;
25  xlabel('Time (t)') ;
26  ylabel('Control (u(t))') ;
27  grid on ;
28  figure(2) ;
```

```
29    plot(t, x, '-r', 'LineWidth', 2) ;
30    xlabel('Time (t)') ;
31    ylabel('State (x(t))') ;
32    grid on ;
33
34    function u = UpdateControl(x, p)
35        global a b ;
36        u = min(max(a*p.*x/(2*b^2), 0), 1) ;
37    end
38
39    function x = SolveStateEquation(u)
40        global a x0 gm bt N h t T ;
41        x = ones(1, N+1) ;
42        f = @(z, v) (bt-bt*z-gm-a*v)*z ;
43        th = h/2:h:T-h/2 ;
44        uh = pchip(t, u, th) ;
45        x(1) = x0 ;
46        for j = 1 : N
47            k1 = f(x(j), u(j)) ;
48            k2 = f(x(j)+h/2*k1, uh(j)) ;
49            k3 = f(x(j)+h/2*k2, uh(j)) ;
50            k4 = f(x(j)+h*k3, u(j+1)) ;
51            x(j+1) = x(j) + h/6*(k1+2*k2+2*k3+k4) ;
52        end
53    end
54
55    function p = SolveCostateEquation(x, u)
56        global bt gm a pT N h t T;
57        p = ones(1, N+1) ;
58        g = @(z, lm, v) -1.0+(2*bt*z+gm+a*v-bt)*lm ;
59        th = h/2:h:T-h/2 ;
60        uh = pchip(t, u, th) ;
61        xh = pchip(t, x, th) ;
62        p(N+1) = pT ;
63        for j = N : -1 : 1
64            k1 = g(x(j+1), p(j+1), u(j+1)) ;
65            k2 = g(xh(j), p(j+1)-h/2*k1, uh(j)) ;
66            k3 = g(xh(j), p(j+1)-h/2*k2, uh(j)) ;
67            k4 = g(x(j), p(j+1)-h*k3, u(j)) ;
68            p(j) = p(j+1) - h/6*(k1+2*k2+2*k3+k4) ;
69        end
70    end
```

The optimal control and state variables are computed iteratively, until at some iteration $k$, the condition $|u^{(k)} - u^{(k-1)}| < \varepsilon$ is fulfilled, where $\varepsilon$ is a small arbitrary constant (selected to be $10^{-15}$ in the example). In this example, it is noticed that fourteen iterations were needed to compute the optimal control and optimal trajectory.

**Example 11.3** In this example we solve an optimal control problem found in [56]. It describes Glucose-Insulin interaction. The optimal control problem is formulated as follows:

$$\min_{u \in \mathbb{R}} \lim \int_0^T (x_1(t) - x_d)^2 + pu^2(t)dt$$

subject to:

$$\dot{x}_1(t) = -m_1 x_1(t) - m_2 x_2(t), \; x_1(0) = x_{10}$$
$$\dot{x}_2(t) = -m_3 x_2(t) + u(t), \; x_2(0) = x_{20}$$

where $u(t)$ accounts for the rate of infusion of exogenous hormone.

Setting $x_d 100, p = 10, m_1 = 0.0009, m_2 = 0.0031, m_3 = 0.0415$ and starting from $x_1(0) = x_{10}, x_2 0 = x_{20}$, the following Python code solves the optimal control problem.

```python
import numpy as np
import matplotlib.pylab as plt

def UpdateControl(p):
    return np.array([max(u, 0) for u in -p[:, 1]/(2.0*r)])

def SolveStateEquation(u):
    from scipy import interpolate
    f  = lambda z, v: np.array([-m1*z[0]-m2*z[1], -m3*z[1]+v])
    th = np.arange(h/2.0, T, h)
    uh = interpolate.pchip(t, u)(th)
    x = np.zeros((len(t), 2), 'float')
    x10, x20 = 300.0, 0.0
    x[0, :] = np.array([x10, x20])
    for j in range(len(t)-1):
        k1 = f(x[j], u[j])
        k2 = f(x[j]+h*k1/2, uh[j])
        k3 = f(x[j]+h*k2/2, uh[j])
        k4 = f(x[j]+h*k3, u[j+1])
        x[j+1] = x[j] + h/6*(k1+2*k2+2*k3+k4)
    return x

def SolveCostateEquation(x):
    p = np.zeros((len(t), 2), 'float')
    th = np.arange(h/2.0, T, h)
    xh = interpolate.pchip(t, x)(th)
    g  = lambda z, lm: np.array([m1*lm[0]-2*z[0]+2*xd, ...
         m2*lm[0]+m3*lm[1]])
    for j in [N-int(i)-1 for i in list(range(len(t)-1))]:
        k1 = g(x[j+1], p[j+1])
        k2 = g(xh[j], p[j+1]-h*k1/2)
        k3 = g(xh[j], p[j+1]-h*k2/2)
        k4 = g(x[j], p[j+1]-h*k3)
        p[j] = p[j+1] - h/6.0*(k1+2*k2+2*k3+k4)
    return p

xd, r, m1, m2, m3 = 100.0, 10.0, 9.0e-4, 3.1e-3, 4.15e-2
T = 90.0
umin, umax = 0.0, 1.0
N = 30000
t = np.linspace(0.0, T, N+1)
h = T/N
u = np.ones_like(t)
```

```
44  x = SolveStateEquation(u)
45  from scipy import interpolate
46  p = SolveCostateEquation(x)
47  uold = u
48  u = UpdateControl(p)
49  Error = np.linalg.norm(u-uold, np.inf)
50  print('Iteration\t\t Error\n')
51  Iterations = 1
52
53  while Error >= 1e-15:
54      uold = u
55      x = SolveStateEquation(u)
56      p = SolveCostateEquation(x)
57      u = UpdateControl(p)
58      Error = np.linalg.norm(u-uold, np.inf)
59      print(Iterations, '\t\t', Error)
60      Iterations += 1
61
62  plt.figure(1)
63  plt.plot(t, u, color='crimson', lw=2)
64  plt.xlabel('Time (t)', fontweight='bold')
65  plt.ylabel('Optimal Treatment (u(t))', fontweight='bold')
66  plt.grid(True, ls = '--')
67  plt.xticks(np.arange(0, T*1.1, T/10), fontweight='bold')
68  mnu, mxu = np.floor(0.1*min(u))*10, np.ceil(0.1*max(u))*10
69  hu = (mxu-mnu)/10
70  plt.yticks(np.arange(mnu, mxu+hu, hu), fontweight='bold')
71  plt.axis([0, T, mnu, mxu])
72  plt.savefig('OCPGI1.eps')
73  plt.savefig('OCPGI1.png')
74
75  plt.figure(2)
76
77  plt.subplot(1, 2, 1)
78  plt.plot(t, x[:, 0], color = 'purple', lw = 2, label = 'Glucose')
79  plt.xlabel('Time (t)', fontweight='bold')
80  plt.ylabel('Glucose (x1(t) mg/dl)', fontweight='bold')
81  plt.grid(True, ls = '--')
82  plt.xticks(np.arange(0, T*1.1, T/10), fontweight='bold')
83  mnx, mxx = np.floor(0.1*min(x[:, 0]))*10, np.ceil(0.1*max(x[:, ...
        0]))*10
84  hx = (mxx-mnx)/10
85  plt.yticks(np.arange(mnx, mxx+hx, hx), fontweight='bold')
86  plt.grid(True, ls = '--')
87  plt.axis([0, T, mnx, mxx])
88
89  plt.subplot(1, 2, 2)
90  plt.plot(t, x[:, 1], color = 'orangered', lw = 2, label = ...
        'Insulin')
91  plt.xlabel('Time (t)', fontweight='bold')
92  plt.ylabel('Insulin (x2(t) mg/dl)', fontweight='bold')
93  plt.grid(True, ls = '--')
94  plt.xticks(np.arange(0, T*1.1, T/10), fontweight='bold')
95  mnx, mxx = np.floor(0.1*min(x[:, 1]))*10, np.ceil(0.1*max(x[:, ...
        1]))*10
96  hx = (mxx-mnx)/10
97  plt.yticks(np.arange(mnx, mxx+hx, hx), fontweight='bold')
```

```
98  plt.grid(True, ls = '--')
99  plt.axis([0, T, mnx, mxx])
100 plt.savefig('OCPGI2.eps')
101 plt.savefig('OCPGI2.png')
```

The optimal infusion rate of Insulin and the corresponding interaction of Glucose and Insulin are explained in Figures 11.2a-11.2b.

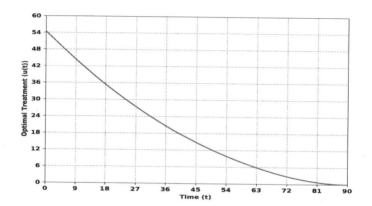

(a) optimal rate of infusion of exogeneous Insulin

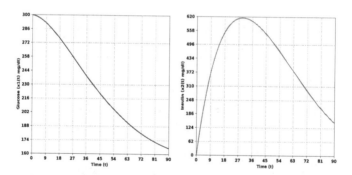

(b) Glucose and Insulin dynamics, left figure: glucose, right figure: insulin.

FIGURE 11.2: Solution of the optimal control problem of Glucose-Insulin interaction.

The MATLAB code for solving the optimal control problem of the glucose-insulin interaction is as follows:

```
1  clear ; clc ; clf ;
2  global m1 m2 m3 x10 x20 N h t T b xd ;
3  xd = 100.0; b = 10.0; m1 = 9.0e-4; m2 = 3.1e-3; m3 = 4.15e-2 ;
4  T = 90.0; N = 30000 ;
5  x10 = 300.0 ; x20 = 0.0 ;
```

```
 6   h = T/N ;
 7   t = linspace(0, T, N+1) ;
 8   uold = 0.5*ones(N+1, 1) ; %p = ones(1, N+1) ; x = ones(1, N+1) ;
 9   x = SolveStateEquationGI(uold) ;
10   p = SolveCostateEquationGI(x) ;
11   u = UpdateControlGI(p) ;
12
13   Iterations = 1 ;
14   Error = norm(u-uold, inf) ;
15   while Error >= 1e-15
16       uold = u ;
17       x = SolveStateEquationGI(u) ;
18       p = SolveCostateEquationGI(x) ;
19       u = UpdateControlGI(p) ;
20       Error = norm(u-uold, inf) ;
21       disp([num2str(Iterations) \t\t num2str(Error)]) ;
22       Iterations = Iterations + 1 ;
23   end
24
25   figure(1) ;
26   plot(t, u, '-b', 'LineWidth', 2) ;
27   xlabel('Time (t)') ;
28   ylabel('Optimal Infusion Rate (u(t))') ;
29   grid on ;
30
31   figure(2) ;
32   subplot(1, 2, 1) ;
33   plot(t, x(:, 1), '-r', 'LineWidth', 2) ;
34   xlabel('Time (t)') ;
35   ylabel('Glucose (x_1(t))') ;
36   grid on ;
37
38   subplot(1, 2, 2) ;
39   plot(t, x(:, 2), '-r', 'LineWidth', 2) ;
40   xlabel('Time (t)') ;
41   ylabel('Insulin (x_2(t))') ;
42   grid on ;
43
44   function u = UpdateControlGI(p)
45       global b ;
46       u = max(-p(:, 2)/(2*b), 0) ;
47   end
48
49   function x = SolveStateEquationGI(u)
50       global m1 m2 m3 x10 x20 N h t T ;
51       x = ones(N+1, 2) ;
52       f = @(z, v) [-m1*z(1)-m2*z(2), -m3*z(2)+v] ;
53       th = h/2:h:T-h/2 ;
54       uh = pchip(t, u, th) ;
55       x(1, :) = [x10, x20] ;
56       for j = 1 : N
57           k1 = f(x(j, :), u(j)) ;
58           k2 = f(x(j, :)+h*k1/2, uh(j)) ;
59           k3 = f(x(j, :)+h*k2/2, uh(j)) ;
60           k4 = f(x(j, :)+h*k3, u(j+1)) ;
61           x(j+1, :) = x(j, :) + h/6*(k1+2*k2+2*k3+k4) ;
62       end
```

```
63   end
64
65   function p = SolveCostateEquationGI(x)
66       global xd m1 m2 m3 N h t T;
67       p = ones(N+1, 2) ;
68       g = @(z, lm) [m1*lm(1)-2*z(1)+2*xd, m2*lm(1)+m3*lm(2)] ;
69       th = h/2:h:T-h/2 ;
70       xh = pchip(t(:), x', th(:))' ;
71       p(N+1, :) = zeros(1, 2) ;
72       for j = N : -1 : 1
73           k1 = g(x(j+1, :), p(j+1, :)) ;
74           k2 = g(xh(j, :), p(j+1, :)-h*k1/2) ;
75           k3 = g(xh(j, :), p(j+1, :)-h*k2/2) ;
76           k4 = g(x(j, :), p(j+1, :)-h*k3) ;
77           p(j, :) = p(j+1, :) - h/6.0*(k1+2*k2+2*k3+k4) ;
78       end
79   end
```

## 11.6   Solving Optimal Control Problems Using Direct Methods

Direct transcription methods are well-known class for solving optimal control problems. Direct transciption methods are also known as discretize then optimize methods, because the problem discretization is a prior stage to the optimization process [10]. Methods of discretization such as the Runge-Kutta methods [50], splines [24], collocation methods [57], etc., are used for the discretization of the state equations, whereas some numerical quadrature method such as the trapezoidal rule or the Simpson's rule is used for the evaluation of the objective function.

One of the efficient direct transciption methods for solving the optimal control problems, is the control parameterization technique [58, 59, 60]. In a CPT, the time domain is partitioned by a number of switching points at which the control variables are evaluated, then each switching interval is partitioned by a number of quadrature points at which the state variables are evaluated. At each switching interval, the control variable is approximated by a constant or linear piece-wise continuous function [58].

At the end of the discretization stage, the optimal control problem is transformed into a large or medium scale finite dimensional nonlinear programming problem (NLP) [11]. The resulting NLP can be solved by using any nonlinear programming software, such as the MATLAB's optimization toolbox [17], the SQP [12], the FSQP [29], etc.

## 11.6.1 Statement of the Problem

In this section we consider an optimal control problem of the form:

$$\min_{u \in U} \varphi(x(t_f)) + \int_{t_0}^{t_f} L_0(\boldsymbol{x}(t), \boldsymbol{u}(t)) dt \qquad (11.25)$$

subject to the dynamics:

$$\dot{\boldsymbol{x}}(t) = \boldsymbol{f}(\boldsymbol{x}(t), \boldsymbol{u}(t)), \qquad (11.26)$$

with initial conditions

$$\boldsymbol{x}(t_0) = \boldsymbol{x}^0, \qquad (11.27)$$

where $\boldsymbol{x}, \varphi : \mathbb{R} \to \mathbb{R}^n$, $\boldsymbol{u} : \mathbb{R} \to \mathbb{R}^m$ and $\boldsymbol{x}^0 \in \mathbb{R}^n$.

The system is also subject to state inequality constraints:

$$\boldsymbol{I}(\boldsymbol{x}(t)) \le 0 \qquad (11.28)$$

and state equality constraints

$$\boldsymbol{E}(\boldsymbol{x}(t)) = 0 \qquad (11.29)$$

where $\boldsymbol{I} : \mathbb{R}^n \to \mathbb{R}^p$ and $\boldsymbol{E} : \mathbb{R}^n \to \mathbb{R}^q$ are differentiable with respect to $\boldsymbol{x}$.

In the two following sections we describe the solution of optimal control problems using `control parameterization technique (CPT)` and implement it in MATLAB. Also, we will discuss the solution of optimal control problems using the Python's `Gekko` package.

## 11.6.2 The Control Parameterization Technique

For the discretization of the control variables $\boldsymbol{u}(t)$, let $N_c$ be a positive integer and $h_c = \frac{t_f - t_0}{N_c}$ then $s_0(= t_0) < s_1 < \ldots < s_{N_c} = t_f$ where $s_{i+1} = s_i + h$ is a partition for the interval $[t_0, t_f]$. Let us assume that, in an interval $[s_j, s_{j+1}]$ the value of the control function $u_i(t)(i = 1, \ldots, m)$ is a constant $(i.e. u_i(t) = v_j^i)$ or a linear function $(i.e.\ u_i(t) = a_i^j t + b_i^j)$. The constants $v_j^i$ or $a_i^j$ and $b_i^j$ are called the control parameters. An interval $[s_j, s_{j+1}]$ is called a `switching interval`. Hence, there are $N_c$ swithching intervals $[s_j, s_{j+1}], j = 0, 1, \ldots, N_c - 1$. For simple discussion, we assume that in $[s_i, s_{i+1})$, $u_j(t) = v_j^i$ where $v_j^i$ is a constant.

For the discretization of the state variables, each switching interval $[s_i, s_{i+1})$ is divided into $Q$ sub-intervals, where $Q$ is a fixed positive integer. The time interval $[t_0, t_f]$ is divided into $N = N_c Q$ subintervals. Let $h = (t_f - t_0)/N$ and $t_i = t_0 + ih$. At a time $t_i$ suppose that the state variable $x_j(t_i)$ is approximated by the value $x_j^i; j = 1, \ldots, N; i = 0, \ldots, N$. We also notice that

$$h = \frac{t_f - t_0}{N} = \frac{t_f - t_0}{N_c Q} = \frac{h_c}{Q}.$$

A point $t_i, i = 0, \ldots, N$ lies in a switching interval $[s_k, s_{k+1})$ where $k = \lfloor i/h_c \rfloor$. Hence,

$$u_j(t_i) = v_j^k \text{ and } x_m(t_i) = x_m^i.$$

Now, the control $u_j(t), t \in [t_0, t_f]$ is given by

The Mayer part of the objective function $\varphi(x(t_f))$ is evaluated at the time $t_N$. i.e. $\varphi(x(t_f)) = \varphi(x(t_N))$. We use the Simpson's quadrature to approximate the Bolza part of the objective function, provided that $N$ is an even integer. Let

$$L_0^i = L_0(x(t_i), u(t_i)) = L_0(x(t_i), v^k), \ k = \lfloor i/h_c \rfloor,$$

then,

$$\int_{t_0}^{t_f} L_0(t, x(t), u(t)) \, dt \approx \frac{h}{3}(L_0^0 + L_0^N) + \frac{4h}{3} \sum_{i=1}^{\lfloor \frac{N}{2} \rfloor} L_0^{2i-1} + \frac{2h}{3} \sum_{i=1}^{\lfloor \frac{N}{2} \rfloor} L_0^{2i}$$

The objective function is approximated by

$$J(u(t)) \approx \varphi(x^N) + \frac{h}{3}\left(L_0^0 + L_0^N\right) + \frac{4h}{3} \sum_{i=1}^{\lfloor \frac{N}{2} \rfloor} L_0^{2i-1} + \frac{2h}{3} \sum_{i=1}^{\lfloor \frac{N}{2} \rfloor} L_0^{2i} \qquad (11.30)$$

The equality constraints (11.29) become:

$$E(x^i) = 0, \quad 0 \leq i \leq N \qquad (11.31)$$

The inequality constraints (11.28) become,

$$I(x^i) \leq 0, \quad 0 \leq i \leq N \qquad (11.32)$$

The classical fourth-order Runge-Kutta method is used for the discretization of the state equations as follows.

$$
\begin{aligned}
k_1 &= f(x^i, u^i) \\
k_2 &= f(x^i + hk_1/2, u^{i+\frac{1}{2}}) \\
k_3 &= f(x^i + hk_2/2, u^{i+\frac{1}{2}}) \\
k_4 &= f(x^i + hk_3, u^{i+1}) \\
x^{i+1} &= x^i + \frac{h}{6}(k_1 + 2k_2 + 2k_3 + k_4)
\end{aligned}
\qquad (11.33)
$$

where $u^{i+\frac{1}{2}} = u(t_i + h/2)$ can be approximated from some interpolation method, such the piecewise cubic Hermite interpolating polynomials.

The continuous optimal control problem described by the equations (11.25), (11.26), (11.27), (11.28) and (11.29) is transformed into the discrete nonlinear programming problem:

$$\underset{u \in \mathbb{R}^{m(1+N_c)}}{\text{minimize}} \; \varphi(x^N) + \frac{h}{3}(L_0^0 + L_0^N) + \frac{4h}{3}\sum_{i=1}^{\lfloor \frac{N}{2} \rfloor} L_0^{2i-1} + \frac{2h}{3}\sum_{i=1}^{\lfloor \frac{N}{2} \rfloor} L_0^{2i} \quad (11.34)$$

subject to:

$$x^{i+1} - x^i - \frac{h}{6}(k_1 + 2k_2 + 2k_3 + k_4) = 0 \quad (11.35)$$

subject to the initial condition

$$x(0) - x^0 = 0 \quad (11.36)$$

subject to the equality constraints:

$$E(x^i) = 0, \quad i = 0,1,2,\ldots,N \quad (11.37)$$

the inequality constraints:

$$I(x^i) \le 0, \quad i = 0,1,2,\ldots,N \quad (11.38)$$

Finally both the control variables $u_k, k = 1,2,\ldots,m$ and the state variables $x_i, i = 1,2,\ldots,n$ are mapped into one vector $Y$. $x_i^j$ is mapped into $Y(i-1+(i-1)N+j), i=1,2,\ldots,n; j=0,1,\ldots,N$, and $u_k^l$ is mapped into $Y((k-1)+n+(n-1)N+(k-1)N_c+\lfloor (l-1)/Q \rfloor), k=1,2,\ldots,m; l=0,1,\ldots,N$. The total length of the vector $Y$ is $n(1+N) + m(1+N_c)$.

The problem described by (11.34), with the constraints (11.35) or (11.35)-(11.37) then becomes:

$$\underset{Y \in \mathbb{R}^{n(1+N)+m(1+N_c)}}{\text{minimize}} \; \Psi(y)$$

subject to the equality and inequality constraints:

$$\bar{E}(Y) = 0,$$
$$\bar{I}(Y) \le 0$$

and subject to the initial conditions:

$$Y(i-1+(i-1)N) - x_i^0 = 0, \quad i = 1,2,\ldots,n$$

### 11.6.2.1 Examples

**Example 11.4** This example is taken from [26].

$$\underset{u(t)}{\text{minimize}} \; J(u) = S(T) + I(T) - R(T) + \int_0^T (S(t)+I(t)-R(T)+C_1 u^2(t) + C_2 v^2(t))dt$$

subject to:

$$\dot{S}(t) = \Lambda - \alpha S(t)I(t) - (\mu + u)S(t), S(0) = S_0, t \in [0, T]$$
$$\dot{I}(t) = \alpha S(t)I(t) - (\mu + \beta)I(t) - v(t)I(t), I(0) = I_0, t \in [0, T]$$
$$\dot{R}(t) = \beta I(t) + u(t)S(t) - \mu R(t) + v(t)I(t), R(0) = R_0, t \in [0, T]$$

where

$$0 \le S(t), I(t), R(t) \le 1$$

and

$$u_{min} \le u(t) \le u_{max} \text{ and } v_{min} \le v(t) \le v_{max}$$

To solve this problem with MATLAB, the model's paremeters $C_1, C_2, \lambda, \alpha, \mu$ and $\beta$ shall be declared as global variables.

Based on the Simpson's rule, a function SIRObjective is implemented for the evaluation of the objective function.

```
1   function OptimalCost = SIRObjective(x)
2       global C1 C2 N NC h Q ;
3       S = x(1:1+N) ;
4       I = x(2+N:2+2*N) ;
5       R = x(3+2*N:3+3*N) ;
6       u = x(4+3*N:4+3*N+NC) ;
7       v = x(5+3*N+NC: 5+3*N+2*NC) ;
8       L = @(j) ...
            S(j)+I(j)-R(j)+C1*u(1+floor(j/Q))^2+C2*v(1+floor(j/Q))^2 ;
9       OptimalCost = S(end) + I(end)-R(end) + h/3*(L(1)+L(end)) ;
10      for j = 2 : N
11          if mod(j, 2) == 0
12              OptimalCost = OptimalCost + 4*h/3*(L(j)) ;
13          else
14              OptimalCost = OptimalCost + 2*h/3*(L(j)) ;
15          end
16      end
```

Write a function SIRConstsRK4 to return the equality and inequality constraints in two vectors ic and ec. The fourth order Runge-Kutta method is used for the discretization of the state equations, where the piecewise cubic Hermite interpolation polynomials are used to interpolate the state and control variables at the points $t_j + h/2; j = 1, \ldots, N$. The MATLAB code of the functionSIRConstsRK4 is as follows.

```
1   function [ic, ec] = SIRConstsRK4(x)
2       global N NC Q h beta mu alpha Lambda t ;
3       S = x(1:1+N) ;
4       I = x(2+N:2+2*N) ;
5       R = x(3+2*N:3+3*N) ;
6       u = x(4+3*N:4+3*N+NC) ;
```

```
7        v = x(5+3*N+NC: 5+3*N+2*NC) ;
8
9        th = t(1:N)+h/2.0 ;
10       Sh = pchip(t, S, th) ;
11       Ih = pchip(t, I, th) ;
12       Rh = pchip(t, I, th) ;
13       uh = pchip(t, u(1+floor((0:N)/Q)), th) ;
14       vh = pchip(t, v(1+floor((0:N)/Q)), th) ;
15       umin = 0.05 ;
16       umax = 0.60 ;
17       vmin = 0.1 ;
18       vmax = 0.9 ;
19       S0 = 0.9 ;
20       I0 = 0.07 ;
21       R0 = 0.03 ;
22       ic = zeros(4+4*NC, 1) ;
23       ec = zeros(3*N+5, 1) ;
24       ic(1:1+NC) = umin - u ;
25       ic(2+NC:2+2*NC) = u - umax ;
26       ic(3+2*NC:3+3*NC) = vmin-v ;
27       ic(4+3*NC:4+4*NC) = v - vmax ;
28       f1 = @(x, y, z, v) Lambda-alpha*x*y-(mu+v)*x ;
29       f2 = @(x, y, z, v) alpha*x*y-(mu+beta)*y-v*y ;
30       f3 = @(x, y, z, u, v) beta*y+u*x-mu*z+v*y ;
31       for j = 1 : N
32           k11 = f1(S(j), I(j), R(j), u(1+floor(j/Q))) ;
33           k12 = f1(S(j)+h/2*k11, Ih(j), Rh(j), uh(j)) ;
34           k13 = f1(S(j)+h/2*k12, Ih(j), Rh(j), uh(j)) ;
35           k14 = f1(S(j)+h*k13, I(j+1), R(j+1), ...
                 u(1+floor((j+1)/Q))) ;
36
37           k21 = f2(S(j), I(j), R(j), v(1+floor(j/Q))) ;
38           k22 = f2(Sh(j), I(j)+h/2*k21, Rh(j), vh(j)) ;
39           k23 = f2(Sh(j), I(j)+h/2*k22, Rh(j), vh(j)) ;
40           k24 = f2(S(j+1), I(j)+h*k23, R(j+1), ...
                 v(1+floor((j+1)/Q))) ;
41
42           k31 = f3(S(j), I(j), R(j), u(1+floor(j/Q)), ...
                 v(1+floor(j/Q))) ;
43
44           k32 = f3(Sh(j), Ih(j), R(j)+h/2*k31, uh(j), vh(j)) ;
45           k33 = f3(Sh(j), Ih(j), R(j)+h/2*k32, uh(j), vh(j)) ;
46           k34 = f3(S(j+1), I(j+1), R(j)+h*k33, ...
                 u(1+floor((j+1)/Q)), v(1+floor((j+1)/Q))) ;
47           ec(j)       = S(j+1) - S(j) - h/6 * ...
                 (k11+2*k12+2*k13+k14) ;
48           ec(N+j)     = I(j+1) - I(j) - h/6 * ...
                 (k21+2*k22+2*k23+k24) ;
49           ec(2*N+j) = R(j+1) - R(j) - h/6 * ...
                 (k31+2*k32+2*k33+k34) ;
50       end
51       ec(3*N+1) = S(1)-S0 ;
52       ec(3*N+2) = I(1)-I0 ;
53       ec(3*N+3) = R(1)-R0 ;
54       ec(3*N+4) = u(end)-u(end-1) ;
55       ec(3*N+5) = v(end)-v(end-1) ;
```

Now, we write a MATLAB script SolveSIROCP.m which uses the functions SIRObjective and SIRConsts to find the solution of the optimal control problem.

```matlab
1   clear ; clc ;
2   global N NC Q h Lambda beta mu alpha C1 C2 t ;
3   Lambda = 0.05 ; mu = 0.05 ; alpha = 2 ; beta = 0.6 ;
4   t0 = 0 ;
5   T = 5 ;
6   C1 = 2 ;
7   C2 = 0.5 ;
8   h = 0.025 ;
9   N = T/h ;
10  Q = 2 ;
11  NC = N/Q ;
12  t = linspace(t0, T, 1+N) ;
13  tc = linspace(t0, T, 1+NC) ;
14  S0 = 0.9 ;
15  I0 = 0.07 ;
16  R0 = 0.03 ;
17  x0 = zeros(3*N+2*NC+5, 1) ;
18
19  Options = optimset('LargeScale', 'off', 'Algorithm', ...
        'active-set', 'Display', 'Iter', 'MaxIter', inf, ...
        'MaxFunEvals', inf, 'TolFun', 1e-6, 'TolX', 1e-6, ...
        'TolFun', 1e-6) ;
20  [x, fval] = fmincon(@SIRObjective, x0, [],[],[],[],[],[], ...
        @SIRConstsRK4, Options) ;
21
22  S = x(1:1+N) ;
23  I = x(2+N:2+2*N) ;
24  R = x(3+2*N:3+3*N) ;
25  u = x(4+3*N:4+3*N+NC) ;
26  v = x(5+3*N+NC:5+3*N+2*NC) ;
27  tc = 0:hc:T ;
28  us = pchip(tc, u, t) ;
29  vs = pchip(tc, v, t) ;
30
31  figure(1) ;
32
33  subplot(1, 2, 1) ;
34  plot(t, us, '-b', 'LineWidth',2) ;
35  xlabel('Time (t)') ;
36  ylabel('Vaccination Strategy (u(t))') ;
37  grid on ;
38  set(gca, 'XTick', 0:0.5:T) ;
39  axis([0, T, 0, 1]) ;
40
41  subplot(1, 2, 2) ;
42  plot(t, vs, '-r', 'LineWidth', 2) ;
43  axis([0, T, -0.1, 1]) ;
44  xlabel('Time (t)') ;
45  ylabel('Treatment (v(t))') ;
46  grid on ;
47  set(gca, 'XTick', 0:0.5:T) ;
```

```
48  axis([0, T, 0, 0.2]) ;
49
50  figure(2) ;
51
52  subplot(1, 3, 1) ;
53  plot(t, S, 'b', 'LineWidth', 2) ;
54  xlabel('t') ;
55  ylabel('Susceptibles (S(t))') ;
56  grid on ;
57  set(gca, 'XTick', 0:0.5:T) ;
58  axis([0, T, 0.0, 1]) ;
59
60  subplot(1, 3, 2) ;
61  plot(t, I, 'r', 'LineWidth',2) ;
62  xlabel('Time (t)') ;
63  ylabel('Infected Population (I(t))') ;
64  grid on ;
65  set(gca, 'XTick', 0:0.5:T) ;
66  axis([0, T, 0.0, 0.2]) ;
67
68  subplot(1, 3, 3) ;
69  plot(t, R, 'm', 'LineWidth', 2) ;
70  xlabel('Time (t)') ;
71  ylabel('Recovered Population (R(t))') ;
72  grid on ;
73  set(gca, 'XTick', 0:0.5:T) ;
74  axis([0, T, 0.0, 0.8]) ;
```

The optimal vaccination and treatment strategies and corresponding ratios of suceptible, infected and recovered populations are shown in Figures 11.3a-11.3b.

### 11.6.3   The Gekko Python Solver

The gekko package can be used to solve dynamical optimization problem at mode 6 (Nonlinear control / dynamic optimization (CTL)). The objective function shall be written in Mayer's form, by adding a new variabe $X$ whose derivative is the integrand part of the objective function, and the problem will be to minimize a function $\varphi(X(t_f))$. The following example shows how gekko can be used to solve an optimal control problem.

**Example 11.5** We consider the optimal control problem of the SIR model described in Example 11.4, and use the Gekko Python package to solve it.

The Python code SolveGekSIR uses the Gekko Python to solve the problem of example 11.4.

```
1  from gekko import GEKKO
2  import numpy as np
3  import matplotlib.pyplot as plt
4
5  Lambda = 0.05 ; mu = 0.05 ; alpha = 2 ; beta = 0.6
6  t0 = 0
```

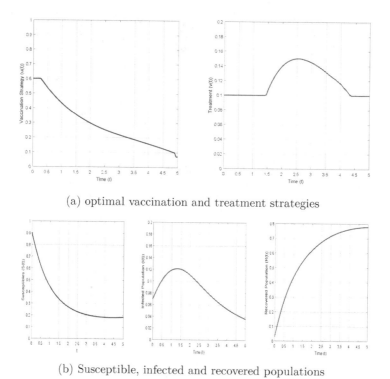

(a) optimal vaccination and treatment strategies

(b) Susceptible, infected and recovered populations

FIGURE 11.3: Solution of the optimal control problem with dynamics governed by an SIR model, using the MATLAB optimization toolbox.

```
7   T = 5
8   C1 = 2
9   C2 = 0.5
10  N = 200 #number of subintervals [t_j, t_{j+1}]
11  m = GEKKO(remote=False) #no remote server is required
12  m.time = np.linspace(t0, T, N+1) #discretizating the interval ...
        [t0, T] by N+1 points
13  S, I, R = m.Var(value=0.9), m.Var(value=0.07), ...
        m.Var(value=0.03) #Initializing the state variables
14  u, v = m.Var(lb=0.0, ub=0.8), m.Var(lb=0.05, ub=1.0) ...
        #Initializing the control variables
15  X = m.Var(value=0.0) #The bolza part of the objective function
16  p = np.zeros(N+1)
17  p[N] = 1.0
18  final = m.Param (value=p)
19  # Equations
20  m.Equation(S.dt() == Lambda-alpha*S*I-(mu+u)*S) #First state ...
        equation in S
21  m.Equation(I.dt() == alpha*S*I-(mu+beta)*I-v*I) #Second state ...
        equation in I
22  m.Equation(R.dt() == beta*I+u*S-mu*R+v*I) #Third state ...
        equation in R
```

```
23  m.Equation(X.dt() == S+I-R+C1*u**2+C2*v**2) #Integrand part of ...
        the objective function
24  m.Obj ((X+S+I-R) * final) #The objective function
25  m.options.IMODE = 6 # Mode 6 is the optimal control mode
26  m.solve()
27  t = m.time
28
29  plt.figure(1)
30  plt.subplot(1, 2, 1)
31  plt.plot(t[1:], u[1:], color='royalblue', lw = 2)
32  plt.xlabel('Time (t)', fontweight='bold')
33  plt.ylabel('Vaccination (u(t))', fontweight='bold')
34  plt.xticks(np.arange(0, T+T/10, T/10), fontweight='bold')
35  mxu = np.ceil(10*max(u))/10
36  plt.yticks(np.arange(0, mxu+mxu/10, mxu/10), fontweight='bold')
37  plt.grid(True, ls='--')
38  plt.axis([0.0, T, 0.0, np.ceil(10*max(u))/10])
39
40  plt.subplot(1, 2, 2)
41  plt.plot(t[1:], v[1:], color='red', lw = 2)
42  plt.xlabel('Time (t)', fontweight='bold')
43  plt.ylabel('Treatment (v(t))', fontweight='bold')
44  plt.xticks(np.arange(0, T+T/10, T/10), fontweight='bold')
45  mxv = np.ceil(10*max(v))/10
46  plt.yticks(np.arange(0, mxv+mxv/10, mxv/10), fontweight='bold')
47  plt.grid(True, ls='--')
48  plt.axis([0.0, T, 0.0, np.ceil(10*max(v))/10])
49
50  plt.figure(2)
51  plt.subplot(1, 3, 1)
52  plt.plot(t, S, color='darkblue', lw = 2)
53  plt.xlabel('Time (t)', fontweight='bold')
54  plt.ylabel('Suceptible Population (S(t))', fontweight='bold')
55  plt.xticks(np.arange(0, T+T/10, T/10), fontweight='bold')
56  mxS = np.ceil(10*max(S))/10
57  plt.yticks(np.arange(0, mxS+mxS/10, mxS/10), fontweight='bold')
58  plt.grid(True, ls='--')
59  plt.axis([0.0, T, 0.0, np.ceil(10*max(S))/10])
60  plt.savefig('SIRConts.eps', dpi=1200)
61  plt.savefig('SIRConts.png', dpi=1200)
62
63  plt.subplot(1, 3, 2)
64  plt.plot(t, I, color='crimson', lw = 2)
65  plt.xlabel('Time (t)', fontweight='bold')
66  plt.ylabel('Infected Population (I(t))', fontweight='bold')
67  plt.xticks(np.arange(0, T+T/10, T/10), fontweight='bold')
68  mxI = np.ceil(10*max(I))/10
69  plt.yticks(np.arange(0, mxI+mxI/10, mxI/10), fontweight='bold')
70  plt.grid(True, ls='--')
71  plt.axis([0.0, T, 0.0, np.ceil(10*max(I))/10])
72
73  plt.subplot(1, 3, 3)
74  plt.plot(t, R, color='darkgreen', lw = 2)
75  plt.xlabel('Time (t)', fontweight='bold')
76  plt.ylabel('Recoved Population (R(t))', fontweight='bold')
77  plt.xticks(np.arange(0, T+T/10, T/10), fontweight='bold')
78  mxR = np.ceil(10*max(R))/10
```

```
79  plt.yticks(np.arange(0, mxR+mxR/10, mxR/10), fontweight='bold')
80  plt.grid(True, ls='--')
81  plt.axis([0.0, T, 0.0, np.ceil(10*max(R))/10])
82  plt.savefig('SIRStates.eps', dpi=1200)
83  plt.savefig('SIRStates.png', dpi=1200)
```

In Figure 11.4 the optimal vaccination and treatment strategies and the corresponding susceptible, infected and recovered populations, obtained by the Gekko Python package are shown.

**Example 11.6** In this example we consider the pest control problem found in [51]. The purpose is to find the optimal spraying schedule to eradicate the number of insect preys of spiders in agroecosystems. The optimal control is given by:

$$\underset{u(t)}{\text{minimize}} \ \ J(u) = \int_0^T (z(t) + \frac{\xi}{2} u^2(t)) dt$$

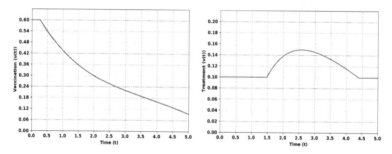

(a) optimal vaccination and treatment strategies

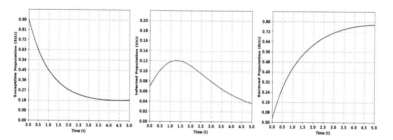

(b) Susceptible, infected and recovered populations

FIGURE 11.4: Gekko solution of the optimal control model with dynamics governed by an SIR model.

subject to:

$$\dot{x}(t) = rx(t)\left(1 - \frac{x(t)}{W}\right) - cx(t)y(t) - \alpha(1-q)u(t), x(0) = x_0, t \in [0,T]$$

$$\dot{y}(t) = y(t)(-a + kbz(t) + kcx(t)) - \alpha Kqu(t), y(0) = y_0, t \in [0,T]$$

$$\dot{z}(t) = ez(t)\left(1 - \frac{z(t)}{V}\right) - by(t)z(t) - \alpha qu(t), z(0) = z_0, t \in [0,T]$$

where

$$u_{min} \leq u(t) \leq u_{max}$$

The values of model parameters are as follows: $r = 1, e = 2.5, a = 3.1, b = 1.2, c = 0.2, \alpha = 0.7, q = 0.9, k = 1.0, V = 1000, W = 5, K = 0.01, T = 50$ and $\xi = 0.05, 0.1$.

The Python code is:

```python
#!/usr/bin/env python3
from gekko import GEKKO
import numpy as np
import matplotlib.pyplot as plt

r = 1.0 ; e  = 2.5; a = 3.1 ; b = 1.2 ; c = 0.2 ; al = 0.7 ;
q = 0.9 ; k = 1.0 ; V = 1000; W = 5 ; K = 0.01 ; D = 0.5 ;
T = 50 ;
t0 = 0
N = 300
m = GEKKO(remote=False)
m.time = np.linspace(t0, T, N+1)
x, y, z = m.Var(value=3.1), m.Var(value=3.7), m.Var(value=2.2)
u = m.Var(lb=0.00, ub=1.0)
X = m.Var(value=0.0)
p = np.zeros(N+1) # mark final time point
p[N] = 1.0
final = m.Param (value=p)
# Equations
m.Equation(x.dt() == r*x*(1-x/W)-c*x*y-al*(1-q)*u)
m.Equation(y.dt() == y*(-a+k*b*z+k*c*x)-al*K*q*u)
m.Equation(z.dt() == e*z*(1-z/V)-b*y*z-al*q*u)
m.Equation(X.dt() == z+D/2*u**2)
m.Obj (X * final) # Objective function
m.options.IMODE = 6 # optimal control mode
m.solve()
t = m.time

plt.figure(1, figsize=(16, 16))
plt.subplot(2, 2, 1)
plt.plot(t[1:], u[1:], color='purple', lw = 2)
plt.xlabel('Time (t)', fontweight='bold')
plt.ylabel('Optimal Spray Schedule (u(t))', fontweight='bold')
plt.xticks(np.arange(0, T+T/10, T/10), fontweight='bold')
mnu = np.floor(1000*min(u))/1000
mxu = np.ceil(1000*max(u))/1000
```

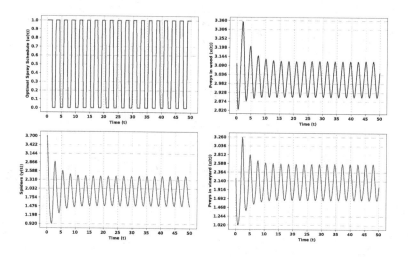

(a) Solution of the pest optimal spraying schedule for $\xi = 0.05$

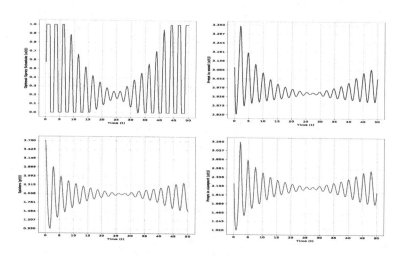

(b) Solution of the pest optimal spraying schedule for $\xi = 0.1$

FIGURE 11.5: Gekko solution of the pest control optimal control problem.

```
37  plt.yticks(np.arange(mnu, mxu+mxu/10000, (mxu-mnu)/10), ...
        fontweight='bold')
38  plt.grid(True, ls=':')
39
40  plt.subplot(2, 2, 2)
41  plt.plot(t, x, color='darkblue', lw = 2)
```

```
42  plt.xlabel('Time (t)', fontweight='bold')
43  plt.ylabel('Preys in wood (x(t))', fontweight='bold')
44  plt.xticks(np.arange(0, T+T/10, T/10), fontweight='bold')
45  mnx = np.floor(100*min(x))/100
46  mxx = np.ceil(100*max(x))/100
47  plt.yticks(np.arange(mnx, mxx+mxx/1000, (mxx-mnx)/10), ...
        fontweight='bold')
48  plt.grid(True, ls=':')
49
50  plt.subplot(2, 2, 3)
51  plt.plot(t, y, color='crimson', lw = 2)
52  plt.xlabel('Time (t)', fontweight='bold')
53  plt.ylabel('Spiders (y(t))', fontweight='bold')
54  plt.xticks(np.arange(0, T+T/10, T/10), fontweight='bold')
55  mny = np.floor(100*min(y))/100
56  mxy = np.ceil(100*max(y))/100
57  plt.yticks(np.arange(mny, mxy+mxy/1000, (mxy-mny)/10), ...
        fontweight='bold')
58  plt.grid(True, ls=':')
59
60  plt.subplot(2, 2, 4)
61  plt.plot(t, z, color='green', lw = 2)
62  plt.xlabel('Time (t)', fontweight='bold')
63  plt.ylabel('Preys in vineyard (z(t))', fontweight='bold')
64  plt.xticks(np.arange(0, T+T/10, T/10), fontweight='bold')
65  mnz = np.floor(100*min(z))/100
66  mxz = np.ceil(100*max(z))/100
67  plt.yticks(np.arange(mnz, mxz+mxz/1000, (mxz-mnz)/10), ...
        fontweight='bold')
68  plt.grid(True, ls=':')
```

The solution of the pest optimal control problem with Gekko is explained in Figure 11.5.

# Bibliography

[1] Eihab B.M. Bashier. *Fitted numerical methods for delay differential equations arising in biology*. PhD thesis, University of the Western Cape, Cape Town, South Africa, September 2009.

[2] Logan Beal, Daniel Hill, R Martin, and John Hedengren. Gekko optimization suite. *Processes*, 6(8):106, 2018.

[3] Amir Beck. *Introduction to Nonlinear Optimization*. SIAM - Society for Industrial and Applied Mathematics, 2014.

[4] M.D. Benchiboun. Linear convergence for vector sequences and some applications. *Journal of Computational and Applied Mathematics*, 55(1):81–97, Oct 1994.

[5] David Benson. *A Gauss Pseudospectral Transcription of Optimal Control*. PhD thesis, Massachusetts Institute of Technology, 2005.

[6] F. Benyah and L.S. Jenning. A comparison of the ill-conditioning of two optimal control computation methods. Technical report, the University of Western Australia, Dept. of Mathematics and Statistics, Nedlands 6907 WA, 2000.

[7] F. Benyah and L. S. Jennings. Ill-conditioning in optimal control computation. In *Proceedings of the Eighth International Colloquium on Differential Equations (ed. D. Bainov)*, pages 81–88, VSP BV, Utrecht, The Netherlands, 1998.

[8] F. Benyah and L. S. Jennings. Regularization of optimal control computation. In *International Symposium on Intelligent Automation Control (ISIAC'98) World Automation Congress*, pages pp. ISIAC 091.1–091.6, Anchorage, Alaska, 1998. TSI Press.

[9] F. Benyah and L. S. Jennings. A review of ill-conditioning and regularization in optimal control computation. In Eds X., Yang, K. L. Teo, and L. Caccetta, editors, *Optimization Methods and Applications*, pages 23–44. Kluwer Academic Publishers, Dordrecht, The Netherlands, 2001.

[10] J. T. Betts. *Practical Methods for Optimal Control Using Nonlinear Programming*. Society for Industrial and Applied Mathematics, 2001.

[11] J. T. Betts and W.P. Hoffman. Exploring sparsity in the direct tran-scripion method for optimal control. *Computational Optimization and Applications*, 14:179–201, 1999.

[12] P. T. Boggs and J. W. Tolle. Sequential quadratic programming. *Acta Numerica*, pages 1–48, 1996.

[13] J. C. Butcher. *Numerical Methods for Ordinary Differential Equations.* John Wiley and Sons Ltd, 2016.

[14] J.C. Butcher. Implicit runge-kutta processes. *Mathematics of Computa-tions*, 18(85):50–64, 1964.

[15] B.C. Chachuat. *Nonlinear and Dynamic Optimization: From Theory to Practice - IC-32: Spring Term 2009.* Polycopiés de l'EPFL. EPFL, 2009.

[16] Stephen J. Chapman. *MATLAB Programming for Engineers.* CEN-GAGE LEARNING, 2015.

[17] Thomas Coleman, Mary Ann Branch, and Andrew Grace. Optimization toolbox for use with matlab, 1999.

[18] SciPy community. *SciPy Reference Guide.* Scipy Community, 1.0.0 edi-tion, October 2017.

[19] Yinyu Ye David G. Luenberger. *Linear and Nonlinear Programming.* Springer-Verlag GmbH, 2015.

[20] Fletcher. *Practical Methods of Optimization 2e.* John Wiley & Sons, 2000.

[21] Sidi Mahmoud Kaber Gregoire Allaire. *Numerical Linear Algebra.* Springer-Verlag New York Inc., 2007.

[22] Per Christian Hansen. Regularization tools: A matlab package for anal-ysis and solution of discrete ill-posed problems. *Numerical Algorithms*, 6(1):1–35, Mar 1994.

[23] William E. Hart, Carl Laird, Jean-Paul Watson, and David L. Woodruff. *Pyomo – Optimization Modeling in Python.* Springer US, 2012.

[24] Tamer Inanc and Raktim Bhattacharya. Numerical solution of optimal control problems using splines, `citeseer.ist.psu.edu/ inanc03numerical.html`citeseer.ist.psu.edu/inanc03numerical.html, 2003.

[25] L. S. Jennings, K. L. Teo, C. J. Goh, and M. E. Fisher. Miser3: A fortran program for solving optimal control problems. A Manual Book, 1990.

[26] K. Kaji and K.H. Wong. Nonlinearly constrained time-delayed optimal control problems. *Journal of Optimization Theory and Applications*, 82(2):295–313, 1994.

[27] D. Kraft. Algorithm 733: Tomp-fortran modules for optimal control calculations. *ACM Transactions on Mathematical Software*, 20(3):262–284, 1994.

[28] Rainer Kress. Ill-conditioned linear systems. In *Graduate Texts in Mathematics*, pages 77–92. Springer New York, 1998.

[29] C. Lawrence and A. Tits. A computationally efficient feasible sequential quadratic programming algorithm. *SIAM Journal on Optimization*, 11(4):1092–1118, 2001.

[30] David C. Lay, Steven R. Lay, and Judi J. McDonald. *Linear Algebra and Its Applications, Global Edition*. Pearson Education Limited, 2015.

[31] J.J. Leader. *Numerical Analysis and Scientific Computation*. Greg Tobin, 2004.

[32] J.S. Lin. Optimal control of time-delay systems by forward iterative dynamic programming. *Ind. Eng. Chem. Res.*, 35(8):2795–2800, 1996.

[33] P. Linz and R. L. C. Wang. *Exploring Numerical Methods: An Introduction to Scientific Computing Using Matlab*. Jones and Bartlett Publishers, Inc., 2003.

[34] David Bau I. I. I. Lloyd N. Trefethen. *Numerical Linear Algebra*. CAMBRIDGE, 1997.

[35] R. Luus. *Iterative Dynamic Programming*. Charman and Hall/CRC, 2000.

[36] R. Marielba. Regularization of large scale least squares problems. Technical Report 96658, Center for Research and Parallel Computations, Rice University, Houston, Texas, 1996.

[37] Ronald E. Mickens. *Applications of Nonstandard Finite Difference Schemes*. World Scientific Publishing Co. Pte. Ltd., 2000.

[38] Ronald E. Mickens. Nonstandard finite difference schemes for differential equations. *Journal of Difference Equations and Applications*, 8(9):823–847, jan 2002.

[39] E. Mickens Ronald. *Applications Of Nonstandard Finite Difference Schemes*. World Scientific, 2000.

[40] E. Mickens Ronald. *Advances In The Applications Of Nonstandard Finite Difference Schemes*. World Scientific, 2005.

[41] Stuart Mitchell, Stuart Mitchell Consulting, and Iain Dunning. Pulp: A linear programming toolkit for python, 2011.

[42] A. Neumair. Solving ill-conditioned and singular linear systems: A tutorial on regularization. *SIAM Rev.*, 40(3):636–666, 1998.

[43] V.A. Patel. *Numerical Analysis*. Publications of Harcourt Brace Collage, 1994.

[44] Jesse A. Pietz. Pseudospectral collocation methods for the direct transcription of optimal control problems. Master's thesis, Rice University, 2003.

[45] Promislow. *Functional Analysis*. John Wiley & Sons, 2008.

[46] Charles C. Pugh. *Real Mathematical Analysis*. Springer-Verlag GmbH, 2015.

[47] J. Douglas Faires Richard Burden. *Numerical Analysis*. Cengage Learning, Inc, 2015.

[48] Lih-Ing W. Roeger. Exact finite-difference schemes for two-dimensional linear systems with constant coefficients. *Journal of Computational and Applied Mathematics*, 219(1):102–109, sep 2008.

[49] Alan Rothwell. *Optimization Methods in Structural Design*. Springer-Verlag GmbH, 2017.

[50] A.L. Shwartz. *Theory and Implementation of Numerical Methods Based on Runge-Kutta Integration for Solving Optimal Control Problems*. Phd thesis, Electronic Research Laboratory, UC Burkeley, 1996.

[51] Silva, C. J., Torres, D. F. M., and Venturino, E. Optimal spraying in biological control of pests. *Math. Model. Nat. Phenom.*, 12(3):51–64, 2017.

[52] James Stewart. *Calculus: Early Transcendentals*. BROOKS COLE PUB CO, 2015.

[53] Strang Strang. *Linear Algebra and Its Applications*. BROOKS COLE PUB CO, 2005.

[54] H. Sussmann and J.C. Willems. 300 years of optimal control: From the brachystrochrone to the maximum principle. *IEEE Control Systems*, pages 32–44, 1997.

[55] John T. Workman Suzanne Lenhart. *Optimal Control Applied to Biological Models*. Taylor & Francis Ltd., 2007.

[56] G.W. Swan. An optimal control model of diabetes mellitus. *Bulletin of Mathematical Biology*, 44(6):793 – 808, 1982.

[57] O. von Stryk. Numerical solution of optimal control problems by direct collocation. *International Series of Numerical Mathematics*, 111:129–143, 1993.

[58] K. H. Wong. A control parametrization algorithm for nonlinear time-lag optimal control problems. *Opsearch*, 25(E):177–184, 1988.

[59] K. H. Wong, L.S. Jennings, and F. Benyah. Control parametrization method for free planning time optimal control problems with time-delayed arguments. *Journal of Nonlinear Analysis*, 47:5679–5689, 2001.

[60] K. H. Wong, L.S. Jennings, and F. Benyah. The control parametrization enhancing transform for constrained time-delayed optimal control problems. *ANZIAM J. The Australian and New Zealand Industrial and Applied Mathematics Journal*, 43(E):E154–E185, 2002.

[61] Dennis G. Zill. *Differential Equations with Boundary-Value Problems.* CENGAGE LEARNING, 2017.

[62] Dennis G. Zill. *A First Course in Differential Equations with Modeling Applications, International Metric Edition.* Cengage Learning, Inc, 2017.

# Index